Embryos, Galaxies,
and Sentient Beings

T0204786

Embryos, Galaxies, and Sentient Beings

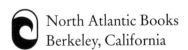

How the Universe Makes Life

RICHARD GROSSINGER

North Atlantic Books
Berkeley, California

RECEIVED

DEC - 5 2003

MINNESOTA STATE UNIVERSITY LIBRARY
MANKATO, MN 56002-8419

Copyright © 2003 by Richard Grossinger. All rights reserved. No portion of this book, except for brief review, may be reproduced, stored in a retrieval system, or transmitted in any form or by any means—electronic, mechanical, photocopying, recording, or otherwise—without the written permission of the publisher. For information contact North Atlantic Books.

Published by
North Atlantic Books   UI Enterprises
P.O. Box 12327   11211 Prosperity Farms Road, Ste. D-325
Berkeley, California 94712   Palm Beach Gardens, Florida 33410

Cover photographs: The human fetus after eleven weeks, © Nestle/Petit Format/Photo Researchers and the Horsehead Nebula, © ROE/AATB/Photo Researchers, photography by David Malin. Photo collage by Paula Morrison. Editor: Kathy Glass. Indexer: Victoria Baker. Cover and book design: Paula Morrison. Printed in Canada.

Embryos, Galaxies, and Sentient Beings: How the Universe Makes Life is sponsored by the Society for the Study of Native Arts and Sciences, a nonprofit educational corporation whose goals are to develop an educational and crosscultural perspective linking various scientific, social, and artistic fields; to nurture a holistic view of arts, sciences, humanities, and healing; and to publish and distribute literature on the relationship of mind, body, and nature.

North Atlantic Books' publications are available through most bookstores. For further information, call 800-337-2665 or visit our website at www.northatlanticbooks.com. Substantial discounts on bulk quantities are available to corporations, professional associations, and other organizations. For details and discount information, contact our special sales department.

Library of Congress Cataloging-in-Publication Data

Grossinger, Richard, 1944–
    Embryos, galaxies, and sentient beings : how the universe make
by Richard Grossinger.
        p. cm.
Includes bibliographical references and index.
    ISBN 1-55643-419-7 (pbk.)
    1. Life (Biology) 2. Embryology. I. Title.
    QH341.G76 2003
    576.8'3—dc21

Printed in Canada   2003008751
1 2 3 4 5 6 7 8 9 TRANS 09 08 07 06 05 04 03

QH
341
.G76
2003

For Bob Bagwell, January 25, 1938, Chicago, Illinois—4 December 2002, Portland, Oregon:

He started my first t'ai chi class (Plainfield, Vermont, 1974); he introduced me to psychobiology and neuro-Darwinism; he was always a reassuring and hopeful voice; his teasing so often saved us from earnestness; he helped me track my own life and the lives of my kids, even when I didn't see him for years; he made exchange of ideas and friendship expressions of each other.

• • • • •

For Richard Handel, 4 January 1949, Springfield, Massachusetts—5 December 2002, Binghamton, New York:

Through his acts Mount Desert Island became sacred ground; he modelled a life of service and playful witnessing of our various plights; he found buddha-nature in everyone, not only swamis and lamas; he was fearless, even daredevil, because he was light as a feather; he taught that "being oneself" is the purest spiritual practice, kindness to others its sole measure.

Selected Other Works by Richard Grossinger

Book of the Cranberry Islands

Book of the Earth and Sky

The Continents

Embryogenesis: Species, Gender, and Identity

Homeopathy: The Great Riddle

The Long Body of the Dream

New Moon

*The Night Sky: The Science and Anthropology
    of the Stars and Planets*

Out of Babylon: Ghosts of Grossinger's

Planet Medicine: Origins

Planet Medicine: Modalities

The Provinces

The Slag of Creation

Solar Journal: Oecological Sections

The Unfinished Business of Doctor Hermes

*Waiting for the Martian Express: Cosmic Visitors, Earth
    Warriors, Luminous Dreams*

As Editor or Co-editor

The Alchemical Tradition in the Late Twentieth Century

Baseball I Gave You All the Best Years of My Life

Ecology and Consciousness

Into the Temple of Baseball

Nuclear Strategy and the Code of the Warrior

Olson-Melville Sourcebook: The Mediterranean

Olson-Melville Sourcebook: The New Found Land

Contents

● ●

Foreword

· ·

John E. Upledger, D.O., O.M.M.

In my opinion the author of this book, Richard Grossinger, is a man far ahead of his time, perhaps a genius. In fact, the rest of humanity may never catch up with him. Richard is also the publisher, through North Atlantic Books, of my own most recent manuscript entitled *Cell Talk* in which, as a clinician (not a biologist, mind you), I put forth the concept that cells—and protein molecules, as well as tissues, organs, and systems—have individual consciousnesses and are able and ready to enter into a dialogue, using human verbiage, with a therapist and/or the "owner" of the body of which they are constituents. In carrying out this dialogue, the "owner" of the body must be willing to allow the cells to freely (without editing or censorship) use his or her vocal apparatus—that is, to speak through a cellular voice. This dialogue can then be—and often is—quite successful in illuminating methods for reversing disease processes, enhancing health, and correcting bodily, emotional, spiritual, and cosmic imbalances.

Since I am a practicing physician, I have included a lot of functional biology, patient perceptions, and patient responsiveness in *Cell Talk*. I speak from that experience rather than from the point of view of a professional cell biologist who might presume he had no reason to look for cell intelligence. My perspective is one of asking cells for their own point of view on this matter rather than imposing a belief system on them. Keeping an open mind on "cell talk" is not that different from keeping an open mind on "interspecies" communication between humans and plants or sea mammals. If you want to hear, you have to tune into available channels, not just those that science deems most likely. You have to trust intuition and not just depend on what the rational mind tells you is possible.

In sum, I have come to believe that every atom has a consciousness, that it contributes that consciousness to the molecule of which it is part. The molecule's consciousness also contributes to the atom's consciousness, and both their consciousnesses participate in every complex of which they are a part. Cell consciousness is a coalescence of the molecular consciousnesses inside it and also influences the consciousness of those molecules comprising it. Likewise, tissue consciousness is made up of all the cell consciousnesses incremental to it as well as all the consciousnesses within those cell consciousnesses. Organs have individual consciousnesses that are aggregates of the tissues comprising them. If we can change tissue consciousness, we can change organ consciousness, and vice versa. Substitute another word for "consciousness" if it pleases you, but it comes out to the same thing.

In CranioSacral Therapy and its offshoot techniques, Somato-Emotional Release and Therapeutic Imagery and Dialogue, I have learned to "blend" with patients and/or clients. This has fostered an exchange of energies back and forth that ultimately mingle according to the intentions put forth by the participants. It is at this time point that we have been able literally to dialogue with body organs, brain parts, tumors and their cells, glands, and cells of the immune system, and so on. Thus far there seems to be no limit. During sessions wherein blending has occurred, we have observed that mutual healing for both patient/client and therapist occurs.

• • • • •

It was Richard's idea that his book *Embryos, Galaxies and Sentient Beings* and my *Cell Talk* were complementary undertakings. With this in mind he invited me to write a foreword to his book. You are reading that foreword right now.

Richard is not a clinician but a visionary and poet. His territories are philosophy and life force. He has some fluency in the sciences, notably biology and physics, from getting a classical education, culminating with a doctorate in anthropology. Most of his knowledge comes from reading, contemplation, imagination, and intuition rather than hands-on laboratory work, although he has taken courses in and practiced CranioSacral Therapy and that clearly informs his viewpoint about the intelligence of life.

In 2002 Richard published an essay called "A Phenomenology of Panic" in an anthology called *Panic: Origins, Insight, and Treatment* edited by Leonard J. Schmidt and Brooke Warner. Its storyline gives me an insight into the orientation of this book. Though Richard's descriptions and ideas are grand in scale and far transcend any reduction of them to psychoanalysis, his family history and childhood provide a window into his act of personal alchemy. Because Medusa gave him a stare early in life, he had to learn to fend her off. He had to live in a universe that had been damaged, in which all hell had broken loose. He had to make it livable. I asked Richard if it was okay for me to offer a brief description of some of the life experiences that have taken him to his present level of understanding. He said I should be my own guide. So here it is:

Richard was born of very pessimistic and oppressive parents who were constantly forecasting gloom and doom. They kept him and his siblings very much up-to-date regarding imagined current dangers as well as those in both the near and distant future. This negative and scary briefing began as early as Richard can recall. It probably began during the first post-partum visit that he had with his mother. Because of these alarming communications, Richard grew up with a high level of anxiety and fear and an inner voice that spoke from his parents' world. As you might imagine, there was a good deal of mental dysfunction in the family, ranging from siblings and cousins in mental hospitals to a mother who committed suicide when Richard was thirty. Rather than allowing anxiety/fear or madness to rule his life, Richard (or some part of him) proceeded to have full-blown, often creative panic attacks. These attacks unconsciously allowed him to blow off some of the anxiety and fear as it was rising to a dangerous level. After each panic ended, the world felt safe again. However, as trepidation began to build again, it interfered more and more with his daily life and created a sense of hopelessness and doom. A threshold would be reached and another panic would bring redemption and relief. This cycle has gone on since Richard was a preschool child and continues at some level into the present. Another way of describing this is to say that when the universe itself is mad, one must find extraordinary means to survive in it. Once Pandora's box is open, you can't close it, and you can't reason with it, so you have to learn to dance with what comes out.

The dogma of science and/or religion is awash in doom and gloom, much of it also mad. Science insists that everything about life is explicable in terms of subatomic particles, atoms, molecules, etc. The belief is that these materials are all governed solely by the laws of thermodynamics. This conviction prevents those who are scientifically oriented from accepting that life is also the energy that is provided by the universe, that a life force/energy drives the material things that science studies. This universal force organizes all things that science thinks of as alive or dead. Once you have seen and felt it, you will never again accept anything as dead. Rocks, fire, wind, ashes, and water are alive.

Several years ago when I was a professor of Biomechanics at Michigan State University, my department chairman, Bob Little, Doctor of Mechanical Engineering, surprised me with the statement that "all things are flowing; even big slabs of granite are flowing. Granite's atoms and molecules are continually flowing. They are just flowing much more slowly than what we call animate things. Extreme slow motion is harder to see." That concept of Dr. Little stuck with me and over time changed my fundamental beliefs about what is alive. In my present view, if it is moving, it is alive. So it would seem that all things are alive, as their atoms, subatomic particles, and molecules are in constant motion. In fact, these so-called inanimate pages may be exchanging bits of material and intelligence with the universe now as you read this.

Richard offers long discussions regarding the concepts of Creationism, Darwinism, neo-Darwinism, and Intelligent Design Theory, but he does not buy into any of them. He describes a constant exchange of ions, molecules, and cells driven by energies between all animate and inanimate objects and creatures. A cell from my liver might become a part of the beloved Norfolk pine tree in our front yard, and a cell from my heart may be moved to my Assistant Alena's new puppy. As our parts blend in this way, all things become one.

From very different starting points Richard and I have arrived at very similar ideas regarding the "conditions" of being objectively alive or dead. It is always nice to have confirmation of your weirdest thoughts come from a totally independent source. Richard and I offer confirmations to each other.

Our civilization is in the midst of a "ceaseless" transformation of matter into "dead" commodity totems that are manipulated in our creature minds for the purposes of indoctrination and disenfranchisement of citizens. We are replacing our very humanity and healing compassion with a kind of mechanical greed fostered by commoditized ambitions. Under such a tyranny, it is fear that keeps individuals and groups clinging to a mechanical or fundamentalist world-view. After all, there is a materialistic aspect to fear and guilt. If you mislead yourself enough about what things are, your fear can be momentarily dismissed and your guilt superficially assuaged. You trade insight into the true nature of life and death for temporary control of imaginary material objects. And even these objects will ultimately overwhelm our feeble attempts to manipulate them, for they are alive in their own terms.

We are left with a sorry choice: would you rather have your brain imprisoned by high-tech science or by old-time religion? Each of them creates God in man's own image; each in its self-righteous arrogance may destroy Life on the Earth.

While science searches for answers in smaller and smaller places, much of organized religion pays little heed to the natural world. John Lilly, back in the 1950s, noticed that "man created God in his own image." This statement has remained with me and has since been confirmed by the arrogance of many organized religions. The oft-cited irony that we have killed in the name of God should be taken seriously as an indictment of our basic morality rather than sadly acknowledged and rationalized. What kind of God would require such behavior? We have also killed for diamonds, for oil, for land, and other commodities and idols.

Science's study of genomics has a strong materialistic bias that shows right through its mirage of humanitarianism. Technology meanwhile has provided us with high-tech televisions, DVDs, VCRs, etc. All of these things imprison original thought areas of our brains.

This book asks how we can free ourselves from the blight of materialism. It shows that, even though firmly materialistic goals are attached to almost everything we humans do, the wonder and miracle of our lives are still available to us. They cannot be explained, and they cannot be reduced to either scientific or religious facts. What makes us alive has its own reality based in the universe itself. And this has nothing to do with any doctrine or belief system.

As I read and digested Richard's book, my horizons widened immensely. I may not agree with everything Richard says, but I have no real grounds for disapproval. Richard has made me think deeply and accept the scope and wonder of the universe in which we have come to dwell.

I was in San Francisco not long after writing this foreword, and Richard and I shot the bull for a couple of hours during my seminar break. He was probably thinking about both his book and my work when he asked me whether I thought people and civilizations on other worlds were doing a better job of things than us here, or whether we were at the forefront of consciousness in the universe and doing it about as well as anyone.

I told him that I thought people were here for a reason and that other worlds had other reasons for people being there. On this world we learn about violence; we learn that it doesn't work by getting to experience that fact directly again and again. We learn how to give and receive love. We learn about sabotaging ourselves and getting in our own way. We learn how everything we really need is inside us; yet we can easily deceive ourselves into not recognizing it or being able to get at it. If we don't need to—or aren't ready to—learn these things, we go elsewhere.

· · · · ·

Thank you, Richard, for sharing many of your deepest thoughts and fears. I must read this book several more times. I'm sure that I will get something more to cogitate on with each reading. My very best to you, and I love your book.

John E. Upledger, D.O., O.M.M.
Palm Beach Gardens, Florida

Preface

· ·

Harold B. Dowse, Ph.D.

I am a scientist. By this I mean I am the real thing. It is what I do all day when I am at work, and that spills over into how I lead my life. Under certain conditions, it would be how I define myself. You can type in my name on one of the scientific literature search engines for biology and medical science and the references will pop up. You can read my papers if you like. I am a cell biologist, and most of the time I work with the ion channels that pass calcium and potassium in the pacemaking tissue of the heart, but I also work on that still largely mysterious entity called the biological clock that times all cellular physiology. The common thread is that both systems oscillate, and oscillators fascinate me. For the most honest among us, science has to be interesting and absorbing first or we could never work that hard on it. We would take early retirement or retreat to full teaching appointments.

Being a scientist makes me a natural skeptic. I tend to question all claims, extreme or not, that fall within the domain of scientific inquiry. You have to do that to be a good scientist because everything demands proof, but with sufficient supporting evidence in hand, most reasonable scientists are forthcoming with approval. Of course, the farther a claim is from commonly accepted paradigms, the more proof is required. You don't initiate an upheaval on hearsay or just a few experiments originating with one lab. There is a redundancy in science that is incorporated to prevent debacles like cold fusion from propagating past a few weeks of sensational press releases and wild talk of free non-polluting energy. But the very process of science makes it seem unexciting, even stultifying to the outsider. It often appears to be designed from the ground up to crush new ideas as they are born, perpetuating a gray and featureless status quo. But this is an unfortunate illusion, not within the heart and soul of the process. If mental telepathy could be proved within the rules of dispassionate, methodical investigation, then after a transition period it would become part of physiological psychology right next to the actions of neurotransmitters, mechanisms of short-term memory, and functions of synapses. But what are the rules of dispassionate, methodical investigation? Lord Kelvin gives us an early clue:

> I often say that when you can measure what you are speaking about and express it in numbers you know something about it; but when you cannot measure it, when you cannot express it in numbers, your knowledge is of a meagre and unsatisfactory kind.

The indisputable proof of mental telepathy would be a major find, probably initiating a true paradigm shift in the field, but it would be accepted. I would welcome it, as would most other scientists, but we demand that the rules be adhered to. Telepathy would be part of Psychology *per se* from that point on. "Parapsychology" would remain the realm of ideas yet to be tested adequately.

This book delineates major paradigm shifts like the acceptance of continental drift and the impact theory of extinctions. It took major work to initiate those shifts, and at first the proposals were received with a lot of skepticism. But their ultimate acceptance shows that science is a powerful enough tool to lift knowledge beyond the hidebound defensiveness of some of its own practitioners.

This book is in search of another paradigm shift and points out where that shift should occur. I have been asked by Richard to comment on this book and have done so twice—once in his kitchen on Mount Desert Island after reading its first version, and again now.

It has undergone a sea change since then. I am not going to take on the mantle of skeptic here, nor shall I argue points brought up in the book. That would be unseemly, as it is neither my book nor my thesis to argue for or against. I may bring up an example or two, but in the service of my basic task. What I wish to do is to set down in a very few words my thoughts on what I think the central issue addressed boils down to, presenting a "scientist's viewpoint." That this viewpoint is orthogonal is a good thing. It makes whatever is at the intersection more intensely illuminated and provides another dimension into which to expand.

Given my scientific background, I tend to pare things down to their most basic form to make them accessible to discussion. I see in the heart of this work a reformulation of a very old discussion. The lines of argument form in a new area, molecular biology, but the opposing forces are very ancient indeed. And it is good to move that discussion into this new space, as any topic becomes more tractable the more quantifiable it becomes. On one hand, we see genetic determinism, the mindset that declares that given 100% knowledge of the genes that comprise an organism's information store, we then know all we need to know about that organism. Across the ancient no man's land is arrayed what must be considered as a new animism, the idea that there is a principle in nature about which the molecules of life dance, and that here is where real evolution and development occur. One no longer must argue teleologically to accept this idea. Modern thermodynamics has presented us with new insights about features of self-organizing structures that are almost certain to occur whenever energy flows through a system. They are fractal eddies in this titanic stream that form, replicate, and become more complex. The book gives several nice annotated examples. So, is life genetic or epigenetic? I can't answer that. I can tell you how I formulate the problem, though.

A tornado is a self-organizing structure that persists for a short while. It is an area of unbelievably low entropy that forms spontaneously between two weather systems. Ludwig Boltzmann tells us that this is vanishingly improbable. The temperature extremes should go to the most probable state: cold plus hot make warm. But what actually happens is that a terrifying "structure" forms in which position and velocity of air and water molecules are tightly constrained. The thermodynamic budget must balance, so energy is dissipated, lost from the system, to keep the entropy change positive overall. What is the principle that drives this? So far, even though we can describe the process and quantify it, we don't know the ultimate "why." A cell is much neater than a tornado in several respects, but the critical one here is that cells carry all the information they need to replicate and continue their lives. But does the information in the chromosomes run the show or is it just a convenient storage area called upon by the cell when needed to get the sequence of amino acids in a protein right, or to come up with nucleotide structures in housekeeping molecules?

Richard chooses well to look at embryogenesis for a field where this question is hotly debated. A genome without a cell is just so much chemically stored information with nowhere to go. But, a cell without a genome is doomed. It can exist for a short while as the chromosomes are packaged up and divided equally during the cell cycle, but the information MUST be accessed soon afterward, or there will be no protein synthesis. When the genes are active, a cell uses them by differential activation processes. This is most evident when a fertilized ovum begins to divide and differentiation starts. Genes are turned on and off like the keys of Bach's organ in a titanic convoluted fugue, information feeding forward and backward, gene products turning other genes on or off and themselves subject to the same controls. One might think of it as a holographic projection of the genome into cell space and time.

Neither the genome nor the dissipative structure that is the cell is dispensable; they must both be in the equations. If you could come up with a generic "consensus" mammalian egg cell, the genome you inserted would determine what you got totally. If you put in a mouse genome, you would get a mouse at the end; if it were a cat, out would come a cat. So in that sense, the genome is the source of "mousehood" vs. "cathood," and all the information is there in the AGCT [amino acid] codons. Here, ultimately, is where changes must occur for evolution to progress, and here is where Darwin made his amazing contribution. If a genome varies, then the cell/genome partnership must change. Perhaps we don't know everything about how that partnership works, but it does work. The changed partnership will interact with the larger higher-dimensioned space that is the environment, and further change will be dictated by the best outcomes. Darwin's theory is really nothing more than this.

But Darwin has been pilloried by groups wielding a frightening variety of opinions. The question is why? I really want an answer to that. Perhaps because Darwin hits too close to home. By the 1850s, it was possible for only the most single-minded zealots to think the Earth's days were numbered by the Bishop of Ussher. Such zealots abide with us in our own time. But by Darwin's time, the sun was clearly seen to be the center of the solar system and was just one star among so many. "Deep Time," which is one of the scarier phrases in science, was demoralizing enough. Hutton blew away our comfort zone and gave us billions of years beneath our feet, "time out of mind." Loren Eiseley adds nothing to how good we feel about that:

> Only James Hutton brooding over a little Scottish brook that carried sediment down to the sea felt the weight of the solid continent slide uneasily beneath his feet and cities and empires flow away as insubstantially as a summer cloud.

And Darwin told us that our ancestors were not human, that we arose through a range of species owing to chance variation and laws of selection acting on offspring. This was (and is, apparently) just too much to take easily. Henri Poincaré puts his finger on the reaction to this cold light of scientific dawn:

> We also know how cruel the truth often is, and wonder whether delusion is not more consoling.

So Darwin takes his lumps; generation after generation he remains the lightning rod for those on the animist side of the battle lines. And general anger at science continues to grow today. Undimmed, however, his theory continues successfully to be the focal point around which work throughout Biology crystallizes. Poor old Charles is not to blame for the excesses of the technocrats. You can't blame him for genetically engineered food. And why the anger against the Celera corporation for sequencing genomes? They succeeded in what they set out to do, a titanic task, and they did it on their own nickel. We know a LOT more about the genetics of a number of critical organisms, including humans, thanks to the efforts of the sequencers, and that knowledge will grow exponentially as new sequencing techniques come on line. You may soon be able to have your very own genome sequenced in a short afternoon.

So, when is knowledge bad? Whenever we learn something new it may work to our advantage in areas we never expected. It is difficult to explain antiscientific bias from an objective look at what is going on today. There are, of course, excesses, but largely the anger ought to be directed at the commercial interests exploiting science. Look rather for the fork in the road in ancient times. We hear the tale of St. Amand, who in the seventh century commanded a blind woman to take an axe to an oak in a sacred grove. She was a pagan, a believer in the animistic spirit of the forest and land. When she took that axe in her hand and felled the sacred oak, she was converted to Christianity and was made to see. With the end of Roman rule, paganism yielded to Christianity in Europe and a bond of humanity with the Earth was severed, occasionally with force. Invention of heavy plows to extend agricultural lands and watermills to grind grain accompanied the destruction of the sacred groves. In this part of the world an aggressive assault on nature to yield wealth and sustenance began. As Lynn White tells us, the blood and spiritual descendants of the peasants of Northern Europe have spread their way of relating to nature across the planet.

But I have become increasingly puzzled that there is a dichotomy at all. If you formulate the problem at a level yet above genetic determinism vs. life force, an interesting thing happens. Douglas Hofstadter tells us that when you look at Escher's remarkable works of art, you must understand one thing about them above all, that they all meet behind the frames. Einstein tells us the same thing about the laws of nature. And, curiously, we seem not to have forgotten this, but the memory lives on a very deep level. In Western society, the most ancient symbol of unity with our planet is the Green Man. His origin lies in the deepest antiquity. He is pictured as a figure covered with leaves, disgorging vegetation. He is botanical nature incarnate, perhaps a dweller within the trees cut by St. Amand. But he did not die under Amand's axes. He persists to this day in art and legend, though he has changed with time as mankind's need for and understanding of him has evolved. The really relevant thing to me is that he appears heavily in Christian art, prominent in cathedral carvings, paintings, indeed every aspect of portrayals of Holy scenes. Why? Was it just superstition on the part of the carvers? Maybe they were slipping these in as relics of an ancient religion. More likely, the Church was consciously taking this image and trying to show that the new religion was now dominant over an ancient pagan power.

But to me these images are too prevalent and too powerful to admit of domination. By Gothic times they were appearing everywhere in cathedrals. I think rather they are an attempt at symbolizing that there really is no "battle." Most telling is an image of St. Michael conquering Satan. On the knee of the leg stomping the devil (who provocatively has a leafy face like that of the Green Man himself) is an image of a Green Man: Christianity and animism in league against darkness and anti-life. So maybe it is not either/or; the genome is not different or separate from the cell. Why is DNA, arrayed carefully in genes and charged with carrying information, qualitatively different from insulin or glutamate or any other inanimate molecule caught up in this fascinating nonlinear, self-organizing, complex dance?

To paraphrase Tolkien's creation Treebeard the Ent, our most recent avatar of the Green Man: "Amand was a Saint, Saints should know better."

Harold B. Dowse
Researcher in Cell Biology
University of Maine, Orono

References:

Anderson, W., and C. Hicks. 1990. *Green Man: The Archetype of our Oneness with the Earth.* HarperCollins.

Eiseley, L. 1958. *Darwin's Century.* Doubleday.

Hofstader, Douglas. 1979. *Goedel, Escher, and Bach: An Eternal Golden Braid.* Basic Books.

Kelvin. 1883. *Popular Lectures and Addresses: Lecture on Electrical Units and Measurement.*

White, Jr., Lynn. 1962. *Medieval Technology and Social Change.* Oxford.

An exchange between a salesman played by Kevin Spacey and his buddy (Danny DeVito) in the movie *The Big Kahuna* serves as a suitable epigraph:

"I believe what I believe."

"Which is?"

"How the hell should I know?"

Introduction/Acknowledgments

....

In general, the theme of this book is: how did all of this get here, and what is it? I address these questions using ten major themes or issues, all of them concerning embryology and its handmaiden, morphogenesis.

1. The decoding of the human genome was disappointing to many because it did not provide a plan for industrial-like repair or future construction of people (as if these were desirable things). The disclosure of the genome did not even provide a comprehensive *Homo sapiens* trait map—or anything close.

But why should that be a surprise? DNA by itself could not assemble a worm or leaf, let alone a crocodile or punk rocker. It could probably not even manufacture promising slime. Genes are not traits. They neither contain blueprints for organisms nor assemble tissues or organs. By themselves they are not even biological information. They code the production of RNAs, which transmit semi-determinate series of amino acids, which program semi-determinate topologies of proteins, which combine in tissues while sending feedback in all directions in a nondeterminative manner. Every gene has multiple, probably limitless cellular and organic expressions. Most if not all of them occur in more than one phylum, and many occur in more than one kingdom.

A caterpillar turns into a butterfly without any alteration of its genes, so its DNA is clearly multipotential. It can break down an earth-bound grub and, from the same database, construct a flying worm. Now we know (as well) that more or less analogous genomes manufacture mice, pigs, and hominids.

Essential at all levels of development, genes paradoxically exist as such at none. In almost all certainty life arose before they did and does not owe its existence or its basic form to them.

DNA is an epiphenomenon of life rather than its source.

2. Strings of genes in coded strands are superfluous until a bio-dynamic event envelops and interacts with them and transposes their information into active structures. Somewhere in this process complexity is generated beyond the message itself. Embryos alone weave something out of the genes encapsulated in them.

The homage paid to the genetics of biology at the expense of its physics (let alone its phenomenology and metaphysics) is misplaced. Epigenesis, not genetics, makes life. The embryogenic, epigenetic process is more originary than the single-target genetic event which alone is deemed now the trigger and activating force for all traits, creature development, and ontology of life.

. . . .

Epigenesis: the theory that morphological complexity develops step by step during embryogenesis by structural elaboration from an unstructured egg as opposed to by simple enhancement of a preformed germ or linear playback of a genetic message. In the eighteenth and nineteenth centuries, epigenetic theorists presumed that complexity was imposed from outside the egg by a vital force or organized intelligence; while the egg contained developmental potential, it did not have the physiological determinants for biological form. In the late twentieth and twenty-first centuries, epigenesis has come to mean multi-tiered, extragenetic levels of control, each one defined by networks of self-organizing proteins and the strategies used by those networks.

. . . .

Genes are our favorite scapegoats because they are conveniently already here and (like us) run esoteric design codes to build complex systems. Yet embryos (distributed nonlinear fields), not genes, are the originless, inexplicable objects that should get the attention and awe of biologists. Embryos are not merely genes' playbacks. Like galaxies they are inherently originating things.

3. *The embryo is the universe writing itself on its own body.* This mantra guides my book. I am convinced that life could not exist unless it was intrinsic in matter, unless it preceded even subatomic particles. It surely could not have come about through matter accidentally arranging in clumps and clusters and arriving haphazardly at organic design principles.

Even if life is implicit, none of its terms are perceptible as such in atoms and molecules. Raw molecules apparently have latent attributes and even intelligence that cannot be specified in their extant chemicomechanical properties.

4. The tissues on which we stake our reality, on which we depend for existing and making it through another day with plans and hopes for yet another—even another second—are mirages, totally tenuous from an atomic or molecular point of view. We cohere and function *only* as emergent principles that arise when atoms, molecules, and cells interact.

We are like the cartoon rabbit who, having sawed off the limb on which he is perched, still imagines he is supported in midair. With each discovery in microbiology and physics, we are standing on less and less. Yet, for reasons that are not entirely obvious, we don't fall. We still have bodies; we proceed through the world, though no physicist, chemist, or biologist can explain why we don't begin unravelling at once, molecule by molecule.

It is not enough to say that we transcend our own chemistry and physics. That alone does not substantiate us.

5. There are two states of matter in the universe: thermodynamic and embryogenic. All matter everywhere is thermodynamic; only life is thermodynamic *and* embryogenic. But life is made of ordinary matter and nothing else—in fact, any molecule anywhere in the universe can become alive if it gets into a biological context.

All matter is potentially both thermodynamic and embryogenic;

some matter develops embryogenically. This metamorphosis occurs only where biological fields arise. The laws generating such fields are unknown but, because bionts are thermodynamic as well as autopoietic, they adhere to regimes circumscribed by heat, gravity, and entropy.

We do not know if embryogenesis represents a special state of thermodynamics or a different law altogether. We do not know when, where, and how it makes matter alive, either in the phylogeny of the first biological fields long ago or their present hourly ontogeneses.

• •

Ontogenesis: the embryological course of development of an individual organism. **Phylogenesis:** *the evolutionary development of a lineage.*

• •

6. The universe, which physics tells us is held together by gravity and thermodynamics and which consciousness tells us is held together by meaning, is also held together by other forces that are impalpable and inextricable. These include—or at least words for them include—sympathy, signature, vibration, and scale. Scalar waves seem to transcend gravity, electromagnetism, and space-time. Thus, phenomena can influence one another across widely divergent scales and dimensionally remote domains. An atom with its orbiting electrons is a planet with moons. A form of consciousness or innate organization flows between levels so that the universe, from end to end, is a single pulse of meta-intelligence encompassing the intelligences of all local systems.

7. Planets and moons are dispositions of ordinary molecules. Some are more innately embryogenic than others, but all have embryogenic potential because all matter has embryogenic potential.

Though we have no explicit contact yet with alien life forms, the basic rules of terrestrial development tell us how life will be made anywhere in the universe, how creatures—whether plantlike, animal-like, fungal, viral, bacterial, protist, or whatever else star-borne elements can churn up—will organize. Our criteria of morphogenesis can be applied to hypothetical life forms on worlds everywhere, even ones orbiting stars in remote galaxies, much as our tenets of thermodynamics can be applied to the stars themselves.

Embryogenesis is how molecules coalesce, form layers, and come alive, and how egoic agency is implanted in them. In this Solar System, Earth, Venus, and Mars, for most of the twentieth century, were considered the lone bodies capable of life. During the last quarter century, however, spectroscopic analysis and satellite exploration downgraded Venus to inanimate and, at the same time, elevated the moons of the Jovian planets to life-possible status. While Venus is ridiculously hot (from the standpoint of our kind of life), worlds like Europa, Callisto, and Ganymede (orbiting Jupiter), and Titan (orbiting Saturn) may harbor water-rich environments under frozen surfaces. Furnaces at their cores—maintained by the gravitational friction of nearby worlds—might melt enough ice to warm subglacial seas. Even Jupiter, which breeds organic molecules in abundance, has gained new respect as a possible biohabitat.

All of these considerations are prejudiced by the provinciality of life on Earth. Since we do not know the rules for establishment of embryogenic fields *per se,* we do not know whether they might apply differently to very hot, dry, cold, or even airless worlds.

8. The relationship between consciousness and matter, e.g. between mind and body—whether algorithmic or epiphenomenal—is established embryogenically according to deep precepts of cell-tissue design. Thoughts and artifacts on worlds throughout the universe are shaped inside bodies inside embryogenic layers. Though thought itself may be irreducible, its material expression in neurons is organized by gastrulation, neurulation, and organogenesis.

9. The only way to understand the transition from particles to atoms to molecules to cells to tissues to organisms to societies to cultures to universal ideas is as a sequence of emergent properties that are indivisible, unique, and spontaneous. They arise vicissitudinously when a phase of matter and energy crosses an invisible threshold. They cannot be hypothesized beforehand.

Each phase of matter and energy nonetheless foreshadows and signifies its successors. Molecules are the only sorts of mega-atoms that atoms make possible, and societies are the sole organizations that could emerge from individual creatures. Embryos ordain cultures and ideas by the way they weave tissues and organisms. This determination, while not absolute, operates approximately at the level of Sigmund Freud's "Anatomy is destiny." The shape of meaning arises from the shape of tissue.

The contradiction between innate and novel runs through nature: emergent properties are both inherent and unconditional, irrevocable and extemporaneous.

Another way to look at this is to imagine meaning spiralling from both ends of the cosmic scale—under atoms and beyond ideas—toward itself. Thus, embryos by their method of organization reinvent the meanings bound in molecules and cells and translate them into denser, higher-level fields—creatures.

10. Consciousness and matter are not opposites; they are meta-states of each other. Every expression of matter is also a form of consciousness—consciousness in its *purest* form, its form least penetrable and recognizable by actual minds.

• • • • •

I have often wondered why I—of all people—should be tilting at the core mysteries of life. My last direct experience of biology (in the late 1950s) was my freshman year of high school, unless I choose to count the watered-down genetics, archaeology, and ecology of graduate-school anthropology. Legions of people are far better qualified to write on this subject than I am. The following rationales come to mind:

1. I have always been interested in origins, from my teenage reading of Sigmund Freud on dream work and the science fiction of Robert Heinlein, Theodore Sturgeon, and Robert Sheckley ("Where in chaos was that hole in time...?"[1]), through my college explorations of traditional magic, Jungian psychology, and American Indian mythology ("The Sun would lay off the gray skin and put on the yellow fox skin, whereupon the bright dawn...."[2]).

2. My early literary elders—Charles Olson, Robert Duncan,

Robert Kelly, Edward Dorn et al.—addressed the nature of being mythopoetically throughout their work: "the actual universe immediate in the body of man."[3] They asked big questions. Olson evoked a pre-Indo-European logos outside linear space-time: "far far out into Eternity ... /the law of possibility"[4]; "the inert/ ... as gleaming as,/and as fat as,/fish"[5]; "the imposing/of all those antecedent predecessions, the precessions of me...."[6] Duncan sought "a map of walls and towers,/painted tents and geometries of distance,"[7] "the spirit of desire ... in all lively things ... /The Golden Ones ... in invisible realms."[8] Kelly drew back theosophical veils to reveal the alchemist in his "robe covered with suns, moons,/motions we call 'planets' and do not know/the green life in their valleys."[9] Dorn pronounced, "We are the end of the leeching which produced us. A spoonful left at the bottom, very refined, pure stuff, the final powder, the dust that lives."[10]

These poets steered my attention to the borders of thought, "the poetic intention itself,"[11] not only as proposition but in the concreteness of each sentence spoken.

3. Graduate anthropology instilled a cross-cultural bias in me—a disposition to view all systems of knowledge as relative, provincial, and contingent.

4. My oldest friend, Charles Stein, with whom my relationship began the afore-mentioned freshman year of high school, has been a student of (in succession) tarot and yoga, alchemy and qabala, Zen and Dzogchen, algebra and complexity theory. As he has encountered the riddles of human inquiry, he has invited me on each segment of his journey. Thus, I have been kept abreast of deconstruction, ontology of mathematics, and non-Western phenomenology without having to work through them myself.

Stein is primarily a poet (sharing the same elders). In the mid-1960s when we were in our early twenties he wrote: "The world (or a world)/is complete/[B]y that I mean/there is no other/world/from this vantage..../I mean/there really is no other world/than that at hand/or way of approach out/of the present/more than/a deepening."[12]

5. In 1978 bioenergetic therapist Stanley Keleman showed me movies of frogs' eggs turning into tadpoles. As he rose from his divan and shook two fingers at the impossible sequence of images unfolding before us, he seemed about to attack the screen. Then he shouted: "Now what in hell is that?" He introduced me to the embryo as the formative principle of human identity and destiny and the foundation of emotional anatomy. I have been thinking about it in that way ever since.

• • • • •

Anyway, my goal is not to write biology but to explore what it means to be alive, to have an existence, to experience visions. I use biology simply as a language and world-view to pry open the box of modernity, as a metaphor for its literalization of life, and as a way to address the consciousness in nature and its pagan gods.

• • • • •

I would like to thank cell biologist and anatomist Stuart A. Newman of the New York Medical College in Valhalla, New York, for his general support, help, ideas, and, in particular, his critical reading of my chapters on genetic determinism and Erich Blechschmidt's embryological theory (which, albeit in the latter case, consisted of figuratively throwing up his arms in frustration).

Richard Strohman, Professor Emeritus in Molecular and Cell Biology at the University of California in Berkeley, carried on a running dialogue with me (mostly over sake and soba at O Chame) *vis-à-vis* the roots, meaning, limits, and consequences of genetic determinism. He made it a daily wonder that we could digest and metabolize our lunch, while he railed against the research establishment for its pig-headed ignorance and arrogant tomfoolery.

Harvey Bialy has been not only a research molecular biologist (founding scientific editor of *Nature Biotechnology*) but a poet with sophisticated aesthetics, a serious student of the traditional occult and African folk medicine, a manager of jazz groups, a champion billiards player, and an actor (with Mick Jagger) in an experimental film by the legendary Kenneth Anger. Presently he is writing a biography of his colleague, cancer and AIDS radical Peter Duesberg. He is one of the few people who would understand the multiple intellectual influences on this book. A patient and modest man however, he is not.

Late in the game (when this book was already in editing) I approached him about it with some trepidation because he doesn't like encroachment on his turf and, though his turf is the universe, cell biology is *really* his turf. After some initial grumpiness, perhaps because I didn't enlist him sooner and also because I was a rank amateur juggling the crown jewels, he guided me through distinctions between gene mutation and aneuploidy and deepened my sense of the role of metabolism in differentiation. I am indebted for both his solicitude and gauntlet.

Though Dusty Dowse was in my class at Amherst, I didn't know him as an undergraduate. In 1991, toward the end of the banquet during our twenty-fifth reunion, a mike was left open for toasts from the floor. After an hour (or maybe an hour and a half) of miscellaneous earnest and clownish pronouncements and cheers and also a good two minutes of silence after everyone thought the speeches were spent and the tables of diners had returned to a decentralized buzz, a large, full-bearded leprechaun strode to the podium and startled the group by announcing: "I was on the campus the other day and I saw this guy at his 71st reunion." He cast a questioning gaze over the room that was suddenly still. 'Last living member of his class." He paused again. "I wondered, was he afraid?" He paused for what seemed like an eternity; then he thrust his glass dramatically upward: "I propose a toast to the last living member of our class. Don't be scared, buddy; we're all with you!"

I have stayed friends with that guy ever since. A molecular biologist and anatomist at the University of Maine in Orono, he not only read this book and critiqued it from his vantage as a practicing scientist but engaged me in long discussions in and around theoretical physics and biology during the summer of 2002. We continued by email and snail mail through the fall. (When Bialy later realized I had been collaborating with, as he put it, "the one big happy family" of homeoboxers and *ingénieurs* of genetic switches, he warned, "As Ed the one and only Dorn once told me, 'Watch out for guys named Dusty.'")

I would like to express my gratitude to Harold Dowse for his skeptical approach to my text and his generous attempt to bridge the gap between our approaches to science. We both honor the mystery, and we both think it is possible to overcome any propaganda with spirit. It is, in fact, the only way to get out of here alive.

• • • • •

This book takes its form by addressing our lot as simultaneously an evolutionary riddle, a koan daring mathematical determinism, and a sacred song.

We need to account for the inexorably thermodynamic terrain of space-time, and we need to account for ourselves. This is the central dilemma for the men and women of an enlightened scientific time. We need to stay honest with who we are, and we also need to get real.

In honor of Dusty, I propose another toast, this one from his Maine Stein Song, which I first joined in a high-school auditorium the year I met Chuck Stein. It always seemed to me the only "alma mater" that meant anything. And that was long before I set foot in Maine or, for that matter, James Hall with Dusty. It remains a perfect adolescent ballad:

> *To the trees, the sky,/ To the Spring in its glorious happiness;/ To the youth, to the fire,/ To the life that is moving and calling us!/ To the Gods, to the Fates,/ To the rulers of men and their destinies;/ To the lips, to the eyes,/ To the girls who will love us some day.*

Poignancy and wonder bathe the restless romanticism of youth and remain at some level our whole lives.

We arise in a manifestation that does not require jungles and stars or ocean crashing upon the rocks of one of its shores for us to recognize its immensity and implacable texture, to crave comprehension and experience of its depths, yet to flee their demonic profundity. The forests and sea merely add to a sense of mysteriousness and awe.

Our existence is ultimately enigmatic enough that it can ride the molecular stream without being diminished or trivialized by it. Natural selection can be true, and we can also be true. In fact, the profundity of our situation is only enhanced by its intersection with its apparent algorithm.

John Keats observed that brandishers of "modern" poetic metaphors encounter a dilemma of which their predecessors were blithely unaware: Newton stole the rainbow from fairies and locked it in a prism; Galileo scrubbed a stark, arid landscape onto the radiant moon.

Keats ransomed those deeds in his poem.

We must likewise convert molecules and embryos into numinous, metaphysical objects. Otherwise, their tyrannies and hegemonies will corrode all other meanings and leave us without solace or defense before monsters from beyond our skies and assorted barbarians at our gates.

We must be present with our lives, these things that we somehow have and thus hold dear, in fact dearest, and that scientists tell us don't exist, not really. We have to salvage it all: love, the taste of mango, the blueness of sky, sparkling crystals of winter, salmon swimming upstream, the hop in a reggae song, the wonder of intuition inside us, because we live in a civilization that would give it all away, that everyday trades our birthright for a bit more instrumentation and capital toward the sterile molecularization and commoditization of everything that is or could ever be. This relentless, nihilistic march now threatens to give back everything that nature gave us once by some unknown magic. And to what end? To serve what better good?

I am trying to imagine what ways there are to conceive of life without yielding to metaphysical and vitalistic considerations, yet without discarding them as wishful idealisms either. I am looking for a path out of genetic determinism that does not foolishly dismiss genes as bogus. So, once again:

To the life that is moving and calling us!

In fact, we have no other choice.

Richard Grossinger,
Kensington, California, and Manset, Maine,
May 2000—June 2003

1

What Is Life?

. . . .

Evolution, Thermodynamics, and Complexity

> The conception of "heat" arises from that particular
> sensation of warmth or coldness which is immediately
> experienced on touching a body.
> —Max Planck, *Treatise on Thermodynamics*, 1921

How are we made and where do we come from? . . . . . .

At the dawn of the third millennium of Western civilization, who
we are is as great a mystery as it was two thousand years ago, or (for
that matter) two million years ago. It will likely be the same mystery at the dawn of the tenth or eleven-thousandth millennium,
should this inquiry last that long.

Biology has not solved the riddle of life or explained "us."

What is a raccoon, softly levitating his head and forelimbs, sniffing at a garbage can? What is a magnolia, spiralling with branches,
blossoms, and fruit? What is a sea anemone, zoanthid tentacles sniffing salty currents of the deep? What is an eagle, feathered arms outstretched seven feet or more as it accelerates up a thermal, talons bearing
a tortoise? What is a horse? a shark? a snake? a skunk cabbage?

Not, what are their anatomies and roles in the plant and animal kingdoms; what are they absolutely? How do they come into being? What motivates the clustering of molecules that make up their bodies, their habits, their semblances and the continuities of those semblances?

What impels each gondola of membranes to form? What gives it dynamic cohesion and a will? What makes a worm wriggle, a cat pounce?

What do they "think" they are as they carry out the unconscious, monotonous deeds of their templates with such eagerness?

· · · · ·

Life is so much bolder, yet more exquisite, than the massive landscape from which it matriculated. Its self-blossoming puppets and polywogs glide, squawk, do battle; enjoy breezes, rain-bursts, arrases of night and sun, which (by the way) could not "exist" without them.

Face it—our knowledge is cut off from our experience. Hopelessly so. Life revels in its own certainty, frothing out viral, plant, and animal entities, long before its custodians reduce existence to mere heat effects and chemical reactions.

We are (according to our own quite exhaustive investigations) delusion pods, heaps of crystals laced with electrochemical threads, automatons that "think" they think. Our robust, acrobatic fleshiness, our scintillating waves of thought are molecular associations congealed by kinetic attractions of particles—in short, big sludge, clever mud, atomic congeries turned into shuddering flocks pursued by carnivores.

To explain life molecularly as a random product of dispersal and differentiation following the Big Bang reduces animal purpose to mechanical principles, thus validating the omneity of matter and (the "hidden agenda") ceding human destiny to the dialectical synthesis posed by (on the one hand) a violent, exploding universe and (on the other) a rescue squad of progressive technocrats and capitalists. Witness: "The processes which occur in living beings at the microscopic level of molecules are in no way different from those analyzed by physics and chemistry in inert systems."[1] Or: "The human brain is an exposed negative waiting to be slipped into developing fluid."[2] Memory is "a 'temporary constellation' of activity—a necessarily approximate excitation of neural circuits that bind a set of sensory images and semantic data into the momentary sensation of an imagined whole ... a set of hardwired neuronal connections among the pertinent regions of the brain, and a predisposition for the entire constellation to light up—chemically, electrically—when any one part of the circuit is stimulated."[3]

This can lead only to a verdict that reality is empty and mind a hallucination. We have no reason to suppose that we actually exist. What we imagine as our elation, our grief, our fear, our wonder are simulacra of these things, atomic-molecular effluvia spinning an existence trance.

. . . . .

The question of what life is and how sperm and egg confer sentient matrices lies at the core of any system of philosophy or ethics—in fact, any system of anything.

Since the post-Mediaeval Renaissance, the West has uncovered an awesome array of facts about matter and life. These collective dossiers (on electrons, subatomic particles, gravity, gases, membrane biosynthesis, metabolic pathways, DNA, genes, the relationship between biochemistry and physiology, ecosystems, and the like) are impressive flinders of a larger puzzle. Yet, no matter how many such scraps we encapture and certify, no matter how elegantly they (sort of) fit together, the pieces do not match the puzzle. They are provisional and circumstantial—rent with self-contradictions, rigged to support and congratulate one another in artificial environments or idealized contexts. They do not answer the big question.

How did this world happen in such a profound, complicated, and subtle way? How could an overhot universe, an exploding melee of rubbish, of initial and boundary conditions, walled by its own statistical mechanics and calculi, subject also to uncompromising degradation by entropy, go on to forge intricate life forms? How could such a universe individuate microbes and amoebas and amalgamate sentient life out of other life? How did it educate pallid molecules to think and know? How did it break the stasis of unminded substance and manipulate itself back? How could it not only stalk and growl but make epistemological distinctions? How could it develop empathy and passion? How could it compose its own poems and codices? How could it educe surreal facsimiles that it recognized as surreal? How could it store so much joy and recognition and sorrow in such infinitesimal space? How could it pick a lemon in words that are not a lemon? How could it get deep enough "inside" to ponder and track its own mystery?

This is not a biology book. . . . .

It is a book about us—what we mean, who we are, how it is we are "alive," what "alive" is.

Really. Not just the establishment's cover story.

In the modern *yuga,* addressing "us" runs into biology-physics everywhere. Physics tells us that we are culled of hydrogen from stars and use circumstantially concocted, eroding microchips to store knowledge and make merry. Biology tells us that we are the statistical outcome of acts of chance agglutination and predation, washed onto these fabled shores by a cruel, purposeless hierarchy.

Unless one addresses our vacuous, atomic core and violent bestial heritage, it is specious to talk about our hopes, fears, dreams, and the possible future of consciousness.

• • • • •

Despite their failure to root out anything other than molecular machinery underlying chemical reactions, most scientists, cheating at their own game, imagine themselves as real, or as real as it gets.

As long as we hang out here, defiant by our presence, thinking these thoughts, experiencing the lugubrious passage of time, materialistic orthodoxy won't cut it—in fact, has never cut it. Random molecular reactions, given even eternity to find the way, do *not* become agents.

At heart every scientist knows this. In 1946 Loren Eiseley said it about as well as it can be said:

> I do not think, if someone finally twists the key successfully in the tiniest and most humble house of life, that ... the dark forces which create lights in the deep sea and living batteries in the waters of tropical swamps, or the dread cycles of parasites, or the most noble workings of the human brain, will be much if at all revealed. Rather, I would say that if 'dead' matter has reared up this curious landscape of fiddling crickets, song sparrows, and wondering men, it must be plain even to the most devoted materialist that the matter of which he speaks contains amazing, if not dreadful powers....[4]

Philosophy, ethics, art, and quantum mechanics all seem beyond the intrinsic capacity of random assortments of atoms and molecules. These are not just complicated, exotic shapes; they are something else entirely—modes of knowledge and being, crawling things that have no basis in nature, in even the loftiest algorithms of inert matter. They do something that nothing could do, something that is probably their destiny; they invert the cosmos into cosmology. Yet, as forms that reflect on nature and witness themselves reflecting, what are they if not molecules somehow bewitched?

Raw matter cannot be the source of "being." Or matter is something else entirely, is essentially and substantially "being" and requires only the roilings of time and space to get there.

The Big Bang

We emerge in a cavern of glittering wave-lengths of primordial luminescence. Through its deceptively sheer vacuum, its tiara of Stone Age constellations, we look along synapses of our own photon-excited neurons into even finer scud of Milky Way center. Where we are, inside the inside, appears to be but one of billions of such crab-shaped and spiral hydrogen fields, the dissipating debris of the splatter, splendid yet in its pinwheel pirouette, "stilled in a shimmy/of its own distance/ ... with the delicacy of water tension/to avoid dispersal."[5] We float here, in wonder, in terror, at the crest of a wave breaking so slowly its atoms barely move, yet so rapidly we cannot feel its motion. ("Pretty trippy," one twenty-something reader told me. "And there's nothing we can do about it. Just enjoy the ride.")

According to scientific canon, *verso uno,* a subatomic explosion sent a heterogeneous cloud hurtling from nothing/nowhere across eternity, creating space-time.

That existence came into being through a circumstantial rip in the void trumps any alternative explanation. Its primacy is now as compelling and incontrovertible to its priests as the Virgin Birth or Holy Trinity are to the cardinals of Rome.

Reality is a smash of pigments against a void blacker than night. A rude, coin-sized irruption, swiftly scattering under laws of entropy, gave rise to stars, planets, planetary landscapes, and (finally) bionts, all from a dilating pip. It made everything, and it also established absolute principles we identify as thermodynamics (decay of heat, impossibility of perpetual motion, etc.).

Don't ask what was here before it came along.

Don't wonder what lies outside it—or if it has an outside.

Don't try to guess what follows its terminus.

The Big Bang is king of space and time; it decrees a strictly mechanical, algorithmic universe—at least until the next Big Bang.

• • • • •

Nature detonates, then is aimlessly reconstructed following another megagalactic contraction to singularity and spasm back into some other version of space-time (or into something else entirely). There are no fixed landmarks, no shoals anywhere, even within atoms.

Anchored by its own laws, each successive universe is unique. Ones prior to ours may have had no gravity, no temperature, no molecules, no proximity, no distance, no locales, no ranches.

We don't even know how many universes there have been or will be, or if number is an applicable concept across cataclysms—for no trace remains in subsequent universes of any prior ones or their inhabitants (if there were any). All subatomic, elemental tints (the bare conformities of matter and energy) are sucked into the cauldron each time and flung back out into the void, making a new novel landscape, a new regime.

In this present peremptory creation with props scattered across a working stage, there is distance all right, a diabolic span of location, location, location. We are born on a reef of staggering absolute proportions. A journey to nearby gaseous Neptune, and back home, would consume more than a lifetime. Any night star with its wondrous science-fiction worlds, let alone the galactic center and other galaxies, is absurdly beyond our reach. A rotifer is as likely to construct a photon-propelled balloon and launch itself to Mars as we are to zip to Rigel or Aldebaran.

We are a speck in infinity which is itself a speck in a greater infinity, and so on across scales daunting thought itself. As bevies of soot in temporary gases and gels, breeding stochastic thought forms, we make maps of nowhere that look deceptively like some Chicago. In sacred times we may have been the masters and maids of honor at the holy wedding, the stars our covenants. Modernity reduces us to vagrants in slag—"numbers in a cosmic lottery with no paymaster."[6]

Evolution is not the problem. . . . .

Nineteenth-century biologist Charles Darwin fashioned the central paradigm of the twentieth century when he proposed that biological evolution occurs by natural selection comprising differential mortality and fertility. Darwinian life forms are happenstance if highly organized products of discursive movements of atoms and molecules over billions of years, combining in cell nexuses that engender and house a genetic device whereby they reproduce, redesign, and further elaborate themselves. That includes trumpet vines and lizards as well as us.

I am not from the camp that maintains evolution is merely a

working model that has not been proven. Its basic premises have in fact been corroborated countless times through several generations of observation, experiments, and genetic, anatomical, and fossil evidence. It is one of the ways nature works. Breeders of rye grass, canines, and equines have been confirming this for millennia. It has been demonstrated more recently by lineages of mice and flies in labs.

In the early 1950s, British medical doctor and amateur moth collector H. B. D. Kettlewell successfully tested the hypothesis that the proportions of light and dark peppered and other multichromatic moths surviving each life cycle vary with local pollution, as their hues "are differentially camouflaged against light and dark backgrounds ... [relative to] birds [that] eat moths."[7] That is, the sootier the industry, the higher the percentage of random mulattos and darker mutants in nearby moth populations. Fortunate individuals evade avian predation because, while roosting on leaves, they are more concealed against the grime. Living longer, they spawn more offspring—so their cell nuclei come to stock more "dark" genes.

With changes in technology and enactment of pollution regulations, the genome tilts the other way toward lighter moths. This in fact has happened in populations in the decades since Kettlewell's survey.

Likewise, though at a radically different temporal scale and anatomical depth, the development of cardiovascular systems from two- to four-chambered hearts among vertebrates provides an almost pedagogical blueprint of natural selection by more efficient mechanisms, stretching in unbroken design successions from fishes to mammals, and oceans to prairies.

As we shall see, it is not evolution *per se* that is suspect—evolution does its job quite well within appreciable ranges and regimes. The problem is morphogenesis, or—more precisely—various attempts to use evolutionary theory to explain the translation of raw information and matter into organized information and matter— living forms—which is not, despite widespread presumptions to the contrary, a clearcut outcome of natural selection. Life forms arise through series of self-organized morphological transformations of matter itself. Evolution may play a role in this event, but it does not warrant its onset or self-correcting precision. It is an explanation by default.

Morphogenesis is a mechanism with most parts patent pending.

Evolution is fundamentally thermodynamics applied to biology. . . . .

Once the prior century's concept of an invisible, indestructible, uncreatable fluid bearing the secret essence of heat was abandoned, mid-nineteenth-century physicists began to fathom mechanical energy as an uncannily profound relationship between the conservation of energy and heat. In a famous experiment using a horse-driven lathe to turn a rod inside a cylinder, a wooden jacket of water was warmed from room temperature to boiling in two and a half hours. As work was performed against friction (and mechanical energy spent), intrinsic heat was transferred to the water. This is how the universe conducts itself at large, from stars to atoms.

Laws of thermodynamics—energy and heat—set mandates for all substance and movement everywhere, regardless of size (subatomic to galactic) or location. These laws depend on the status of heat as a fundamental property of matter, a random flow of energy without exception.

Heat is simple, gross, and pedantic. It would seem to be a poor building block for anything with more initiative than a cyclone or solar flare. All heat's predilections are inimical to life, not perversely so, but because heat is wanton and literal. It doesn't tend to want to better itself or anything around it. Yet the life we savor is a form of heat and must abide and prosper on its terms.

• • • • •

No matter what philosophy or belief system we espouse *vis-à-vis* nature and God, we need to track thermodynamics in order to hinge life to the ingredients of which it is made, i.e., in order to segue operationally from inanimate to animate stuff. The quantum leap across that gap must be underwritten. While thermodynamic laws clearly tyrannize inanimate events, something changes between there and the domain of life, and that "something," though not necessarily nonthermodynamic, is different from ordinary thermodynamics. It is impossible to grasp either conceptually or physically.

In the equations that follow, life should not occur, but life also does not violate them in any actionable way. The result is a paradox. Being alive means being thermodynamic—being subject to the order of the universe—and also means something else.

• • • • •

The first law of thermodynamics states that, as energy is conserved, one form of it is always converted proportionately into another. Heat supplied to a system (including a metabolic one) must equal the change of internal energy within the system combined with the external work done (mechanical energy released) by the system.

The second law, governing energy flowing between objects of differing temperatures, states that, of its own accord, heat only travels to a colder region. As a state of random motion of particles, it must always degrade toward disorder and become (ultimately) unobtainable energy.

The measure of the internal state of a system at any point is its entropy (its relative order or disorder, or to what degree it is coming apart—something it will inevitably do anyway, no matter how robust at a given time). Entropy can thus only do two things: remain constant or (most often) increase. From the standpoint of beings who rely on beating Murphy's Law, this is like saying things can only go from bad to worse.

Although a machine, whether horse-drawn or petroleum-fired, may be doing useful work for now by the means at its disposal—changes in pressure, volume, and temperature—it always consumes more energy than it converts toward these goals. Living machines, i.e., developing embryos, are no exception. Usually some energy is squandered in friction or drag from the wearing out of parts (in locomotion or metabolism) but, even if it were not, the available mechanical energy to a machine cannot exceed that supplied to it from its heat source, i.e., from the directed random motion of particles into it. In fact, it must always be less, even if—in the case of the absolute best machine created somewhere in the universe—only fractionally so.

Heat cannot be converted to even merely an exactly equivalent amount of mechanical work. This is why a Golden Age, or even a surviving industry, is impossible. *Aprés moi, le deluge* is the motto of the universe.

Entropy is the ultimate arbiter of energy becoming inaccessible and systems degrading. By its essential nature it decrees that no nexus has limitless energy to draw on; every machine will run itself down. No matter how many water-slides turn wheels to roll "frictionless" balls to raise water back to the beginning of a slide, without external energy pumped into it, the system will stop. Nothing maintains its own motion.

So how do organisms develop from germ cells? Why do their parts not grind and wear out or bolt the regime well before they engage in higher-order functions such as digestion and thought?

The third law of the universe's thermodynamics adds the impossibility of cooling any substance, even a gas with a zero temperature, to absolute zero (−273.16 degrees C.). Once molecules slow upon the approach of absolute zero, there is no energy left to cool them further.

An additional so-called "zeroth" law states that, as hot and cold objects are brought into contact with each other, absorbing and emitting energy, even in unequal quantities, their temperatures equalize: two objects in thermal equilibrium with a third object must be in thermal equilibrium with each other.

These laws may be summarized as: 1) the conservation of energy within systems and the change in their internal energy solely as a result of the sum of work done on them and the heat they absorb in the process; 2) the natural increase in entropy of any system left alone, plus the change of entropy in that system in ratio of the heat absorbed by it to its temperature; and 3) independent of all other parameters, the inevitability of a constant value for entropy as the temperature of a system approaches absolute zero.

. . . . .

Laws of thermodynamics guided Darwin, both implicitly and explicitly, toward his theory of natural selection. He tried to maintain a credible balance of energy in his models of how living systems evolve, though of course he could not demonstrate it by real-time experiments the ways physicists could with vials of gas particles or pans of water. He assumed that a pigeon cannot do anything with heat that a furnace can't. Thus he limited pigeon etiology to conservation and dissipation of mechanical energy.

Darwin's successors came to argue compellingly that energy and matter alone could account for everything—initially everything inanimate, every breeze, lake, and rock formation, but ultimately (with fluctuant fertility and heterogeneous survival of molecule complexes) every life form too: every creature of every size and scale, every degree of intricacy and circuitousness of design and behavior. It could even account for the concatenation of life itself from barren chemicals.

No form in nature—no river or drumlin, no eagle or orchid, no plant, animal, or tissue network, however exotic, however delicate, however devious and unlikely its manifestation—requires a supernatural explanation or has a meaning or identity apart from its arbitrary, gradual occurrence and adhesion through trillions of separate material events. In fact, Darwin defied anyone to find an organ possessing degrees of complexity "which could not possibly have been formed by numerous, successive, slight modifications."[8]

Darwin's creatures evolved as the collective incidental characteristics conferred on them by random events gave them advantages over other creatures in their vicinity in a ceaseless contest for nature's energy (mostly in the form of prey). One heat machine consumed another, though both ultimately degraded into the same chemical constituents. More successful tissue complexes flourished temporarily and were perfected and amended as environments changed. By degrees of fertility and mortality transferring chance alterations (or not), horses, gazelles, zebras, llamas, cows, goats, caribou, and yaks all galloped from blueprints of the same quadrupedal prototype along with hundreds of other extinct and extant grazers. Cheetahs, foxes, pigs, and jackals propagated along its side branches. Pigeons and doves likewise arose once upon a time as the same bird and diverged from each other solely by slight variations of energy and incremental retention and loss of features.

The first life forms adhered once upon a time in a legendary tepid pond because their incipient metabolism (heat) was its own *raison d'être,* and they converted neighboring molecules into the compounded bubbles of their emergent biochemistry.

Genes ....

Despite the inherent statistical underpinnings of his theory, Darwin did not provide a *mechanism* for natural selection or for the retention, archiving, and inheritance of hard traits that lock in and reify "numerous, successive, slight modifications," nor did he did know how to quantify their selective and reproductive process. He assumed that some sort of tiny nutlike germ archiving templates of potential forms was passed from creature to creature, though the precise shape, etiology, and location of it and the manner in which it conveyed heredity through its infrastructure were unknown. It was also unknown whether it was directly reprogrammable by environmental influences during life—that is, whether its plasm was susceptible to an organism's life experiences in such a way that utilitarian (and nonutilitarian) offshoots of some of these were converted into new traits in that organism's offspring.

Over the next half century heredity was fully mathematicized. German monk Gregor Mendel, a contemporary of Darwin, had demonstrated that characteristics of organisms were determined by stable units (genes) that were donated from parent to progeny. Darwin kept a copy of Mendel's published work in his library but apparently did not realize its implications—no one did until forty years later when the focal plane of biology had so shifted that the mathematical basis of heredity had become obvious. Its "discoverer" was then posthumously annointed.

Genes, however, were still hypothetical objects represented by numbers and a trope. The birth of a genetic concept of inheritance before anyone had seen or measured a gene is a legacy that microbiology is still trying to overcome. A gene is now both an object (a section of a chromosome) and a series of phenotypic effects, but these identities do not coincide any better than they did forty years after Mendel's tentative demonstration of their ghost.

• •

*Genotype: the entire genetic makeup of an organism, as opposed to just its physiological appearance. **Phenotype:** the observable appearance and expressed traits of an organism, as opposed to its genetic constitution.*

• •

With a background in statistics and probability theory, Mendel sought modestly to apply the same range of mathematical rules to organic inheritance. Sexually reproducing round peas provided an excellent test population. Working primarily with their colors and shapes, Mendel calculated principles for the proportional inheritances from pairs of parents. Alleles are contributed by each progenitor pea to a zygote such that either the differing characteristics of two parents are combined or the trait of one parent—a dominant allele—is chosen unilaterally over that of the other.

While some inherited factors combine and blend, others stay particulate and discrete. Co-dominant (blending) alleles contribute intermediate effects, such as brown fur when one allele codes for black and the other for light. Dominant (particulate) alleles, however, result

• •

*Allele: one of two or more alternate forms of a gene. **Zygote:** the fertilized egg, e.g. the diploid fusion of haploid gametes from two parents.*

• •

in fundamental shifts—black *or* light. The replacement of "b" in ball with "t" changes the core meaning of the word. Replace the "a" in either ball or tall with "e," and another absolute change occurs. An "o" instead of an "a" makes a chair into a choir.

If it were not for particulate selection, evolution could not proceed past a certain level of complexity because single innovations (mutations) would be diluted in genetic fields. Changes would evaporate; life forms would be too deep-seated to morph. This doesn't happen. Speciation makes use of sudden, discontinuous, and irreversible variations that maintain their integrity in transmission. They insert novel forms into established sequences, shuffling the gross expression of heredity itself. New plants and animals are synergized.

While only dominant alleles give rise to attributes when both dominants and recessives are inherited from parents, the unexpressed recessives remain in the genome and many return in future generations. If a mother and father each carry recessive alleles for a particular nondominant trait, approximately twenty-five percent of their offspring will express that feature phenotypically—i.e., inherit the same recessive trait from each parent because there is nothing to override it.

Sickle-cell anemia is carried recessively and transmitted by matings of two healthy organisms, each of whom bears a sickle-cell gene; their children include potentially some twenty-five percent sick individuals. Because the heterozygous condition (one dominant healthy gene with one recessive sickle-cell gene but no anemia) provides a matrix of immunity against malaria (unrelated as such to sickle cell) while two "healthy" genes do not, sickle-cell genes continue to be favored selectively; they are not bred out because, while dormant as actual anemia, they protect against premature death from malaria. This explanation shows how genetic meanings are submerged, multivariate, and interdependent in more than one context simultaneously.

Statistical heredity laid the basis for the neo-Darwinian synthesis, a fusion of the "law" of natural selection (thermodynamics) with the algebra of trait inheritance. Yet the Mendelian solution, though regarded after the fact as a historic breakthrough, was in truth restricted to a range of isolated traits in idealized situations under controlled environmental conditions. Mendel ignored any inheritance or loss of characteristics that did not fit his model—a bias that has persisted in biotechnology.

He never found "genes" either. He left it that an unseen entity confers and distributes biological traits. And he traced that entity's mathematical shadow.

. . . . .

Though the exact physical locus for hereditary transfer remained unknown for a while after the rediscovery of Mendel's experiments, evincing it required only more powerful microscopes. Obviously some mechanism, likely much more minute than Darwin estimated, had to store and sort shapes, sizes, and colors of peas and other living things.

During the 1940s, scientists tracked the long-speculated agent for the transmission of traits to the previously undistinguished DNA molecule in the nuclei of cells. In 1953, using Maurice Wilkins' and Rosalind Franklin's x-ray diffraction of its molecular structure, Francis Crick and James Watson derived a model for how DNA could harbor and perform a hereditary role. The molecule's double-helical structure was, in fact, its method of trait repertory and replication: information-indexing genes arranged spirally in molecules copying "letters" onto proximate molecules.

DNA: deoxyribonucleic acid, a long, twisted and twin-stranded nucleic-acid molecule that bonds sequences of base pairs to replicate the inherited structure of a cell's biological information.

Early genome mappers not only identified ostensible genes by phenotypic traits but extrapolated the geometry of their placement and linkage from statistical inheritance of those traits. The closer genes are to each other on chromosomes, the less likely they are to become separated during zygote fusion and thus the more likely to be inherited or transferred together. Conversely, genes on different chromosomes are more liable to sort and redistribute their factors independently.

Information . . . .

Despite the apparent discovery of a real locus for Mendel's genes and their positions, it was unclear initially what they did physico-mechanically to bring about the transfer of actual traits. This puzzle dovetailed with another, far older predicament: how did the pristine universe of heat effects into which life protruded accommodate and express it without changing its abject operating principles? By the middle of the twentieth century, biologists had articulated these two puzzles into a clear challenge to their physicist colleagues, and to themselves: "Figure out if life forms, which do not decay spontaneously and immitigably toward equilibrium (as they should), provide additional principles of thermodynamics that operate outside known laws. If biological activities introduce exceptions to the equations governing stars, gases, meteors, liquids, and the like, identify and measure these. Replicate their effects."

No one looked very long or tried very hard. As soon as DNA was discovered and lodged in the collective imagination, a third member was vested in the prior exclusive club of matter and energy — information. The advent of information theory (as a by-product of the invention of the telegraph and development of linguistic analysis) contributed a cover story whereby integers transmitted instructions in discretely coded binary digits (bits) grounded atomically in thermo. Genes comprise DNA molecules and, through a sequence of molecular activities, transcribe amino acids, thereby regulating the chemical reactions necessary for protein assemblage and metabolism.

• •

Bit: *A unit of information or information storage capacity, or a unit representing equal likelihood of choice of one of two states within an information-containing system.*

• •

This proximal mechanism is now seemingly viewed daily in laboratories across the world and elicited with outcomes as reliable as those achieved while operating a dam or thresher according to its manufacturer's manual. Yet its explanation rests on some unexamined presumptions regarding the nature of information. According to biologist Barry Commoner:

> Crick's crisply stated theory ... hypothesizes a clear-cut chain of molecular processes that leads from a single DNA gene to the appearance of a particular inherited trait. The explanatory power of the theory is based on an extravagant proposition: that the DNA genes have unique, absolute, and universal control over the totality of inheritance in all forms of life.
>
> In order to control inheritance, Crick reasoned, genes would need to govern the synthesis of protein, since proteins form the cell's internal structures and, as enzymes, catalyze the chemical events that produce specific inherited traits. The ability of DNA to govern the synthesis of protein is facilitated by their similar structures—both are linear molecules composed of specific sequences of subunits. A particular gene is distinguished from another by the precise linear order (sequence) in which the four different nucleotides appear in its DNA. In the same way, a particular protein is distinguished from another by the specific sequence of the twenty different kinds of amino acids of which it is made. The four kinds of nucleotides can be arranged in numerous possible sequences, and the choice of any one of them in the makeup of a particular gene represents its "genetic information...."

Crick's "sequence hypothesis" neatly links the gene to the protein: the sequence of the nucleotides in a gene "is a simple code for the amino acid sequence of a particular protein." This is shorthand for a series of well-documented molecular processes that transcribe the gene's DNA nucleotide sequence into a complementary sequence of ribonucleic acid (RNA) nucleotides that, in turn, delivers the gene's code to the site of protein formation, where it determines the sequential order in which the different amino acids are linked to form the protein. It follows that in each living thing there should be a one-to-one correspondence between the total number of genes and the total number of proteins. The entire array of human genes—that is, the genome—must therefore represent the whole of a person's inheritance, which distinguishes a person from a fly.... Finally, because DNA is made of the same four nucleotides in every living thing, the genetic code is universal, which means that a gene should be capable of producing its particular protein wherever it happens to find itself, even in a different species.[9]

This is a précis of the modern biological world-view. Creatures, whether they be plants, animals, viruses, funguses, or future species, are interpolations of events originating from molecular, thermodynamic shifts in DNA-based code, and then delegated to protein foundries.

• • • • •

As to how entropy was outfoxed, it was presumed that bits of information at a subcellular level temper the undisciplined romp of gravity and heat, generating closed metabolic cycles of them, imbedding these interdependently and seamlessly in one another, and running them efficiently in rigorously coordinated designs. The upshot on primordial worlds is conversion of matter and energy, through full biological evolution, into bodies packed with information-bearing nervous systems.

Once it is established, DNA inscription dramatically augments raw thermodynamics in life forms, though it must itself be primordially invented, fabricated, and operate only by thermodynamic principles. Whether life once summoned and now draws on additional laws of thermodynamics, it is presently running and refining itself on the superfuel of information without expressly violating the known three. We know all this because we exist, but try explaining it to a universe of pure matter. Try getting it to discriminate without a mind. Try building intelligence from random motion. Tell chaos to find a thread out of itself.

. . . . .

The conveyance of information in binary molecular streams of "either/or," "0/1," "on/off," etc., according to thermodynamics was demonstrated in a gedanexperiment (thought scenario) by nineteenth-century physicist James Clerk Maxwell. Maxwell proposed a "wee creature" (later called "Maxwell's demon") who perched inside a box of gas particles in front of a valve that he manipulated in order to funnel faster, hotter molecules into a chamber at his right and slower, colder ones into another at his left. Thereby a temperature gradient was introduced, information was created, and mechanical work induced, such as driving a piston (by heat) or turning the blades of a tiny windmill. This would seem to permute entropy into information (energy) in challenge to the second law of thermodynamics. Yet Maxwell's demon must work even harder than that, work harder than this bonus would intimate, in order to carry out his dull task. Although he is creating information that can later be extracted from the system, he is dissipating it at the same time.

Maxwell was saying: hypothetically at least, we can reverse entropy. All we need is to have a wee creature sitting at an orifice between two chambers of equal temperature. If he lets high-energy molecules go one way, stopping them from returning, and lets low-temperature molecules go the other, also impeding their return, clearly this will separate hot and cold *using information*. By information alone, we can potentially do useful work and increase the free energy of the system. After the fact, we can turn matter into ethics.

Even so, the equation must be balanced in the end. Its energetic surplus can be gained only at the expense of the loss of free energy in the surroundings. In fact, if you know how much entropy is removed from the system, you can assign a value to information used

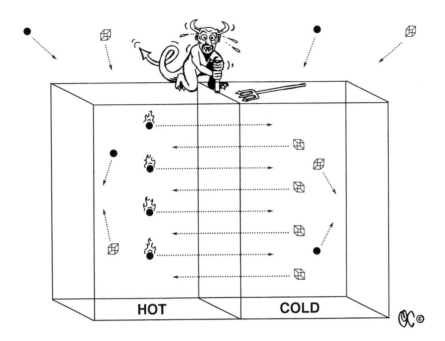

to separate its molecules, so many Joules per bit.[10]

• • • • •

Maxwell's fable of a demon separating gas particles and using thermodynamics to establish information flow is something we need not go into very deeply to understand that the connection between knowledge (in the abstract) and the physical world (as molecules) lies in actual physical systems (nascent DNA helices) turning entropy and energy into smart bits, toying with the Second Law and somehow seeming to circumvent without circumventing it. At the hypothetical crossroads of matter, energy, work, and information, the conversion of debris into life occurs. There is no other conceivable (or inconceivable) mechanical site.

In a world at or near equilibrium, a world in compliance with our Second Law, matter resists life with every grim morsel of its mere existence. Heat flows from source to sink. A meteor crashes into the surface of some moon or other, making a whole bunch of noise and kicking up far too much dust. If entropy could be reversed, you might collect all that sound and vibrational energy, turn it into a wave under the dust, and put it back together in a rock, perhaps even send it off into space. That can never happen; the collision is irreversible. Any attempt to reverse it—to collect the dust and re-form the original rock—will only create more entropy, i.e., "kill" more heat. In the words of German physicist Rudolf Clausius, who first characterized entropy and delimited it mathematically: *"Die Energie der Welt ist konstant, die Entropie der Welt strebt* [strives] *ein Maximum zu."*[11]

The terms of life's origin cannot be predicted into the past any more than Newton could calculate the future of the trajectories of three or more planets orbiting around each other. We do not know how Maxwell's demon got extricated.

• • • • •

The demon exemplifies a predicament at the crux of information theory and thus of consciousness. If information (life) is operating in the physical universe, it cannot be anything other than entropy in another form. Yet how can entropy take measurements, make discriminations, and transmogrify itself from a box of gases into a computer (or trillions of tiny computers)? How did the Earth escape its initial equilibrium landscape effectively enough to sort ludicrously huge bunches of molecules into not only hot and cold chambers but stable morphogenetic strands?

A mathematical contradiction, a reversal, lies at the root of matter. The statistical version of the second law of thermodynamics turns out to harbor a universe far different from its own literal rendition: the classic nineteenth-century machine running down. Entropy is (yes, at first glance) ignorance, mayhem, the opposite of information, dead heat—but information is itself conveyed in pure molecular streams; hence entropy must also not quite be the opposite of information. In distinguishing itself from entropy, information must be another face of entropy, not neg-entropy, as one might expect; not dead heat only—or dead heat at just one level. Information can literally ride on nothing; in fact, it rides far better on "nothing" because "something" would always get in its way and clutter its ultra-sleek track. Information can ride on "nothing" as long as "nothing" means irrepressible atoms rather than "void and without form."

In other words, entropy is more than just a bad guy.

On the one hand, information depends on thermodynamics for a body at its hardware core; on the other, thermodynamics naturally turns chaos into information. It sorts because it exists. It exudes intelligence because we who emerge from our own such paradigm employ its sorting as—call it whatever you want. Maxwell's demon trapped at a fateful conduit between hermetic chambers sends molecules one way or the other according to a stamina and austerity it cannot rightly have—but don't tell it—at least not till after the party begins.

* * * * *

Somehow the constituents of DNA molecules were tamed by enough "wee creatures" among heat and gravity to arrive at their forms in the first place. Their successors now incorporate astounding amounts of data by relatively straightforward thermodynamic, algebraic methods of bit flow. They are repositories of so much intelligent chatter that we think of them as transcendent of ordinary matter.

Yet genes began as matter, are continually made anew from matter, and operate only by matter-energy flow. Their precursors were once rills in streams, mud in suspension, gas in snares. The origination of their art of fooling entropy and turning inherent laziness into work occurred long, long ago in a circumstance beyond our capacity to conceptualize or reconstruct, so we imaginarily reinvent the kinds of molecular, signal-transferring dynamics that could concoct knowledgeable helices in the rough from water and silt and then continue to discriminate them.

The Neo-Darwinian Synthesis . . . .

Putting Darwinian selectic dynamics and DNA information transmission together provides a contemporary version of the evolutionary tale: Organized physicochemical structures arose here on

Earth because an open-ended flow of incident energy from the Sun in the form of light and other electromagnetic radiation was absorbed, transformed, reemitted, and stored in discrete packets (quanta) by tiny structures held together by cybernetically coordinated chemical bonds. The solar stream was thereby subliminated, ultimately to be released as different (lower-energy) quanta to carry out the kinds of physical and chemical tasks that living systems make useful. These came to include information networks with distributed controls.

Elemental configurations—matter/energy in chance designs—capture other, higher-energy quanta, elicit new excited states in them, and step them down into biogeochemical cycles of planetary organisms and ecosystems. We are "energy specified and congealed." This conspicuous breach of thermal equilibrium is systematized and systematizes itself by pure information flow from DNA molecules that arose from disturbances in sea foam by laws of thermodynamics, hence forfeiting free energy elsewhere, and continuing to forfeit it anew as they reproduce themselves and transcribe their ambitious plan.

Information is accumulating, but only temporarily and statistically, not as real stuff. But then nothing else is any more real or any less temporary. Through the mirage we experience as self, we see a slightly different mirage, the natural world. Both events are impossible, but both seem quite at home in the universe.

• • • • •

Matter-and-energy, given the opportunity to roam free and amok forever, instead gathers in hierarchies. From then on (here on) we assume genetic regulation of life. Once nucleic-acid chains bearing information and ciphering strings of protein molecules were winnowed out of cosmic debris on Earth, they set themselves in full transaction in steady-state programs. They could program events (heat conversion and metabolism) and run projects (germ cells and embryos) without "further" violating thermodynamics. Whenever explanation was needed (i.e., where thermodynamics seemed insufficient), agency was assigned to inherited codes that earned their spurs by evolving once upon a time in rigorous thermo regimes. All other questions were referred to Maxwell's demon or his stand-ins.

Nucleic prototypes milled by Precambrian thermodynamics now use everyday thermodynamics unfolding at blinding speed in informational matrices to maintain their kingdoms. They flourish just about everywhere, on at least this planet. In fact, this ultra-sophisticated operation covers the Earth in mites and jellies, gazing indifferently from their sanctum into the smoldering entropy of night from which it came.

To pretend this is okay—that biology supports its own weight—is a sham.

* * * * *

Since thermodynamics is apparently all that can order atoms and molecules to do what they do, and since entropy attends to every movement of every atom, there is no ultimately credible explanation for neg-entropy, for why embryos organize and why they so flawlessly now assemble flamingos from flamingo eggs, mice from mice, giant squids from tiny squid germs, oaks from acorns, and so on; why they manufacture such acorns and other DNA-bearing germs, or even exist as devices, in the first place. These are ostensibly the kinds of highly sophisticated things that matter and energy do when given billions of years of unmonitored time. They make not just complicated patterns but existential objects. And the objects stick.

Yet the paradigm holds fast. No one believes it. Even those who say they believe it don't believe it. It holds because civilization runs efficiently by its rules—all the factories, vehicles, hospitals, power grids, computers, economies, weaponries, and laboratories. The seasons, tides, winds, crops, and elemental cycles are also docilely compliant.

Information from DNA is the great catchall, the saver of appearances. Everything nonentropic and inexplicable in morphogenesis is assigned uncritically to bits stored long ago in the nuclei of cells and drawn on to pay all current debts—even though using information and life to explain each other requires constant unexamined extrapolation from ontology to ontogeny and back.

This burlesque ultimately sinks all otherwise-seaworthy ships.

Biology's initial challenge to physics remains substantially unmet: we have not begun to bridge the real gap between inanimate and living systems. We cling to an informational, material model so tenaciously because we have no other paradigm—or hope of one—that could replace it.

Biochemistry to Physiology . . . .

Needless to say, scientists waste little time worrying about the gap between matter and life or letting it get in their way. Assaying factors they *can* pin down and using Darwin's elegant logic, they correlate genes, chemical codes, levels of physiology, and behavioral traits in ancestral sequences and known survival patterns among living, extinct, and hypothetical bionts; they establish lineages of life shapes and behaviors. They don't dismiss the prior chasm between chemistry and biology; they simply presume that matter, energy, and information (in sum) have fairly played their assigned roles in getting us to where we are. They know that present and accessible systems cannot be untangled or analyzed at an ontological level. There are certainly no tools or autopsies for reconstructing original biogenic states; we see everything at a very late stage of either fossil remains or DNA activity.

Substantial lacunae between matter and DNA, and early genes and complex organisms, must have been filled long ago stepwise and evolutionarily by relationships between the components of emerging genomes in environments, leading to body plans and culminating in phylogenetic pathways, anatomical systems, heredity, and motif divergence. These diverse events and structures are now rationalized by assigning millions of years of chance effects and incremental selective regimes to their assemblage. What occurred in a nonlinear fashion and was condensed in lineal series can be excavated more or less by comparative studies of extant biochemistry, genes, and mechanistic, behavioral adaptation.

The primal imbedding and synopsizing of information in life can be reconstructed from the archaeology inherent in contemporary structures and activities.

One such decipherment is the evolution of heat production in mammals by brown-adipose-fat (BAT) conversion. Creatures migrating into cold-challenge environments need to generate and maintain increasingly more internal warmth. Norepinephrine, a neurotransmitter originating in sympathetic nerve terminals, excites BAT heat production by promoting its cell proliferation. It differentiates and controls a gene for uncoupling a form of protein thermogenin such that some cells get "filled to bulging with mitochondria and lipid droplets."[12] In these, thermogenesis supplants energy production.

This explanation is both genetic and thermodynamic.

All in all, about three hundred genes collaborate to run the complex system, including its oxidation, ion channels, ion exchange, metabolite movement, electrochemistry, control cascades, and signal-transduction pathways. Recalling the horse-drawn lathe (above), heat literally wastes ATP while warming tissues. Genes get into the loop by programming enzymes to permute cell metabolism and trigger the cycle.

Using thermogenin as a marker for the system, molecular biologists have traced it to the evolution of mammals two hundred million or so years ago, and they have also located most of its distinct structural, regulatory, and catalytic components in other settings. Genetic analysis by sequence comparisons tells them that the brown-fat complex likely originated from the thermogenin gene's recruitment of a mitochondrial carrier protein in a preadipocyte cell. True adipocytes were then naturally selected as genomes reproduced in increasingly colder regions.

Biologists have similarly been able to reconstruct the evolution of thermogenic functions by a brain-heater organ among fish, a hierarchy involving over two hundred separate genes and effecting morphological and chemical changes in the mitochondria and Golgi bodies of fish cells. The structure *per se* likely arose independently more than once from slow-twitch muscle fibers in mackerel-like creatures.

Other studies have traced the diving response of sea lions and seals and electric organs in fishes to a variety of "cardiovascular metabolic, neuronal, and endocrine responses"[13] related to tissue-specific biochemistries.

. . . . .

All these evolutionary series involve relatively approachable organs in known reaches of geological time, organs whose precursor cells can be tracked by genetic sequence analysis and whose genes were recruited from other uses that are still active. Their patterns of organization suggest how most genetic, anatomical, environmental, behavioral loops might work and how all thermodynamic-morphogenetic links originated.

Yet we are working with tiny, interrupted segments of a massive supersystem (the biosphere) made up of intricate mega-networks (bionts). We can observe specific genes, metabolic pathways, and thermodynamics in play, but we cannot guess how so many discrete elements among them got so entangled and efficient at the same time at so many levels of structure and function and then were maintained in such states to form denser, more sophisticated networks. There is certainly no way to ferret out how macromolecular entities and core tissue complexes came into being in the first place. Most biochemical systems "were 'invented' nearly four billion years ago when the three basic branches of life (archaea, bacteria, and eukaryotes) were being established."[14]

We also don't know who lit the first campfires during the Stone Ages or how monkey yelps became hieroglyphs. We have virtually no census for the soldiers of Hannibal or those who fought at Agincourt and Thermopylae. Those genetic, anatomical, and behavioral vectors that are accessible to us do not explain the methuselan origin of cells or basic body plans. Famous assignations of molecules and genes are infinitely beyond us. Yet it is specious to hold any of that against evolutionary theory.

Complexity as a Method of Organization . . . .

The best candidate for preserving thermodynamic integrity across the matter-life chasm is the innate, infinitely complex dynamics of matter itself. Raw turbulent clusters of molecules—churning in thermal convection on stars and planets, billowing in atmospheres, oscillating in electromagnetic waves, and swirling in viscous flows— for no good thermodynamic reason seem to develop mathematically consistent topologies, infinitely complex shapes that project self-similarity across scales. These patterns arise spontaneously out of zillions of randomly moving particles. When they are disturbed, they don't just disperse randomly (entropically) as they should; they rebound to their peculiar motifs. Matter and energy, by themselves, using their own molecules as canvases, paint multidimensional landscapes. Neg-entropy proves itself to be just as "entropic" as entropy.

Jupiter's weather, basically a gigantic gaseous ocean under a strong Coriolis spin, keeps returning to characteristic multicolored bands highlighted by a giant red spot. As the planet is sheared globally by hurricane forces dwarfing any we can imagine, the configurations of its atmosphere continue to self-organize back to their standing states. Jupiter is not quite as stable as a zebra's bands or a jellyfish's mesogloea, but it is semi-stable over eons during which hundreds of thousands of zebras and jellyfish die and decompose, trusting their motifs to germs cells alone.

The same intrinsic inanimate patterning occurs more ephemerally in weather systems here on Earth; they are turbulent and coherent at the same time.

Even random information from disparate sources (for instance, cotton prices, fluctuations of wildlife, the rise and fall of rivers, distributions of earthquakes, and noise in telephone lines) displays this uncanny clustering behavior.

The persistent geometry of chaotic (discrete nonlinear) systems remains the single most mysterious and bizarre "wild card" in the

Newtonian universe. It impels crystallization in rockbeds, three-dimensional scroll waves in gels, striated bands and cyclones, and molecular clumps and layering in bogs and pools.

The conclusion is inescapable: physically and statistically, order progresses relentlessly out of disorder. It does so unrehearsed, unsummoned, without any demonstrable mechanical basis. Something is driving inherent organization, something no doubt thermodynamic but with a strikingly impulsive tilt.

• • • • •

The Belousov-Zhabotinsky (BZ) reaction was discovered more or less accidentally when B. P. Belousov was trying to simulate the reaction-sequence loop of the Krebs (citric acid) cycle *in vitro* (see p. 127). Instead, he induced a different cyclical chemical reaction, one simulating Krebs but using its own self-organizing wave patterns and component reaction pathways to oxidize an organic substrate. Numerous such self-propelling nonbiological loops have since been recognized: a generic recipe is sulfuric acid + malonic acid + cerium ammonium nitrate + sodium bromate = carbon dioxide + dibromoacetic acid + bromomalonic acid + water. "The system will oscillate in color until equilibrium is achieved—in typical demonstrations ... [regularly] for more than thirty minutes ... setting the system up for the beginning of a new cycle."[15]

In some BZ reactions a wave of oxidation rotates around an inhomogeneity (its so-called pacemaker) such that each set of spirals induces another, displacing remaining concentric rings in the medium insofar as they are more efficient mechanically.

That most of these sorts of self-generating reactions do not breed the kind of chemistry needed for terrestrial metabolism as it evolved is why they didn't interpolate themselves within organisms and build cellular loops—but the point is: citric-acid and other energy cycles following the same principles *did*. Simple life at its beginning was probably little more than pathways of photosynthesis and glycolysis imbedded in primordial membranes among enzyme-protein triggers.

Phase States and Attractors . . . .

Complexity analysis began with systems-level or general systems theory in the 1920s. Scientists began to focus on an obvious but easily overlooked facet of nature foreshadowing chaos' innate complexity—the accession of new properties of matter at inexplicably critical jumps in scale. Although separated by powers of tens to thousands, atoms, molecules, cells, and organisms interact together along their inherent gradient in a single thermodynamic framework (the universe). Rules that govern each exponent of that gradient pertain to that exponent alone. The behaviors of molecules cannot be reduced to aggregate atomic traits any more than organisms can be inventoried in cumulative cell activity. The fluid dynamics observed at large on Jupiter is not inherent in any single water molecule. The characteristics of life that cannot be anticipated from coalescences of purely atomic or molecular traits are emergent properties, underivable from their components. Along this invisible ladder, mind is an emergent property of brain.

Machines demonstrate the same attribute in a less hierarchical fashion: a washer or computer enacts quite unique capacities when its various wires, circuits, power source, and other parts are brought together, but the elements by themselves lie dead on the work bench.

Relationships among atoms clearly bring molecules into being, and feedback among cells knits and stabilizes organisms. In addition, nature as a whole introduces emergent properties at each tier that are not locatable at any one threshold *per se* and cannot be attributed to the collective or additive properties of the tier under it.

While extemporaneous traits do not violate thermodynamic principles, they confer behavior that can in no way be predicted from simple thermodynamic interactions. They also keep complex systems in theoretical compliance with thermodynamics because they serve as an alternative to a purely metaphysical or divine force adding new qualities. In fact, emergent properties, though vaguely conceptualized, may betray the very laws that generate and maintain distinct levels of matter.

• • • • •

For much of the twentieth century scientists toyed with chains of integumentary events but didn't truly grasp their nature because they didn't need explanations for their experimental proofs or models. Atoms did their atomic things in the land of atoms; molecules did molecular things where appropriate; cells did cellular things; and creatures provided the statistics of neurological and psychobiological waves—each safely in its own realm. Correspondingly, each academic department—physics, chemistry, biology, ecology, psychology—developed its own course of study and licensed its own researchers. Most scientists don't have to address why cells have traits that individual molecules don't give them, because molecules and cells otherwise scrupulously obey the laws regulating their respective domains.

However, since the 1960s with new interest in the extrinsic origin of life and the physical basis of mathematics, systems theory has blossomed anew under shorthand like "complexity" or "chaos." As such, it has been routinely applied to self-organizing networks and given rise to a set of laws that sound as grave and authoritative as those of thermodynamics except that they are not laws; they are metaphors derived from statistical analyses and subjective presentiments—working corollaries that provide a modern, post-thermodynamic idiom for events that change states and flout entropy without any purely thermodynamic excuse.

They are harbingers of a separate branch of physics that has not yet been integrated into the reigning paradigm.

• • • • •

In the language of complexity, nonlinear dynamic systems are self-maintained and convoluted by multiple layers of interconnection and feedback within themselves, the hypothetical number of levels increasing exponentially with each further complexification in their matrix. Each system contains vastly divergent, convergent potential states—possible behavior combinations. Every one of these semi-stable arrays of interlacing parameters is a "phase state."

A multiplicity of variables affects single phase states such that thermochemical analysis of them or translation of them into linear vectors is futile. A solution to such a system must be statistical rather than thermo—though, don't forget, since Maxwell's demon (and grandchildren of Maxwell's demon), a lot of thermo is already little more than statistics:

> Phase state gives a way of turning numbers into pictures, abstracting every bit of essential information from a system of moving parts, mechanical or fluid, and making a flexible road map to all its possibilities.... In phase space the complete state of knowledge about a dynamical system at a single instant in time collapses to a point.... At the next instant, though, the system will have changed, ever so slightly, and so the point moves. The history of the system ... can be charted by the moving point, tracing its orbit through phase space with the passage of time.[16]

In phase state we confront more than meted thermodynamic higgledy-piggledy careening with gases or bubbles; the collective maelstrom comprises a series of alternative positions—discernible whole shapes—colliding and dispersing among one another, while potentiating themselves and each other by an unknown principle of interactive sorting to yield regular irregularities.

With the help of an attractor, a complex dynamic system evolves from any prior phase state to a next "preferred" semi-stable disposition. An attractor is basically anything that holds a complex system, in the absence of other attractors, in its nonlinear configuration. It could be a planetary orbit, a fence collecting debris along a highway, the bottom of a bowl, a river-bed, or a medley of subtle and gross influences so complex and multivariate itself that it is for all intent and purposes infinite and indefinable (e.g. the metabolism of a cell).

Each attractor in its own way specifies the dynamical status of its system while maintaining it in a fixed sequence or structure, a so-called basin of attraction: "The ratio of the volume of the basin to the volume of the attractor can be used as a measure of the degree of self-organization present."[17] A basin includes itself and all its sub-states, pre-states, and emergent states—a simultaneity of what it once was, what it is doing now, and what it is reshaping into. One system can have many attractors, i.e., be meta-stable with alternative phase states. As it oscillates among attractors and various basins that codetermine its position, it mutates and evolves over time.

A configuration generating an infinitude of simultaneous vectors is called a strange attractor. "Did you ever ask God whether he created this damned universe?" retorted physicist Floris Takens when science writer James Gleick asked him whether *he* had coined the term.[18] This exchange rests on a darker parable: there *is* a thin line between the state of infinite attraction and the emergence of a universe as delicately vast and locally dense as this.

．．．．．

When a feedback system becomes canalized into a set of reactions such that a consistent cycle of phase states develops, it is known as an autocatalytic set, i.e., a simple, protean organism, perhaps the first on its planet. Within its life cycle, molecules catalyze one another's states in a consistent fashion as in BZ reactions:

> [F]or some distances from equilibrium, the system will behave as a chemical clock, its dynamics governed by a limit cycle attractor. For other distances from equilibrium, qualitatively new behaviors, chaotic behaviors, appear as the dynamics become governed by a strange attractor. . . . In such a dynamic regime, the chemical system can show emergent properties not explicable in terms of the properties of the components of the system.[19]

This could explain not only the origin and stabilization of a brown-fat heating system but far more ancient lineages of morphology and metabolism underlying BAT's very cells and physiologies and bridging their gap between inanimate chemistry and biology.

• • • • •

Chaos "laws" provide both a systemic and physical basis for protoplasm coalescence, organelle and cell differentiation, and diversification of species, in part because they allow an infinite range of *possibilities* to be run through a mesh so fine that *no actual organism* has to fit. What is being selected are basins of attraction engendering semi-animate crystals, then multiple genes and protein complexes—plus their networks and progressive associations—rather than just the gametes of single victorious creatures. This means that natural selection is more another level of chaotic complexity than chaste feedback from actuarial equations. It moves by pattern jumps rather than—or as well as—census shifts.

Whether because of decreasing food supply, lowered temperatures, new predators, an organizing catalyst, or some hidden spur, complex living as well as preanimate systems inevitably reach critical edge-of-chaos points at which they radically transmute. Bacteria-like entities become protists, protists become organelle colonies, organelles coalesce into cells, cells regerminate in layers of tissues, and tissues metabolize into organs. All these are strung together in speeded-up time in ontogenesis. At any level of each of these phase states, variant attractions can be "selected" in place of actual genes or phenotypes. Quantum leaps are made when chaos is threatening either to return to—well, chaos—or to organize at ever higher, more integrated levels. In this way Darwin's evolution achieves a molecular thermodynamic basis which escorts it through its two most serious crises—the emergence of life from inanimate chemicals and the discontinuous leaps from species to species.

Natural selection in complex systems is more than a zero-sum game; it represents unceasing series of interrelated pattern choices made within meta-stable fields, each with emergent properties. Over epochs, multiple attractors (selective pressures) fluctuate among predator, climate, and prey, giving rise to not just isolate discrete organisms but the interdependent functions of their digestive, excretory, and immune systems, as well as whole ecosystems, societies, and sociopolitical hierarchies of them, at each level drawing on complex nonlinearity as a screen against random perturbations, hence maintaining and returning to new semi-stable states.

Perhaps this is how subatomic particles once forged atoms, atoms spawned molecules, and eventually molecules catapulted into organic configurations—although it is difficult to imagine what premetabolic environmental events constituted phase-state crises for inanimate particles. The atomic-molecular trigger was more likely intrinsic; either that, or exogenous and innate factors meet in a hyperdimensional space, and whatever is driving a system from within "coincidentally" finds factors outside itself that allow it to fulfill its destiny.

• • • • •

Once state cycles begin matriculating through edge-of-chaos metamorphoses, they radiate self-similar, scale-independent shapes through all their modes of expression. Molecules and cells organize in semi-congruent motifs at different calibers in both directions, so puddle ripples repeat in the atmospheres of whole planets, organelles reconstitute as tissues and organs, and tiny darting monads become dense lumbering organisms. Along the way, substances meld, bifurcate, and shatter in regularly irregular patterns.

The fluff and fuzz—the inevitable rough edges of matter-energy—represent not only experimental error and gratuitous drift (as pre-chaos scientists presumed in consigning them to the garbage pail); they are the unevenness and ruggedness of stuff itself, stubbornly defying fake smooth edges and symmetries. For all their bristliness, crags and squalls throughout nature are deeply and mysteriously patterned in other ways.

When assuming physical forms under thermodynamic regimes, self-imbedded objects create and occupy their own fractional dimensions, or fractals, through which they crowd seemingly infinite length into finite volumes. Becoming asymmetrically symmetrical allows them to fit into actual space, perhaps even to incubate their emergent properties.

The rough edge of a piece of metal or the porous pattern in pumice embodies this same fractal suffusion. Intricate in-between dimensions define tree bark, the boundaries of coastlines, the Mississippi to even the smallest of its tributaries, and the scrolls of Neptune's atmosphere. Gargantuan stellar and minute snowflake forms occupy the same dynamic fields—and use the same algebraic lexicons. Shapes are coiled and buried in other moving shapes *ad infinitum*—all the way up to entire galaxies.

Complexity is inherent, emergent, or both, and the thermodynamics driving it is tied to a transdimensional etiology or encompasses an infinite, incalculable sequence. Either way, the universe has a penchant. "Alive" and "dead" are no longer meaningful distinctions.

· · · · ·

Chaos-based cycles provide a rubric for emanation of life as an emergent property in a self-organizing phase-reaction system in any friendly environment on any planet regardless of native specifics. Interacting autocatalytic sets engaging each other in homeostases over time become self-creating and also novelty-generating, so emergent phenomena swell fractally and pandemically out of prior emergent phenomena. Autopoietic states maintain their equilibria against noise, returning to homeostasis even under external assaults and inherent wobble. As patterns mature in complexity (more cells, more tissues), they produce further complex, equally semi-stable patterns. If an inanimate red spot three and a half times the size of the Earth can do it, why can't an amoeba?

Through chaos and complexity, evolution advances to a universal cosmic law.

· · · · ·

An autocatalytic matrix all by itself, the putative earliest life form, turns out to be a more effective mechanical, metabolic engine than its molecular predecessor; thus, it arose spontaneously, a chemically selected phase state. Then it translated its fundamental motif over millennia to successors, generation by generation through fractals.

We see a dramatic demonstration of an instantaneous transformation of single cells into composite physiques when, during times of diminished food supply, members of a species of slime mold amoebae *(Dictyostelium discoideum)* assemble into bulky slugs. The monads "signal to each other using waves of cyclic AMP that propagate by a diffusion-autocatalysis process analogous to that found in BZ-waves, and with the same dynamical properties."[20] As cyclic AMP is created by one enzyme and destroyed by another, these enzymes integrate themselves into the signalling pathway. Heralding by molecule triggers the metamorphosis of a single-cell organism to a metazoan state and then back again. The physical (informational) processing of the signal depends on a cell surface receptor and the pathway's downstream target, which could be stimulatory or inhibitory. "The slugs are able to exploit the chemical dynamics for aggregation and differentiation, leading to dispersal, and subsequent replication. BZ-type dynamics are an integral part of their survival strategy."[21]

• •

Cyclic AMP (cAMP): this is cyclic adenosine monophosphate, or ATP with only one pyrophosphate group. A triggering molecule in cells, it turns pathways on and off (see p. 127).

• •

Amoeboid clustering is many strange attractors and fractal levels of organization beyond cloud formations and lava flows. It dances on the borderline between simple chaos thermodynamics and chaos systems funnelling thermochemical events into self-organizing phase states that kindle "stranger" meta-stable blobs.

Fourth Laws of Thermodynamics . . . .

Stuart Kauffman, a biologist grounded in mathematics and physics, attempts to use his own rendition of complexity theory to provide a working model for the types of biological structures that classical thermodynamics cannot easily explain. Even when taking into account possible self-organizing complexity, strange attractors, meta-couplings, and layers of emergent properties, Kauffman concludes that the most astute and tenacious Darwinian purists would still have trouble explaining the occurrence of viruses and plants (to say nothing of full personalities in bodies) without bending if not breaking the most basic established laws of matter. Even the most rudimentary pieces of the simplest bionts could not evolve in a strict thermodynamic, Darwinian regime.

Through mathematical and thermochemical calculations, Kauffman shows how improbable it is for primitive autonomous monads to have arisen *ex nihilo* from matter solely by interactions of particles subject to natural selection. Elemental distribution of galactic matter and natural selection on planets in the context of heat and entropy are, by themselves, unlikely ever to construct anything so notable. In fact, a variegated bubble replicating itself precisely in a second bubble, and then another and another, is out of the question under such puritanical conditions.

> In short, the known universe has not had time since the big bang to create all possible proteins of length 200 once. Indeed the time required to create all possible proteins at least once is at least the ratio of possible proteins to the maximum number of reactions that can have occurred in the lifetime of the universe, or 10^{67} times the lifetime of the universe.
>
> Let that sink in. It would take at least 10 to the 67th times the current lifetime of the universe for the universe to manage to make all possible proteins of length 200 at least once.[22]

In establishing its method, life somehow eluded random equilibration of all possible different molecules. It had a better algorithm. Not only does each developing embryo break clear of entropy, it does so in such a totally clandestine, even sneaky, way that it is dutifully observant of entropy (and thermodynamic laws) at every other point (i.e., except the one that makes it alive).

. . . . .

In his book *Investigations*—published in 2000 and named specifically to recall Ludwig Wittgenstein's philosophical journey into structures and meanings beyond logical atomism—Kauffman testifies:

> Unlike the well-defined and formal transformation rules of an algebra or a calculational process, the transformation rules of the biosphere enlarge and change in ways that cannot be prespecified.... [C]onsider the evolution of the genetic code and consider the structure of eukaryotic chromosomes, whose ... coordinated behaviors underlie both normal mitotic cell division and the astonishing sequences of meiotic reduction cell divisions in which maternal and paternal homologue chromosomes synapse, undergo recombination, and separate such that the final sperm or egg cell receives, at random, only one homologue of each parental chromosome. The emergence of these complex macromolecular systems has altered the way evolution itself unfolds....[23]

This is a telling admission. In morphogenesis, the incumbent system regulators and transistors turn out to be more chaotically and nonlinearly complex than any, even infinitely entangling strings of computer-driven mathematics. Given the one-of-a-kind talents of energy in membranes and the subsequent novel behaviors of autonomous agents made of them, we need more than just our own nonlinear dynamics to map the real universe.

No one stands any chance of calculating the historical phase states in an actual living system. The representation of so-called

• •

Boolean algebra: a mathematical system using binary variables (AND, OR, NOT, IF, THEN, EXCEPT) to generate symbolic-logic elements and propositions, i.e., in computer science. Genes can be considered elements in a network such that each gene receives Boolean inputs from other genes at more than one level of function.

• •

state cycle attractors and edge-of-chaos metaphenomena culminating in feedback/phase loops of genes in cells and embryos in a search-algorithmic system requires something like a series of four- to thousand-dimensional binary (Boolean) hypercubes each bearing equivalently exponential numbers of vertices (sequences). Imagine 80,000 structural genes in humans, each of which can be activated or suppressed. That leads to $10^{24,000}$ possible states of gene activity to be explored in the 10^{17} seconds since the Big Bang.[24]

> In the evolution of a biosphere, the emergence of systems such as the genetic code and meiosis seems rather like the emergence of new laws.[25]

• • • • •

Kauffman chooses to devise speculative antecedents of a possible new physics via four different versions of a fourth law of thermodynamics, a kinetics that synergizes effects out of the more evident three. This law could be some previously undiscovered principle that "governs biospheres anywhere in the cosmos or the cosmos itself," that explains "living entities—bacteria, plants, animals [which] manipulate the world on their own behalf: the bacterium swimming upstream in a glucose gradient that is easily said to be going to get 'dinner'; the paramecium, cilia beating like a Roman warship's oars, hot after the bacterium; we humans earning our livings ... [all] 'autonomous agents,' able to act on our own behalf in an environment."[26] Looked at from a purely chemical, atomic perspective, this is the monster mash.

• • • • •

Entropy is the final arbiter of a statistical universe, the second law of thermodynamics requiring all matter to slide inexorably, irreversibly into disorder. But entropy requires a closed space. What if the universe is neither bounded nor confined? What if light and matter are two utterly different, co-impinging domains? What if information systems featuring genes are neither linear nor nonlinear? What if information creates genes rather than the other way around? What if thought is a metaphenomenon rather than an epiphenomenon of electrons along cell membranes?

If the universe is cosmologically open, entropy has no antecedence, let alone status as an absolute gradient.

• • • • •

While not accepting the profligate speculations above, Kauffman assumes there must be hidden routes of system formulation and maintenance, "general laws for thermodynamically open, self-constructing systems such as biospheres"[27]—or at least one other overriding principle, one meta-property of matter and energy: "a law in which the diversity and complexity of the universe increases in some optimal manner … [in which] the nonergodic universe [a universe that has some predilective inclination] as a whole constructs itself persistently into an expanding … workspace. This is in sharp contrast to the familiar idea that the persistent increase in entropy of the second law of thermodynamics is the cause of the arrow of time. But the second law only makes sense for systems and timescales for which the ergodic [randomly equilibrating] hypothesis holds. The ergodic hypothesis does not seem to hold for the present universe and its rough timescale, at levels of complexity of molecular species and above. Perhaps we are missing something big, right in front of us."[28]

Although I am not on the same playing field as Kauffman, the intimation that "something big is right in front of us" is the point of this book.

• • • • •

Even a devout antievolutionist would have to concede in Kauffman's behalf that he appreciates the epistemological dilemma of life's autonomy, an aesthetic that mere physicalist scientists do not share. He writes of old egrets, "long-legged, knowing how winter is thin, when the first phase transition of water to ice forms slight solidity across meadow streams, [as] small creatures of flesh and concept tiptoe gingerly to some far side where, perhaps, something new is to be found." How do they get this way, he asks—independent, curious, proud?

"We have lacked a physical definition of an autonomous agent able to manipulate the universe in its own behalf," he continues—"the egret whose foreboding of winter leads to lifted wing and steady, powerful flight."[29] How do random particles accrete and then cull one another to produce the bird's remarkably feathered and neuron-thick body? The creature's gesture likewise cannot be dismissed as a mere temporary congery of molecules; yet it must be. Nature, though highly organized where alive, oddly has no way of getting there except by creative mechanical chaos—and then barely. "The egret is as much a part of physical reality as the atom, and perhaps more than the vaunted quark [which is only an intellectual construct]. But autonomous agents carrying out work cycles, we who daily manipulate the world on our own behalf, we to whom 'intentionality' and 'purpose' are so inevitably attributed by our common languages, we are by definition of autonomous agents, also nothing but physical systems with a peculiar organization of processes and properties"[30]—the aftermath of gusts, rain, sunshine, and congestion of sea minerals.

. . . . .

In the course of his trans-thermodynamic excursion Kauffman manages to construct topologically elegant entities—as equivocal and weird as anything drawn by Maurits Escher or narrated by Jorge Luis Borges. Yet, in order to do so, he relies on Boolean hypercubes, logarithms, reaction graphs, phase transitions, quantum mechanics, and other creative statistics without taking into account that these topologies and their numbers are stiff semantic constructions, not building blocks of actual reality. His creatures are compelling, but they are not mercurial enough. They must finally be cartoons, earnestly proposed fictions. The three big laws are so simple and absolute and any fourth-law candidates (thus far) so problematic and convoluted, they end up sticking about as long as snow in Carolina.

Kauffman cannot break into the epistemology of the embryo. Far more sophisticated than most reductionist scientists, he is condemned to making arguments of the same order, splicing new emergent thermodynamics out of known thermo-algebraic pieces, while succumbing to the malady of believing in calculi as fair measure of the universe, though he can see right through them because he can see at all, i.e., because he experiences the irreducible complexity of his own calculus-shattering act of discernment and being. Of course, without such calculi, he has no place to go.

A game that does not play by its own rules is not winnable, not for Kauffman, not for Darwin, not for Einstein; not now, not ever. It is not even a real game. Life is neither a calculation nor an algorithm. It is not "a physical system with a peculiar organization of properties." That is its modern alias. It is qualityless motion of essence itself, which occurs almost independent of the forms and agencies it takes. The gradual accretion of mechanism is an illusion masking the sudden, encompassing reality of existence. Science can effectively analyze the independent sources and algebra of moving parts, but it cannot get at the character of life's vital element or autonomy.

2

Is There a Plan?

Creationism, Cultural Relativism, and Paraphysics

> The lord whose oracle is at Delphi neither speaks nor
> conceals but gives signs.
>> —Heraclitus, 540–480 B.C.

Darwin's Opponents

The three most serious epistemological challenges to neo-Darwinism
have been: 1) creationism, intelligent design, and other pseudosciences
grounded in religious fundamentalism; 2) cultural relativism and
symbolic deconstruction, i.e., the totemic basis of all thought; and
3) paraphysics and vitalism—reconsidering the universe by hyperspace, telekinesis, and events posited outside closed systems of thermodynamics and ordinary space-time.

Religious fundamentalists make formidable adversaries for scientists
insofar as both groups are trying to establish their beliefs as a religion; i.e., a universal cosmology and moral system. Conservative
priests, rabbis, and mullahs allege that a Supreme Being, whether

God, Allah, or some other demiurge, planned, mapped, mysteriously created, and now rules the entire universe and its creatures. They regard evolution as a wicked blasphemy, one that threatens to render any such deity obsolete. Plus, if the modern universe has no divine ruler, it can have no intelligent, let alone ethical, basis: existence is rudderless.

For evangelical spokesmen in seminaries as well as *madrasas* the transfer of creationary power to governments and objects is a sacrilege, a crime of hubris against the Boss. Though they are convinced that evolution is mere dogma and dead wrong, they still fear its charismatic power over thought, as natural selection and survival of the fittest provide a satanically rational explanation for the presence of life and diversity of species on Earth.

They don't realize that theirs is not the real "God," but a totem serving equally secular and materialistic goals. Remember Emile Durkheim's "elementary forms of religious life." It hardly matters if the setting is the Australian outback, a church under Jerry Falwell's ministry, or a mosque in Qum:

> The totem is their rallying cry; for this reason ... they design it upon their bodies.... Since they are emus or kangaroos, they comport themselves like the animals of the same name. By this means, they mutually show one another they are all members of the same moral community and they become conscious of the kinship uniting them.... The Australian seeks to resemble his totem just as the faithful in more advanced religions seeks to resemble his God....
>
> The traditions ... express the way in which society represents man and the world; it is a moral system and a cosmology as well as a history.[1]

Intelligent Design . . . .

At first glance it would appear that "intelligent design biology" (ID) provides a credible third option, reconciling creationism with molecular dynamics. Its proponents argue simply that life and mind are too complex to have originated from random selection; instead, they must be products of some extrinsic biological plan. Just as the faces of George Washington and Abraham Lincoln at Mount Rushmore patently could not have been formed by wind and water so, they reason, a cell, by its sheer intricacy of function and design, is beyond agglomeration by chance events in nature.

Stuart Kauffman argued on a similar basis (and for equivalent reasons) that a universe engaged solely in random molecular sorting has not been around anywhere near long enough to assemble the proteins that make up rotifers and fleas, let alone the beasts themselves—but he did *not* go on to claim that either animals or their macromolecular components are so irrevocably complex as to require a Supreme Being.

* * * * *

Also known as irreducible complexity, specified complexity, and designed complexity, ID challenges the very concept of *innate* complexity. Its premise is that tides and precipitates cannot concoct living structures no matter how long they swish, sputter, and brew, for alive creatures are synopsized or epitomized in stages.

For an object to be irreducible by the tenets of intelligent design, it must consist of multiple components, each one of them irremovably critical to its function. Any precursor entity along a hypothetical lineage of cellular organization, necessarily lacking some essential device, would never achieve systemic function and thus could not have been selected according to evolutionary principles.

Even relatively primitive biological forms, for example, require two hundred or more genes to designate the proteins underlying their configuration and biochemical function. If a single locus among these were missing, the ID argument goes, none of the other interdependent structures would work or, if one somehow lurched into operation, it would be out of context and, at best, inefficient. Thus, in Darwin's version of nature, it could not have evolved.

How, for instance, could cilium structure (and function), the genetic code, and innumerable, diverse metabolic pathways have emerged in gradual stages by evolution if none of their pieces have any selective value in and of themselves without the others? DNA missing a section cannot function as DNA; it is irreducible. Like a watch without a spring, separated helical strands—the more likely products of random oceanic sorting—are functionless.

• • • • •

Lehigh biochemistry professor Michael Behe is the most well-known present spokesperson for ID. In *Darwin's Black Box,* Behe tries very hard to convince us that, purely as an open-minded (albeit Roman Catholic) scientist, he has applied advanced microbiology and biochemistry in the laboratory to uncover systems that, on their own merits at the cellular level, are irreducibly complex.

According to Behe, a "bacterial flagellum is literally an outboard motor that some bacteria use to swim.... [It] requires dozens or even hundreds of precisely tailored parts" that must be realized together for the "motor" to function. "Nobody has ever proposed how something like that could be put together step by step."[2] Other irreducible devices include: the eye; blood coagulation; the Krebs or citric-acid cycle; the relationship between messenger RNA and protein synthesis; and myriad neurons, ganglia, and immunity mechanisms in beetles, birds, and rodents, as well as other cellular structures. These consist of parts that, if Darwin (as Behe reads him) is correct, must have evolved independently before their cohesion into a functional whole.

Yet they did not, for, if they had, they could not have been acted upon by natural selection, as they had no function prior to their final assembly; thus they "could not" have been fabricated without a plan.[3] They are impossible objects unless someone made them.

To support his teleology, Behe quotes first Darwin as acknowledging that the very existence of such complex assemblies, if it were demonstrated that they could *not* have been formed by gradual selection, would cause his own theory to "absolutely break down"[4]; and then popular Darwinian Richard Dawkins as "confirming" that organs as complicated and designed as eyes must either represent a gradual "coming into existence" or be totally inexplicable and a "miracle."[5] The implied verdict is obvious.

Trying to wheedle evolutionists' arguments to turn against each other, Behe engages in shameless sophistry, imputing by his juxtaposition that even the redoubtable Dawkins doubts Darwin. Like an editor from *Forrest Gump* cutting and pasting scenes from different clips, he does not call attention to the fact that the two individuals in his example, writing nearly a century and a half apart, are talking about quite different aspects of evolution (organismal and molecular, respectively). In the context of evolution, "gradual" means "stepwise; in stages" as opposed to "in a giant leap." It does not mean, as Behe implies, "taking a long time" as opposed to "rapid."[6]

This is a red herring anyway, for (as readers of the previous chapter know) neo-Darwinian theorists do not posit life and speciation as the outcome of chance progressions forging machinelike parts (Behe's claim). Nature has shown that it can be intrinsically designed without our either assigning its instrumentality to an exogenous agent or fully grasping its working principles. Darwin's living designs arise algorithmically from molecular organizations and adaptations, their indispensable parts coalescing by probabalistic, nonlinear determinism. They mature wholecloth, phase state by successive phase state, not by independent evolutions of different device-like parts at different speeds. The latter would be an anthropomorphism of nature, viewing it through a machine-shop metaphor.

· · · · ·

Contrary to Behe's implication, evolution *does* employ preexisting structures in new uses. It achieves homeostasis by alternative interlocking biochemical pathways to the same tissue plexuses. A biont is not a mousetrap because mousetraps do not have inventories of surplus springs and back-up platforms. That the embryo continues to develop even when damaged means that it was created in the first place by redundantly complex—though not necessarily supernatural or immutable—processes.

By contrast, a watch or outboard motor is manufactured economically and nonredundantly. Crack a part or clog a valve, and the machine instantly stops running. It has no alternate pathways.

The removal of a critical genetic or enzymatic component does not shut down a biological system's flow of energy (as it would with an engine). (See also pp. 99–100.)

> [B]iochemical processes frequently do not involve simple, linear sequences of reactions, with function destroyed by the absence of a given component. Instead, they are the product of a large number of overlapping, slightly different and redundant processes [that turn out] to lie at the heart of the stability that these processes manifest in the face of perturbations that ought to catastrophically disrupt systems from the standpoint of Behe's central interpretative metaphor of the well-designed, minimalist mousetrap—the absence of any component of which should render the system functionless.... It is a hallmark characteristic of evolved biochemical systems that there are typically multiple routes to a given functional end, and where one route fails, another can take over.[7]

The phenotypes of an evolving organism can remain functional even after severe genetic perturbation. Loss of some hereditary elements usually leads to compensation by other genes such that the task of an absent or displaced module is subrogated by another one flawlessly taking up the slack (e.g. different pathways from species to species to manufacture identical lens crystallins). Fundamental motifs are applied in new anatomical and behavioral contexts—as thermogenin for brown-fat heat when mammals moved to colder climates, float bladders for lungs in the first amphibious fish. Nonlinear dynamics restores essential design and biochemistry.

Living systems in fact *must* make use of preexisting components because they have nothing else; they cannot order parts by UPS. A basin-of-attraction model wins out over two guys repairing vacuum cleaners in a Torrance, California basement.

•••••

In a Darwinian universe, an evolving, self-organizing, interactive design does not rely on aimless shuffles of matter and information. It internally sifts through itself, through its multiple attractors and complexity feedback networks, merging them irreversibly into unique states of organization, as a drop of food coloring kneaded into a pile of bread dough can never be retrieved (even if time were magically reversed). Randomness and entropy are represented as much by their irrevocability as their blind, desultory wandering. This is a subtle but crucial distinction.

Complex structures such as eyes plausibly developed step by step along independent trajectories, sometimes slowly and sometimes rapidly, through ordinary chemical processes with countless intermediate steps—from primitive photo-sensitive spots in marine invertebrates to iris diaphragms, movable lenses, and retinas in squids, and so on. Each light-intercepting and -resolving phase of an emerging ocular organ could collect vision-enhancing oscillators and trigger the next basin of attraction, as each tissue nexus complexified and its beneficiaries came to detect their predators and prey more acutely, thus survived in greater numbers to have more offspring.

A nascent organism plays with the edge of chaos in order to achieve order, dancing among levels of strange attractors, triggering one phase state out of another. This is natural, unspecified design; it is design that is not "designed" but elicits and reveals itself through autopoiesis. Its complexity is inextricable but not irreducible.

•••••

Behe disdains all such attempts to explain ordered systems by complexity theory—as idle speculation, abstract algebra aloof from real biochemistry. It is hard to see how BZ reactions (see p. 30), to take one example of nonequilibrium complexity, can be dismissed as lacking in biochemical specificity. BZ chemistry fits Behe's definition of irreducible complexity—remove or alter one part and the organizing oscillation fails—yet it develops intrinsically and "defends" its own cycles.

But Behe's game is solely to embarrass "Darwin" and his followers; otherwise, he would acknowledge complexity theory as at least partial explanation for apparent design. Instead he singles out Kauffman's Boolean hypercubes as an indication that evolutionary theorists are hopelessly out of touch with life, "trapped in the mental world of mathematics."[8]

Evolution and intelligent design are irreconcilably antagonistic explanations. . . . .

It is possible (though not likely) that the new creationists do not understand what a bad idea intelligent design is, and not because Darwinism is unassailable but because it has real vulnerability at a much deeper level from which we are distracted by the ID tautology.

And a tautology it is. The fact that the precise evolutionary origin of biological designs cannot be demonstrated experimentally or confirmed on the spot by molecular biology is an incidental dead end of about the same ilk as the possibility that Francis Bacon or someone else wrote plays under the name "Shakespeare." We cannot substantiate the exact history and mechanism used by evolution to achieve genetic and tissue structures because we have no access, either direct or forensic, to the actual relationship among biochemistry, physiology, and heredity at the time of system "invention": "Macromolecular structures and complex metabolic pathways do not make good fossils, so biologists must simply accept that we will never be able to rise to Behe's challenge by proving to him how each of his microbial models of irreducible complexity evolved some 4×10^{19} microbial generations ago (assuming a division rate of about 20 generations per day). For blood clotting and the immune response the same considerations apply, although their origins do not penetrate as far into the past."[9]

We can extrapolate the kinds of mechanisms involved and their required phase depth, but we cannot track their precise evolutionary blueprinting (see also p. 28).

* * * * *

Some ID advocates actually see the tautology in their own "irreducible complexity" argument and so try to have it both ways by embracing key aspects of neo-Darwinism, while substituting supernatural agencies for molecular dynamics. In the universe of intelligent design, morphogenesis is a preprogrammed seed planted in the soil of differential selection: "[T]he divine designer infused ... complexity into his creatures ... programming the very first cells with the entire repertoire of genes needed for every successor species...."[10] After an inscrutable first cause is allotted to them, designs can be irreducible without being inexplicable.

Creation myths handle breakthroughs with totems that precede their own existence. That is how the Eskimo stormy petrel caught seals before there were any seals and how the mother of all caribou manifested already wearing breeches of caribou skins.

Though the neo-creationists clearly intend an omniscient deity to be their only candidate as the grand molecular designer, in order to retain credibility as scientists they leave open the logical option of any outside architect, including, for instance, an alien mage.

The blatant inefficiency of nature and the wastefulness of random selection and its staggering cruelty are rationalized as God's testing of his creatures or—grasping at straws—his mysterious ways opaque to humans: We don't see the whole picture. If we did, it would all make sense.

Some modern IDers concede points that would have horrified 1920s Scopes-era creationists: "Yes ... our planet has been in orbit for billions of years. No, Earth's ten million species weren't crammed into Eden together. And yes, the extinction of some 99 percent of those species through eons preceding our own tardy appearance is an undeniable fact. Even the development, through natural selection, of adaptive variation within a given species is a sacrificed pawn."[11] But they cannot allow life to arise and diversify harum-scarum from interactions of particles in kinetic systems without the intercession of outside wisdom.

Their real quarrel, they contend, is not with evolutionary science, which is an honest attempt at deciphering of God's labyrinth, but with the replacement of their *bona fide* religion (metaphysical biblicism) by a fake one (metaphysical naturalism, i.e., extrapolation from natural section to species creation). Claiming to be allies of modern biology, many of these Christian philosophers insist—perhaps disingenuously—that evolutionary science not only runs well but "doesn't imply that there is no God, or that God has not created human beings in his image, or that the second person of the Trinity did not become incarnate, or that there aren't any souls, or that if there are, they are in fact material processes or events of some sort, or anything else of the kind. It is only evolutionary science *combined with metaphysical naturalism* that implies these things."[12]

By posturing that evolution and the origin of life are two different working paradigms aimed at different targets, they promulgate the double illusion that science is not concerned with the origination of systems and that intelligent design is filling that gap without getting in the way of evolution. In fact, while a Supreme Being is not in explicit conflict with natural selection, his intrusion as system-designer is superfluous to nonequilibrium dynamics. The relation of organized shapes to a creator *per se* is irrelevant, as life forms can be probed and understood in their own terms (even given the present impossibility of getting to their molecular bottom).

Any putative intrinsic "intelligence" of nature (archetypal or deific) is so deeply imbedded by now anyway that it hardly matters what kind of entity—transpersonal or personal—"put" it there in the first place. It is a metaphysical not a scientific question, so it has nothing to say—pro or con—about the existence of God or the immaculate conception.

Why anyway is an omniscient deity preferable—or less fake—than autopoeitic sorting by strange attraction?

· · · · ·

Metaphysical bioengineering events, like the pre-Copernican planetary epicycles of an earlier generation, can fit anywhere and be argued backwards (reverse-engineered) at leisure by those proposing them, thus are truly tautological. No supernatural wand igniting Earth's biosphere or *deus ex machina* in nature can be either confirmed or invalidated by empirical evidence, since any omnipotent agency could override local cause and effect to effect a burning bush or whatever else he or it wanted. He could rudely break into the laws of physics with any whim or miracle or, more reasonably, disguise his activities in a vast, immaculate, nonlinear mechanism. A real God would hardly employ patchwork laws or devices that exposed his meddling. For Him, a sacred tablet or parting of a sea must be the same as a genomic alphabet.

Even engineers "of unknown identity and methods, be they cosmic, or merely alien"[13]—an eligible creator for most ID partisans— would be impossible to locate from within his handiwork. And he would require a creator himself.

Furthermore, what is the ultimate functional implication of intelligent design as a substitute for evolution? Does a designer specially create each organ, each gene, or each creature *in toto*? Or does he merely set in motion systems that create genes, organs, and organisms? If so, how (again) is he different from inherent natural complexity fluctuating from attractor to attractor? Is it that he mostly lets things go on their own, intervening only at key moments to help creation along?

These various possibilities suggest the puerility and ridiculousness of specified design.

ID gets rid of the algorithm but not other, equally mechanical euphemisms and anthromorphic metaphors. In fact, ID is nothing more than a metaphor masquerading as an explanation. In this case the metaphor has taken on such a life of its own that its originators have forgotten how they confabulated it.

Orwellian Doublespeak . . . .

Intelligent design didn't arise as a concept from legitimate scientific inquiry and discussion, nor does it have any lineage in intellectual history. It appeared full-blown, much like an advertising slogan launched on Madison Avenue by Ogilvy and Mather. The client was the Discovery Institute of Seattle, a conservative think tank founded in 1991 by Reagan-era politicians with the goal of influencing scientists and academics to abandon Darwinian materialism and the Big Bang and to accept a creator of the universe, likewise to ram ID down the gullets of school children in order to restore the great white Christian totem with its sanctimonious political agendas, including *Pax Americana* and global capitalism.

The strategy of those associated with the Discovery Institute has always been to make it seem that they are debunking Darwinism solely as flawed science while advancing irreducible complexity in its place as a more optimal biomolecular interpretation of the facts. Their main spokespersons are, in fact, scientists, many with Ph.D.s, some of them well-funded to affiliate—and they don't even have to share the evangelical agenda. They merely need opine about the depth, wonder, and impenetrability of biological complexity—a sure bet by any standards. A Christian God is the subtext but, artfully, never the text.

• • • • •

Appearing on National Public Radio's "Talk of the Nation," February 13, 2002, in connection with Darwin Day to chronicle—ostensibly impartially—scientific alternatives to Darwinian theory, Behe employed the insincere techniques of a politician more inclined to sound bites than transparently characterizing his platform for the audience. Routinely denying not only any connection to creationism but any knowledge about it, playing dumb when anyone brought up a possible ambition to back-door the Bible into the classroom, he made it seem as if he were talking only about his own research: "He represented himself as a scientist persuaded by the evidence, not a creationist with an agenda."[14]

The science he promoted was mush, its distinctions purposely conflated. He did not acknowledge, for instance, the clear categorical difference between artificial objects of demonstrable design like wristwatches and Mount Rushmore (whose history we know) and natural objects of apparent design. "Real biological systems are quite unlike economically designed engineering artifacts such as mouse-traps"[15]; they merely resemble them, as noted, in some features.

He confounded "proximate usefulness and current interdependence of parts" with "historical modification and change of interdependent,"[16] multiply redundant parts.

He implied that intelligent design and natural selection are on the same footing in terms of both their scrupulous application of scientific method and the amount of research that has gone into each one. What research? What data-gathering? What testable or falsifiable hypotheses? What science? In truth, *at least* several million times more laboratory research has gone into evolution than ID.

• • • • •

Traditional science is unique because it proceeds by strictly enforced rules. It is pretty much an honest broker; a mostly objective, self-correcting series of theoretical models based on repetitions of peer-reviewed experiments. What can't be demonstrated in a lab and then redemonstrated is provisionally discarded. Before advancing any law or agency, science runs it through mathematically certified protocols.

Holding to rubrics like this makes it difficult to say much, or anything that is not simultaneously discrete, concrete, and repeatable.

But creationists feel justified, no doubt buoyed by the sanctity of their cause and inflamed by theological zeal, in saying anything they want, hence to mimic science's rigor by using rigorous-sounding metaphors in place of hard-core experiments. Theirs is the *Reader's Digest* or *Classic Comics* version of microbiology.

In another drama-laden performance of pseudoscientific sophistry, ID embryologist Jonathan Wells supposedly trounced the iconic "proof" of moth evolution by camouflage, on the basis that light moths supposedly "did not increase in frequency after air pollution was reduced..., [while he also feigned] righteous indignation about 'fraudulent,' 'staged' textbook photographs of light and dark moths against light and dark backgrounds ... photographs [that] merely illustrate the differential camouflage that field experiments tested."[17]

He conveniently failed to mention "the role of migration and gene flow among populations, or that the light colored morph has now recovered in all populations."[18]

Squelching unimpeachable evidence subversive to his position, he put his energy into rearranging the original text to look like dissembling and propaganda, which was of course his own m.o. Strange tactics for a spokesman of God! Odd that a preacher would think that God needed a rigged defense! Even odder that he would not see how this deceit trivializes God!

This same author revealingly provides a bookmark and an on-line template sticker to download (www.iconsofevolution.com) entitled "Ten questions to ask your biology teacher about evolution"—for students to anneal to any of their textbooks that support evolution (e.g. just about any traditional scientific text) and then to use as a prompter in class to protest against "materialistic claims that we are just animals"[19] and to quote from during Q&A following lectures by evolutionary biologists. He urges "Congressional hearings to stop 'supporting dogmatic Darwinists that misrepresent the truth to keep themselves in power.'"[20]

Talk about enforcement of beliefs by mob instead of reason and persuasion! Political operatives flying in to steal meetings and vote counts!

But who could deny that this is how elections are won these days? (And, by the way, *who* exactly is in power?)

• • • • •

Behe, Wells, William Dembski, Philip Johnson, and their Discovery Institute colleagues are additionally disingenuous in: 1) claiming that science's recent discovery of the greater complexity of cells and proteins confirms ID and makes it impossible for molecular biology to delve any further into ineluctable phenomena—when it actually more dramatically confirms complexity theory and opens microbiological experimentation to a new, dimensionless universe of research through further parsing the previously irreducible (and sacrosanct); 2) playing "Gotcha!" by overblowing mistakes one or another IDer fakes having discovered in a scientific textbook, errors that are minor, incidental, and have already been widely acknowledged or corrected; 3) taking advantage of a large number of well-meaning, scientifically illiterate citizens who can be swayed to support creationism after being confronted with the seeming paradox of highly designed life forms with minds confabulated by random assortments of atoms and molecules (what could be more absurd?); 4) appropriating stature they have not earned within the scientific community or by peer-reviewed work but which comes solely from political pressure generated by extracurricular publicity—which has the deleterious effect of making it look as though aggressive p.r. is more gainful than good science; 5) using funding from the religious Right to foist their writings on textbook adoption committees, if not as required reading then as supplementary volumes—they make these demands on the basis that including both sides of the debate is only fair despite the confusion caused among students by blatantly biased and unscientific misinformation [of course, when supporting their own interests (e.g. opposing abortion for some or promoting unfettered gun ownership and the death penalty for others) they no longer push the motto that "fair" means equal opportunity for what is "right" and what is "wrong"]; and 6) drumming it into the public awareness that evolution has not been experimentally proven when obviously no one can conduct an experiment by time machine or over hundreds of thousands of generations.

This brouhaha is the inevitable consequence of science's conflation of materialism with meaning colliding with religious fundamentalism's conflation of literal biblicism with meaning.

What will always be—because of the incomplete nature of conscious inquiry and the shifting cultures and languages in which it occurs—a transient mirage of irreducibility should not get frozen ideologically as a condition of theory itself or a dogmatic requirement for thought police to control. It should not become a pretext (in the short run) for outlawing Darwinism from public schools or, even worse, for teaching intelligent design along with evolution as two equally unproven, mere hypotheses.

Irreducible complexity has no standing at all outside the Darwinian logos it assails; it is not a theory on its own merits. It is an end run to supplant evolution with theism, a modern version of creationism, and an attempt by the Christian right in America to retry the Scopes case.

This truth is well characterized by a 2002 letter to *Scientific American*. Would that Behe, Wells, Plantinga et al. could drop the bogus kid gloves and speak as sincerely. When not posing as gentlemen scientists, this is about what they have to say:

> As a young-Earth Christian, I find all the answers to the
> meaning of life in the Bible. Even if I were not a Christian, I
> would find the theories of evolution insane. God gave men
> the brains to develop computers and all the amazing inventions we enjoy today. . . . It seems that the more we learn, the
> more hardened evolutionists become in their rebellion against
> God. If the genetic code discovery does not prove intelligent
> design, nothing will convince evolutionists.[21]

Nature is more profound than our thought processes. . .

The deconstruction and refutation of orthodox Darwinism is a worthwhile undertaking for reasons that should not have Christian theologists in their vanguard.

Materialistic science offers a grimly jaundiced view of the universe, one that is based on a fundamental objection to spirit. Everything, both inanimate and alive, is subdivided and pigeonholed, robbed of essence. Every possibility (other than technological) is subjected to cautionary arguments of "this will only lead to that," so every trail out of nihilism is as fatal as the condition itself—or even worse, for wasting hope.

We are so afraid of ourselves that we have expended centuries of monumental labor trying to prove in theorems and laboratories beyond a doubt that we don't exist, that meaning is not meaning; life is not really life.

We tell ourselves we don't have to evolve beyond our present lot because the world is in such a terrible state—and only getting worse—that it wouldn't do us any good anyway. People trapped from childhood in a cynical world-view think, "Why bother?" We have an ideal alibi, a foolproof excuse for not trying. We nurse a built-in bias against ourselves—that, as spiritless, transitory objects, we are fundamentally and decisively flawed—and then we zap ourselves with it continually. We are self-validating only in a negative way that gives us a negative identity.

Our ingrained, individual tendencies then recognize themselves in a group creed. After all, the culture seems to thrive while it confirms our worst prejudices against ourselves. (The current ecological crisis has its roots in another crisis. Not existing ourselves and with nothing inside us real, we have no conscience, no restraint, no reason to cherish or protect our world.)

We are good at knowing what everything is in its most limited, menial sense, not good at being alive. We are stuck in a tragic consciousness we have stamped upon ourselves. We assume the entire universe is grounded in that reality, that it is what everyone experiences. At the same time, we don't think we are being indulgently nihilistic. Science tells us that we are realistic and hard-nosed, not negative.

We have become, to our own proud judgment, the world's most destructive species, maybe even the cosmos' most destructive creature, and now we think that we can get away with anything.

All this is because we see ourselves only from the outside, not from within as spirit sees us. No wonder priests and clerics are trying to reinstate humility, to get our tail between our legs before it is too late. Unfortunately the fundamentalist pundits are equally arrogant and external; they are merely another symptom of our negative identity.

You don't get kids to stop shooting heroin and aborting fetuses by imposing a joyless, despotic figurehead on them, by telling them, "Christ died for you, and you are nothing and you better know it." They see right through that, and they can find plenty of other gods and voodoo masters to support acts of sacred disobedience.

You must encourage them to find their own magic. After all, Christ died so that people would be as he was, find their origin in God as he did, not so that they would feel terrible because he died for them.

·····

In March of 1999 an internal white paper was leaked from The Center for the Renewal of Science & Culture at the Discovery Institute. It addressed the social and spiritual damage wrought by Darwinian materialism. The first four sentences are right on, but from there it is partisan politics. The whole statement is irredeemably corrupted by its last four words:

> The social consequences of materialism have been devastating. As symptoms, those consequences are certainly worth treating. However, we are convinced that in order to defeat materialism, we must cut it off at its source. That source is scientific materialism. If we view the predominant materialistic science as a giant tree, our strategy is intended to function as a "wedge" that, while relatively small, can split the trunk when applied at its weakest points. The very beginning of this strategy, the "thin edge of the wedge," was Philip Johnson's critique of Darwinism begun in 1991 in *Darwinism on Trial,* and continued in *Reason in the Balance* and *Defeating Darwinism by Opening Minds.* Michael Behe's highly successful *Darwin's Black Box* followed Johnson's work. We are building on this momentum, broadening the wedge with a positive scientific alternative to materialistic scientific theories, which has come to be called the theory of intelligent design (ID). Design theory promises to reverse the stifling dominance of the materialist worldview, and to replace it with a science consonant with Christian and theistic convictions.[22]

I too think that materialism is corroding civilization while threatening the biosphere. I too think that we are shutting out our higher selves. I too think that life is too complex to have been formed by random events, probabilistic determinism, and survival/fertility regimes. But I do not believe that that points to the teleological strokes of a supreme architect in the Judaeo-Christian or Muslim sense.

Ultimately nature is more profound than any of our thought processes, scientific or theistic, which should be no surprise given that consciousness is an effect within nature. We were never meant to replace living the riddle with solving the riddle. And only living it will approximate solving it anyway.

Darwin came to a parallel conclusion during his journey through matter, though his own sole alternative to willy-nilly selection was likewise stern figureheads from the Old and New Testatments. In 1870, at the age of sixty-one, he wrote:

> My theology is a simple muddle; I cannot look at the universe as the result of blind chance, yet I can see no evidence of beneficent design, or indeed of design of any kind, in the details.[23]

He wanted desperately to find something other than godless predation and consumption. He was actually both the first creationist and the first Darwinian. At heart he was closer to some forms of creationism than the inconsolable neo-Darwinism his career launched.

As elegantly and with as much precision as he proposed modern laws for biology, that explicitly and paradoxically close did he come to the great labyrinth of human existence. There is no way the convictions of the human mind "developed from the mind of the lower animals are of any value or at all trustworthy. Would one trust in the convictions of a monkey's mind?"[24]

"How can we escape this recursive paradox," added his intellectual heir Stephen Jay Gould, "that our brains, as biological devices constrained by the history of their origin, must be enlisted to analyze history itself?"[25]

Darwin could not deduce evolutionary laws without meeting, perhaps one starry night between dreams, the mystery of his own power of deduction in seeming contradiction of so many diverse "random" events—the wonder of his own consciousness among orchids, tortoises, and pigeons in their same esoteric world. Though he saw no divine instrument or basis for one, there was also far too much order, beauty, and design for this all to be expressly what it seemed to be. It was a problem so profound that Darwin could neither have conceived its true depths nor foreseen how meekly his successors would do away with it by consecrating his aphorisms.

• • • • •

Darwinian science fingers the truth without necessarily recognizing it: forms actualize their own complicated depths and unfold from cosmic debris without any outside directive. They manifest intrinsic, transpersonal intelligence.

Biblical creationists misunderstand the axe they grind with Darwin. The physical fact of evolution among moths, grasses, voles, cats, cattle, catfish, and the like is undeniable, but it is only one form of creation. The goal should not be to censor or pigheadedly deny the thermodynamics of speciation but to meditate upon it deeply enough to see what it *really* is and where in it the Spirit of Waters dwells.

The way out of the present trap is to imagine the full scope and entanglement of what we are, and see that it is vast enough to include anything, even the divine.

Gaia . . . .

James Lovelock was a biologist working with NASA during the 1960s' development of the *Viking* spacecraft program. His assignment was to search for, among other things, life on Mars. While devising models for a Martian biosphere, he recognized that life on any planet required first and foremost an entropy-reversing system all the way from its atmosphere down to its geosphere. Maintaining a biosphere meant keeping the entire landscape, inanimate as well as animate layers, in a quasi-nonentropic state and maintaining it there by moment-to-moment adjustment of its homeostases.

Eventually Lovelock proposed a theory of the Earth as a single living entity including not only its biota but its near-surface rocks, oceans, mountains, atmosphere, etc., in a unified, self-organizing, self-regulating system with feedback unconsciously flowing among its levels and stabilizing its biosphere in such a way as to support a habitat nurturing life and intelligence.

Overall systemic control of our planet occurs through global oscillations, ostensibly keyed by a transbiotic intelligence, a designer without a brain and extrinsic to any particular organism (or deity, for that matter). This master operation covers maintenance of numerous supersystems, including atmospheric balance by life forms removing some gases while replenishing others—a normally unstable mix at 79% nitrogen, 20.7% oxygen, and .03% carbon with trace amounts of methane and other elements; the mineral equilibrium of oceans at precisely 3.4% salinity for several billion years—the ratio most suitable for the emergence and survival of cells; and a temperate surface over geological times during which there has been a 25% increase in solar heat.[26]

. . . . .

A living Earth in the spirit of Gaia transcends a biblical deity, or at least encompasses Him, because it embraces all cultures and species and has no axe to grind. It is pagan, nontragic, resourceful, and closer to the spirit of the real God of Genesis or Matthew than any theocratic totem.

If life is to be found irreducible, that should not be in order to assign a personalized creator or designer to it; it should be in order (first) to understand that complexity in its own terms so as to continue the millennial inquiry into matter, consciousness, and nature; and (second) to keep the case open because what is irreducibly complex today may yield to analytical and laboratory methods of the future.

The one place where I might deviate from the position of complexity theorists is on the matter of whether intrinsically developing, self-organizing designs have roots in systems that are more like the phase states and oscillators found in chemical experiments or more like the archetypes that Jung intimated in the transformational projections of alchemists (both Taoist and Western) who found God and consciousness immersed in matter, and matter immersed in consciousness. But that is somewhat like asking, "How complex is complex?" or "What role do our own minds play in our coming to be, both before and inside the fact?"

Macho Marxism

Editors of serious mainstream publications (like *The New York Times*, *The New Yorker*, and *The New York Review of Books*) continue to act as though the debate between neo-creationists and neo-Darwinians is substantial and juicy, the single most serious one between religion and science. This inflation serves both factions (as well as their journalistic allies): Darwinophiles can summarily write off IDers as uneducated and brainwashed sophists, and creationists can dismiss strict Darwinians as ideologues blind to the holes in their own theory. Each side gets to posture vaingloriously against a straw horse. It is a staged match, like G. Gordon Liddy against Timothy Leary.

Writers for the same "New York" periodicals quash or ignore all other challenges to neo-Darwinian liturgy, whether those be grounded in particle physics and molecular biology or come from ostracized subcultures. They dismiss such "heresies" much as Christian fundamentalists rebuff gauntlets to the Bible—they are sacrileges hardly meriting consideration, let alone rejoinder.

In both cases it is a hardball authoritarian prejudice. They do not want to yield a nano-unit to Mother Nature herself and her pagan, mysterious, synchronistic, primordially manifesting, incipiently telekinetic, tantric universe.

Under science's martial regime, discussions of alternatives to neo-Darwinism and alternative cancer therapies run into the same ideological brick wall, the same unyielding censorship, because both inquiries affront the literal, hard-and-fast commoditized nature of cells—what they are, how they organize, where they come from, what they do with information.

It is perhaps a carryover of faddish ideologies arising from Darwinian-based Marxism that not only neo-Marxists but Marxist renunciates, anti-Marxists, and Marxian illiterates promulgate not even so much Marxist beliefs or their antitheses but the shadow of prudish and macho materialism cast by the whole Marxist liturgy and enterprise over civilization. To put it another way, it is perhaps the modernist assumption that the great debate of human society involves only the relative roles of labor and capital; of free markets and communism; of matter, energy, and the valuations and prerogatives generated by their transformation into products and ideas. Since politicians—in the guise of corporate executives, union leaders, generals, lobbyists, and guerrilla chiefs—run everything, physicists and biologists want most of all to be politicians, to be players in the ruins of the recent Marxian wars, because there is still no other recognized battlefield.

To address the riddles of the universe and to call them such by name is foppish to them. They would rather be taming broncos at the Commodities and Products Rodeo. To be hip these days is to subscribe to the tough-love, compassionate conservative, atheistic theistic, terrorist, preemptive-strike world-view of genes, humans, tumors, and societies—to accept toxic, global-market commodification like a man. Even female, gay, and lesbian scientists buy into it.

· · · · ·

The clash between "Darwin" and God is a sterile and solipsistic one. The dichotomy is apocryphal. There are many admirable theories in which evolution, self-organization, nonequilibrium thermodynamics, embryogenesis, and consciousness come together without an outside agent. It is far more fruitful to consider how a God who is inseparable from nature works through secularized matter to generate autonomous creatures. In Buddhist and Hindu cosmologies, divine presence manifests itself through multidimensional universes of atoms and polar energies, which are both alive and thermodynamic, infinitesimally material yet phenomenologically evolving.

Are science's objects real? . . . .

While scientists endow ostensibly airtight laws, originating in the Big Bang, with control over all events (including self-organizing ones), this is a lot to lay at their feet without first tracing them empirically back to their own cultural and linguistic origins to see if they are true unbiased witnesses or mere clan objects.

Claude Lévi-Strauss' structural mythology, Maurice Merleau-Ponty's proprioceptive phenomenology, and Benjamin Lee Whorf's and Noam Chomsky's phonemic genealogies—each in its own way—show that language and its depictions are psychoneural arrangements of sounds and syntaxes in hyper-strings and not mediums in which absolute truth or pure things can be inscribed. Pea soup, salami, and lime soda (to take one example from a cafeteria) disintegrate into "s's," "l's," "m's," and other phones. Words defined only by other words are talismans. Numbers, terminologies, and kindred signs are morphophonemic conceits; they are not *really* real things. They do not deserve coronation. When they are long forgetten, the inner world from which they arose will endure.

• •

Morphophonemics: The variant sound relationships and patterning, including addition, loss, and stress shifts, of indivisible morphemes ("meaning" units) in languages.

• •

Biology and astrophysics knit priceless but flimflam medicine bundles for Western civilization—hives of tribal totems gathered mythohistorically and sewn together as if facts. "Cells," "genes," "black holes," and "Big Bangs" are more clues to how our lineage addresses "being" and nature than they are components of actual living or cosmic things. Their laboratory credentials notwithstanding, they are adumbrations of conceptual things. They generate business for science; they raise scientists to levels of esteem and social reward; and, as "glorious souvenirs,"[27] they address the beginning of time and the origination of plants and animals. They "render the mythical past of the clan present to the mind."[28] Corn and kangaroo ceremonies and tales of Spider Woman and Dream Time do about the same among the Hopi and Warramunga.

Scientific objects ultimately hold together no better than marriage rules of the Witchetty grub clan at Alice Springs, Aranda moieties, or totemic operators linking animal, fish, and bird sodalities among the Ojibwa. They are no more *real* than the Yellow Eel or Mud Turtle phratry, though they speak to our own (not an Algonquian) crisis of meaning.

Ideological reductionists like Stephen Jay Gould, Carl Sagan, Richard Dawkins, and Steven Weinberg have particularly bristled at such critiques of science because these come from sources that are intellectually at their level (so, by their standards, should know better).

· · · · ·

Yet the notion that science is relativistic and subjective, underwritten by Jacques Derrida and Bruno Latour (among others), is, by comparison to militant Judaeo-Christian or Islamic imputations, respectful (though cultural relativists think scientists are smart enough to know better, too). Philosophers and cultural historians define scientific modernity as the *deluded* quarantine of physical objects and technological things from human things (e.g. those of morality and aesthetics) in such a way as to elevate science to an abstract status alone unpolluted by human egoity, alone able to explore the universe impartially: "the rational progress of science, in potential league with progressive politics, patiently unveiling a grounding nature."[29] Every bird, every molecule is dragooned.

Treating their measurements and inscription devices as facts designating final events in which action and power are uniquely located, members of the Sagan-Dawkins klatsch annoint their own mode of knowledge as the closest thing we will ever have to absolute truth and "speak as if they were the mouthpiece for the speechless objects that they just shaped and enrolled as allies in an agnostic field called science."[30] Yet natural objects have an ancient and pagan right to exist on their own, apart from being detained or taken into custody by science.

Despite the present technocracy, science's girders are affixed to a mirage, i.e., scientists' own unacknowledged and undiagnosed totems. We maintain our Brobdingnagian edifice upon the shallowest of foundations. What we don't see is that the bare mechanism of nature (lurking in electrons, viruses, rabid coyotes, and dying stars) presents few insoluble predicaments for mankind. Real danger attends scientific and social objects configured in symbols. Our relations to one another and to things, while tribal and generic, are contrivances, forever in disquiet and incipient anxiety. Culture barely holds together its own paradoxical rules and acts. It cannot obviate its own crimes or protect its civilians and innocent wards from crossfire. It can hardly claim to incorporate atoms, molecules, cells, and their technologies as neutral or solely beneficent guests.

Ignoring this flaw, liberal modernists still think science is not only correct but, by the righteousness of its correctness, uniquely humanitarian and just. They peg our survival as a species to its future advances. Science's critics regard such beatification as precisely the delusion that clouds our vision and in fact *threatens* our survival.

The resolution (or further pathologization of this crisis) marks a crossroads for the nine-billion-plus humans of the twenty-first and twenty-second centuries.

Paraphysics, Science, and Professional Skepticism . . . . . .

Despite fastidious policing by scientific and skeptical authorities, strange things apparently do happen in ordinary places in normal times. Fortean frogs fall from the sky. During the late 1960s Ted Serios, a Chicago-based elevator operator sequestered behind lead in a Denver lab, seemed to cloud film with images he received from the mind of a man staring at photographs in another room.[31] This same basic experiment (remote viewing) was later conducted between Argentina and Manitoba and other sites.[32] Children occasionally remember former lives in faraway villages, travel there, and find lost objects; this is also how lamas traditionally are identified in Tibet. Other people begin to speak ancient languages they were never taught. Electrodes register the hormones of startle on skin an instant before any actual surprise.

Urban myths or true blue, information keeps showing up where it *couldn't* be.

• • • • •

Most real scientists would whoop with a child's glee if some "supernatural" event like a spontaneous crop circle or telepathic message repeated itself in a consistent, analyzable way. They live for adventure and astonishment.

But they don't move on hearsay or hit-and-run events. They stick to rubrics and rules; they don't know any other way to play, and they have no terms for admitting that they just witnessed the impossible. They'd rather not report UFOs or ghosts to their peers.

The engineers and interdisciplinary experts who viewed Serios' feats in the morning, to the consternation of their host forgot them over lunch. Without peer-reviewed, replicable experiments, they'd rather discuss the weather, their vacations, hoops.

Sticking to the rules is the only way science can protect humanity from being taken over again by religious blackmailers and ideological terrorists.

Good for science!

• • • • •

This culture works not because we are clever and have beaten the cosmic game but because there is a relationship between the structure of the universe and the human endeavor characterized by empirical experiment. In critiquing science, we must not forget this.

Physicochemical research, Dzogchen Buddhism, Taoist internal alchemy, and shamanic transfiguration are among the best attempts of our species to meet the universe head-on, to aspire to the intelligence that made us.

Where would we be without science? If we didn't have it, we would have to start inventing it from scratch at once. The forerunners of Anaximander, Thales, and Aristotle took care of that. The goal is not to lose their thread at this uncertain juncture in human history.

. . . . .

Professional skeptics and practitioners of ideological scientism, by contrast, seek a future controlled by computers, chromosomes, and neuropharmaceuticals (hardwired connections only) and—more than that—by their own kind of folk. Eggheadtopia or revenge of the nerds. They are *against* spirits, synchronicities, telepathies, and alchemizations of all kinds, just on principle. They would be appalled by an actual, irrefutable UFO, free-energy machine, or homeopathic cure. If silicon chips can accomplish the synthesis of mind outside bodies or microwirings someday feign telekinetic action, then great—let's do it! Let's make all love and entertainment virtual and patentable.

Repressed yuppie technocrats want to run their own utopian world of cyborgs—no unconscious intrusions, no gods or spirits, no vital energies, nothing irrational or supernatural.

Just don't let anyone else do it by yoga or astral projection. Block all escape hatches. Stomp out all mind-directed molecular effects. Close the X Files. Kill mutant ninja turtles.

The goal of a ghostless, spiritless universe has led to an inquisition far more devious than anything Spanish or Roman clerics could have dreamed of. Our modern constabulary bribes and tampers where holy cardinals once bludgeoned and burned at the stake. The message of both, one papal, one anti-papal, is the same: "I am ... and thou shalt have no other. . . ."

It is uncanny how closely the ID crowd resembles the professional skeptics in both strategy and mood.

. . . . .

Something has been missed that is simultaneously so incongruous that it could be an elephant in a tea shop and yet so intangible that physics and biology can shun it entirely and make off like thieves.

With expanding domains of uncertainty, quantum information, and chaos-based order, a tight mechanical universe is outmoded, and the Darwinian synthesis is breaking down (without any push from neo-creationists) in every aspect except its ritual hold over an ideologically reductive imagination (fans of all-powerful genes).

Science is not wrong; it is just that it has limited its explorations to one province in a single dominion—physical objects and events. It mistakes things for meanings, content for context, trees for forests. With no mature context for its thermodynamic laws, galactic formations, or bionts, it is susceptible to any new red or blue shift, black or white hole, supersuperstring, charmed quark, or transdimensional quantum error-correction. It locks onto transient piecemeal views and then employs intellectual tactics to enforce them.

Most practitioners feel that, in order to be rewarded, they have to deliver permanent cause-and-effect to the corporate office and then enforce any issued patents.

Mainstream science has dismissed or ignored the paraphysical, the epiphenomenal, and the proprioception of "being" as if these were not part of nature. Its aspiration is, in fact, to bury phenomenology and provincial morality.

It has birthed an epistemology of dissociation, unleashing demonic forces that rival dragons and ghouls of mythological worlds.

Most citizens, whether they know it or not, would rather be ruled these days by atoms and slogans than their own restless natures and unmentionable desires. We have such a desperate need to be accounted for, to be "real," that we have composed a facile algebra for the fathomless forces that birth and validate us.

What has once been imagined cannot ever be unimagined. . . . .

Established doctrines and ideologies lull people into assuming our essential condition is known or accountable. This could not be more fallacious. Whenever someone asserts a dogmatic fact—the universe is this or that; life is this or that; "meaning" is this or that—they may be interesting (or not), but they are always wrong. The dilemma of being unable to tell the truth is far more basic than any theory or proposition.

In fact, we know absolutely nothing about what this is, except what our being tells us it is. Phenomenology is alive and well.

Pleistocene hunters butchering a caribou, slobs dumping beer on each other in a club, bus-drivers following routes, latrine-cleaners in India, and jugglers keeping balls in the air are all experiencing near syllogistic truths about the nature of reality.

Until we know what reality is, we are confined to what reality does, what reality feels like to its habitants. And I mean *everything:* no one thing closer to the mark than any other—physics no better than bowling, microbiology no more empirical than necking.

. . . . .

When I mistyped "Bing Bang" above, I saw how easily concepts are altered—and that's all any of these "things" are. As the poet Gary Snyder remarked to me in 1971 (re: dime-sized universes, Big Bangs, et al.): "Those guys are paid to sit around all day, smoke expensive dope, and think things like that up."[33]

You can't stop mentation from going anywhere it wants, mind from conceiving promiscuously. Murder, decapitation, pink elephants, genocide *(kill the bastards),* blasphemy, XXX, masochism, zombis, suicide bombs, time travel, endtime, nonexistence, prayers to every imaginable gnome and god, lie ripe and irresistible in layers of consciousness.

The damage has already been done, in fact long ago. Obsessive compulsive disorder is a modern epidemic. Though we designate it as a disease (OCD), it is in fact a burgeoning civilization, covered by a working press and broadcast to itself in staged episodes via satellites.

Mao Tse-tung said, "Science is simply acting daringly."[34] And he made it stick. Anyone who lived in China during the second half of the twentieth century understood the lethal power of abstractions like "red guards."

That the universe might never have happened is a thought. That we are constructed of cells by genes is a thought. To pretend thought doesn't exist anywhere in a physical universe is a trick of thought, a failure to identify where thought hangs out. Here comes Miss America is a thought. God is a thought. Evolution is a thought. Allah is a thought. Intelligent Design is a thought. Self is a thought. A talking duck is a thought. Death is a thought (that neither a goat nor a spider has). The Big Bang—or Big Gang or Big Gong—is a thought.

These thoughts create lives, ethics, whole belief systems. They have enormous, far-reaching consequences. It hardly matters that they are also not real.

Our consensus world is testimony to the compulsive worship and transmigration of thought forms, most of them semi-random, all for their own sake.

What has once been imagined cannot ever be unimagined. It continues to flutter through designants like so much mince. It becomes habit, precedent, basis. It *is* the universe.

Quantum Tygers . . . .

New Age cosmology—with its assorted claims of hyperdimensional energies, chreodes, galactic humans, quantum this and quantum that—is even more provincial than science in that it makes things up without any mathematical or ontological basis. Few of its practitioners recognize that the stuff they contrive, though passionately and super-sincerely conceived and attested to, does not constitute solutions even to the problems for which they are proposed. They are metaphors or coincidences preening as facts—or they are on the level of stopped clocks which are right twice every day.

Everything that is not understood, such as synchronicities or faith healing, can be delegated to a "quantum" realm wherein some Dutch or French physicist is ever about to discover, say, the physical basis of prayer or healing neutrons among subparticles. When you give yourself permission to say anything at all, what does it matter what you say?

How does cosmic energy become psychic and personalized? How does a thing called a "chreode" send messages by "morphic resonance" through time and space? How do thirteenth-dimensional wave forms create life ... life in its raw tuberousness? Might as well go back to guys pouring hops on one another's heads.

I am guilty of plenty of "quantums" myself—assorted decoherence waves, extended phase states, poetic uncertainty principles, transdimensionalities, and other meta-babble throughout this book. I just hope that no one takes these ganders as serious convictions of mine or proposals of fact.

• • • • •

There is little intelligent dialogue between proponents of a supernatural vital force (on the one hand) and supporters of autocatalytic molecular sets (on the other). Few readers of this book would accept a co-consideration of these riddles.

My New Age audience valorizes telekinetically channelled energy, cosmic consciousness, a nonhuman intelligence designing crop circles, hyperdimensional cities, and astral beings. An equally charming gang is loyal to deconstruction, artificial intelligence, neo-Marxism, set theory, and abstract topology. These groups don't talk respectfully to each other. In fact, they generally don't talk. They munch away in coccoons of their own enthusiasms, vigilant against contamination from the other.

• • • • •

Indigenous hunters in the Canadian Arctic gainfully track caribou using scapulimancy, divination by the charred scapula bones of prey. Their success is probably because the cracks that are scorched by flames into the bone map randomize search patterns and defy any incipient caribou savvy, while the rite itself invokes transcendent group ideals in the manner of a football rally.[35] This is how a lot of fortuitous outcomes occur. Magic (like life) is just system depth and intuitive dowsing.

Quantum information and its uncertainty states are relevant to life and mind in some fashion. Clearly we are rooted in a universe of language depicting emergent chaos-based, nonequilibrium fields. We are generated inside their reality, and our minds and energies must be expressions of it. There must be uncertainty events at the basis of life and consciousness; otherwise, we would not exist. Scapulimancy works for a reason. Yes, we are quanta more than bricks.

Yet the actual if elusive "quantum" that arises from Max Planck's and Werner Heisenberg's atomic experiments and has been rediscovered in mathematical factoring and discrete logarithms is not kin to the New Age "quantum" applied to everything mysterious or inexplicable. The expression of a quantum domain among organisms, symbols, and societies cannot correspond to quantum effects among subatomic particles. Despite the organization of the universe by emergent properties, mathematically "hard" versions of quantum theory, superconductivity, and superstrings cannot be dragged roughshod across scales from quarks and electrons to atoms to molecules to cells to organisms to societies. Each system, passing up or down fractal levels of existence, is sufficient unto itself since knowledge at that level is uniquely required to predict systems-level behavior.

It is neither necessary nor sufficient to understand iron at the atomic level to predict the path of an inelastic collision between two steel ball bearings, yet the bearings' fundamental nature is derivable from knowledge of those atoms. The interactions of children playing with those ball bearings in a game of marbles is removed yet another level. There are levels below that of the atom and above the group playing marbles. These layered sets are fuzzy at both extremes.

At the outer limits, noise rules; down the microcosmic tunnel, quantum effects blur details excruciatingly.[36]

The universe roars across all these scales, oblivious to noise, oblivious to uncertainty. At least it gets down into the mess where life itself is—the thicket from whence hunters and caribou arise.

"I can feel my scales growing already." . . . .

After being out of touch with each other for more than thirty years, Andrew Lugg and I are now (2001) exchanging emails between Ottawa (where they are ice-skating to work) and Berkeley (where the plums are in blossom). He was my graduate-school friend in Ann Arbor, and he warns me he hasn't changed much since ("save for the grey hair and weight"). He is still, in his own words, "on the side of the woodenheads," adding: "I am always intrigued by the fact that whereas people like you are very interested in science, it does virtually nothing for me. I figure things have to work some way and I could care less which way they do."

Great point, Andy! The Wittgensteinians deserve to demolish the technocrats, even over the latter's Internet. And "some way" is just about how things work.

There is not a physical, scientific universe and another, metaphysical one—a natural domain and a supernatural one. Everything that happens is natural and physical both. My belief is that reality is complex and messy enough that neither Yaqui shamans and Da Free Johns nor Lacanians and chaotecians have the whole pie, have anything other than paradoxes needing totally opposite paradoxes to explain how four and twenty blackbirds got in.

• • • • •

My text takes place not only in gaps of intellectual cultures (scientific, linguistic, and psychospiritual) but in stand-offs of countercultures (hierarchical priesthoods and anarchic street rappers).

Academic Buddhists preach on matters of souls and destinies, how beings transmute between lifetimes and get reborn ("one life

of being a government official results in nine lives of being an ox"[37]).
Hardcore artists and urban clowns surf the same wave with a sense
of prankishness and the absurdity of it all.

At a book-signing for the publication of my earlier writing on
this topic (*Embryogenesis,* in a classroom at the California Institute
of Integral Studies, San Francisco, April 2000), I read some passages,
then responded to questions. Charlene Spretnak, a faculty member
and author of much-admired tomes on the politics of women's spir-
ituality, good-humoredly but imperiously objected to my presentation of spirits travelling through bardos and reincarnating down the
line as humans: "According to my Theravada teacher," she advised,
"a human form is very hard to attain. Most people do not get to be
reborn in a human body again."

The event was being videotaped by veteran beat poet and vaga-
bond chronicler Kush, founder of the Cloudhouse Poetry Archive
and more used to open mikes and slams than karmic exegesis. Charlene's comments aroused his startled attention.

Not noticing his rustle in the corner, she paused and smiled
smugly: "We are more likely to come back as cockroaches."

He stared at her, mock aghast. Then under his breath he mut-
tered, "I can feel my scales growing already."

3

Biogenesis and Cosmogenesis

. . . .

Cells, Genes, and Planets

> It's lovely to live on a raft. We had the sky, up there, all
> speckled with stars, and we used to lay on our backs
> and look up at them, and discuss about whether they
> was made, or only just happened—Jim he allowed they
> was made, but I allowed they happened; I judged it
> would have took too long to *make* so many. Jim said
> the moon could a *laid* them; well, that looked kind of
> reasonable, so I didn't say nothing against it, because
> I've seen a frog lay most as many. . . .
> —Mark Twain, *Adventures of Huckleberry Finn*

Embryos uniquely herd molecules into animate systems.

Embryogenesis—life out of lifeless matter—is the ghost in the celestial machine, the oracle of nature. The shaping construct is axiomatic. It surpasses relativity, charmed quarks, and star nurseries, for it directly confronts the mind-matter interface, hence, the condition of "mindedness" from which we emanate.

Morphogenesis (morphing) is so contrary to galaxies, comets, asteroids, and Mercury-like/Pluto-like planets that it, viral and crab-like, spurns all their physics and chemistry, making nature far more fecund and enigmatic than it should be. "Indeed, with the sun shining beatifically upon [it]," Stuart Kauffman notes, "... the biosphere may be one of the most complex things in the universe."[1] From where he is dealing, that indeed seems a droll understatement.

The last 4.8 billion years or so, the Earth has been taking in and converting solar energy by agile, diaphanous structures that are not entirely equilibrium-dissipative. Because of them life is now its own epidemic with private rules—flowering, polyping, gastrulating, ciliating, starring, cocooning, calving—creatively exploring the basis of organic design in near blissful disregard of entropy.

If entropy is the tyrant it is rumored to be, this here is a full-scale rebellion. Highly refined information comes pouring out of the unpromising maelstrom; not only information, but predation, community, desire, text.

It is in truth outrageous that so much sentient activity, including everything we think and do, is anomalous and flies in the face of ordinary equations and their classic degradation.

According to Kauffman, the biosphere "is doing something literally incalculable, nonalgorithmic, and outside our capacity to predict, not due to quantum uncertainty alone, nor deterministic chaos alone, but for a different, equally, or more profound reason."[2] That profundity lies at the heart of our sense of ourselves and of all things. The ignited universe exists as two contradictory domains.

The outcome of insurrection on Earth is a range of phenomena that defy their own initial and boundary conditions. Quahogs don't just bubble in place. They are aggressive, accessive whorls. All animals are immeasurably opaque, cognitive events—not only phenomena but phenomenological. Even paramecia and amoebas have autonomy and agency. Mountebanks of their domain, rabbits and bears celebrate their existences and, within the limits of their corpora, seek meaning ... and dinner. Ultimately the descendants of worms name gods, establish lineages.

• • • • •

From this perspective, two distinct states of matter define creation: raw molecules and the embryogenic field. Their separate domains are addressed, respectively, by astrophysics and morphogenesis.

Physics comprises interactions of particles, radiations, gravity, vacuums, stellar objects, quasi-stellar objects, and space-time in their most universal and rarefied states as well as coarser *entremêlements* among stones, fluids, gases, tides, shear and compression forces, etc.

What has not been fully acknowledged, or perhaps even recognized, by biologists is that embryogenesis is more than just a special case of general thermodynamics; it is a radically different method of organizing matter. Swarms of molecules under heat and gravity alone don't ordinarily turn into vermicular metabolizing shapes. They don't become cells, let alone snakes.

Embryogenic fields herd commonplace dust into self-organizing, self-plicating, replicating forms. Their fetuses and hatchlings introduce an entirely unique expression of physics to a universe that, because of it alone, generates symbols.

An embryo-mass is beyond binary functions of quantum mechanics; it is only figuratively a vigesihedral spin network—or a host of other abstractly transgeometric objects. It is, in truth, an absolute and complete break in mechanics—a state of innate and probably inevitable de/coherence. Its vintage includes (thus far) cells, tissues, fungi, plants, animals, sambas, spears, huts, factories, languages, totem poles, coyote myths, cars bumping along on roads, etc. One way or another these all come out of cell-based orchestrations of molecules. None of them could exist in a bare heat-and-gravity universe without the intercession of an embryogenic field.

Yet embryogenesis as a system of meaning is astonishingly ignored. While the public pays homage to genes, pulsars, electrons, hyper-strings, fractals, black holes, etc., as clues to reality and reigning mysteries of creation, it overlooks the seething, gyrating, self-mutating writ of molecular cobble collecting in tissuing sheets that alone animalizes matter. The embryo is mistaken as a trademark, a staid chemicomathematical event—at best, somewhat less determinate than gas molecules—elapsing mechanically protein by protein, even if in probability states within the nuclei of its cells and nonlinear flux of its macromolecules.

Yet life is weirder than any contingent state of matter, for it is in life forms, not electrons, not quarks, that uncertainty principle and the space-time continuum take on their ultimate—in fact, their only—expression. Life brings "our" circumstance to recognition, as creatures literally turn the universe inside-out.

An inanimate universe is an empty universe. Without life, the universe is a dimensionless void.

Nothing can get itself alive and aware without being embryogenic in some fashion. . . . .

Embryos on Earth tell us what "being" is and how it originates, metabolizes, reproduces itself, and evolves *anywhere on any world.* Terrestrial cell fusion and design, our local mode of embryogeny, are a clue to all others across creation. As we extrapolate outward from local embryos to the hypothetical virulence of "plant" and "animal" organizations on other worlds, exobiology can be derived from biology without having to encounter a single alien life form in the flesh.

We know this for sure: where life exists, it will have come into being by invagination, microbial layering, and a quantal step-down of solar rays through membranous papyruses—even life that is totally foreign to Gaia. The universe cannot manufacture animate forms out of molecules without an intermediate process for convening and organizing them, molding a permeable inside and outside, aggregating membranes in layers, linking them to one another metabolically, and capturing and conducting energy for their individual uses. This will hold on planets throughout the physical cosmos, though it may not always involve mitochondria, chloroplasts, and DNA-like helices, or their analogues. The shelf fungi-like intelligences, spiderfish, and flying snakes of the Andromeda system, if they exist, are composed of aggregate nonequilibrium, meta-stable folios packed with arterial infrastructures that ravel from a brood of archetypal ova and then turn themselves, as they heterogeneously snowball, inside-out.

If leviathan creatures the size of Earth's mountains whip flagella-like halyards to soar through the core-heated depths of Jupiter's ocean like an endless Caribbean, they too must be embryogenically whelped. They could not have been forged from scratch without some sort of incremental enveloping and indexing, without dense, recombinative eggs and giant curling gashes. A rolling seam or series of rolling seams must put stuff inside and outside itself. Life plasm must condense in monad-filled sheets, layer by layer, fractally and geodesically, until they achieve enough tunnelling and truss to withstand their planet's crunch. And this will be true whether such creatures are made of cells or some other particle-form.

Two-kilometer Jovian jellyfish trailing rows of tentacles hundreds of meters behind them, their regalia beating a minute and a half per undulatory cycle in stately unison, may spurn terrestrial modes of morphogenesis, but they will still be materialized by some form of radial cleavage, Boolean distribution, lamination, foliation, squamation, cavitation, incremental thickening and thinning, vacuolization, outpocketing, and neural filamentation. Their ontogeny will reenact and condense phases of their phylogeny. This is atomic, molecular certainty.

The same condition will pertain to metabolic crystals budding in Jupiter's atmospheric turbulence and raining down into leviathans' food filters.

Even sessile, rooted plants throughout the universe arise from seeds and have deep binary, embryogenic structuring.

* * * * *

Incorporating geometeorological events from the gaseous fields of their pebble planets, embryos complexify to the point where, despite the miniscule size of even the largest of them, they rival their local suns. They x-ray and digitalize their own atoms and subatomic particles; they peer into their cells and cavities. They build megascopes and launch satellites to find where they are. They make weapons that replicate solar fusion. Hypothetically, they could grow powerful enough to destroy whole solar systems.

The infinitude of the universe is captured in the nucleus of the cell. . . . .

Long ago, here and elsewhere, through the complicacy of the cell (or some other sporing and hiving bud), the termless exponents of galactic geography grounded themselves in life forms. Stardust and mineralized muck were crenulated by molecular machinery. Their flakes agglutinated and came alive.

The process of microcellularization has used the breadth and girth of the universe to encapsulate its own nondeterministic domain— to scratch, scallop, laminate, and mold carbon-phosphate clay; to seal sequential eskers in recombinative meiotic bundles; to establish hierarchical organization and replicative machinery; in short, to fashion its own nucleocosmos. Gossamer microbes, diatoms, and minds have been spun from intermediate patterns into the great hollow zone among stars such that the catacombs of space are now rendered hologrammatically inside boxes of trillions of soft rhomboids.

One night on a raft slipping down the lower Mississipp', Jim got it: Stars and frogs are germs of the same order. Conception of life as cells is a precise corollary to the synthesis of stars from hydrogen. Cosmogenesis was once, and has again become, oogenesis.

An egg in a nest precisely countervails and defies the Big Bang.

• • • • •

The physical universe may be vast beyond measure or conception. Yet paradoxically, the nucleus of the cell is equally vast. The cell is the full explication of galactic vastness at another scale in another dimension. Blistering infinity, it makes the heavens truly complex.

Cells are a warp in the otherwise unbroken membrane of linear space containing sortless debris. They are what turns the universe both inside-out and inside-in and lights it with its own visceral radiance. There is no other trajectory along which scurrying fantoccini could be fleshed—no other vector for space and time to display their substance and subtlety.

Above us, the milky explosion of night. In local moonlight, a rustling tree. These two entities—the cosmos and the cellular nucleus—are in perfect, excruciating, impossible balance.

We are compiled of other creatures because nature had no other way to make us. . . . .

Plants and animals are fundamentally modular—quorums of pulsations networked into transformers driven by chemical reactions and solar particles. Our tissue is varmints thronging in the dissipating fields our existence imposes.

It is *beyond* incredible that unprodded, inanimate waters bundled in batteries and wrote a nonlinear, revampable script for reproduction of their own molecular designs; it is equally beyond incredible that pairs of wee germ units collaborate daily to construct skinks, skunks, and ursine philosophers. This could not have happened without long, discursive series of prior evolving matrices—phylogenies.

First, the intelligence and integrity of "cell" had to coalesce from ocean broth and "be"; had to precipitate stage by stage yet whole-cloth from the solar system's wash; then plants, animals, and other runts had to be syncretized from thickening clusters of these monads. They had to hold together long enough to be worth making—to swallow other organic molecules, to assimilate what they ate, to accrue infrastructure and expedite energy cycles, and then to make themselves over.

Since apparently wolverines and elephants cannot be woven *ex nihilo* out of molecules, nature has to amass greater integrities out of masses of much tinier ones, first sealing energy in grids of them and then subsuming and translating their innate autonomy and agency across scales. (On alien worlds, likewise, buds and gemmules fission and meld to generate living designs.)

• • • • •

Through their interactions long ago, independent cells (for whatever reason) gave up selfhood, sank beneath the threshold of the organs and ego-minds they wove, and abandoned their separate lives and self-reliant agendas to assemble consortia in which each one of them had to function as an indivisible part of a whole.

This is a remarkable circumstance. One kind of thing, a cell, a monad, a ceramic-like panel, transubstantiated every kind of thing necessary for forging animals. If a car were a cat, identical mosaics would be turning into rubber for tires, synthetic fabric for seats, transistors for the radio, wiring for the odometer, steel alloys for the engine blocks, other metals for the muffler and gas tank, more pliable metals for the shell, paint for the exterior, plastic for the steering wheel, coils and glass for the light bulbs—all *in medias res,* in coordination with one another and in response to one another's changes.

Component cells are both sovereign and lackey. Atavistic descendants of free-living protozoa inside us (i.e., mesenchyme, lymphocytes, and spermatozoa) continue to express aspects of ancient autonomies. They are wards who are not totally free and whose ancestors were never fully enslaved. Their continuing independence within tissues is a crucial factor in the assembly and replication of life. Combining autonomy with tenancy, they send out immune patrols, lay wide-ranging neural networks, heal wounds, and project holograms of themselves. An extended organism uses these renegade zooids to feel and connect itself, to police its milieu, to explore the world outside its shell, to rebuild its shape after damage and disease, and to sort the dies of its prototype. Totally obedient gonidia would congregate in gooey, palpitating lumps. Feral cells coalesce into motivated, replicatable creatures. A cell as much *becomes an organism* as colligates with other cells into an organism.

• • • • •

While alliances among cells weave tissues, cells are composite beasts too. Welded out of primitive zooids—captured protists and microbes that generate their own proteins, enzymes, and nucleic acids—they refashion and incubate these prisoners into organelles (microscopic organs) which then collaborate in the breakdown of other molecules as digestive and locomotory energy. These exiles from chlorophyll sea-jungles also align to mint their own hereditary versions.

Ultimately all the conscripted microbes—Golgi bodies, mitochondria, microtubules, microfilaments, lysosomes, endoplasmic reticulum, and their kin—cohere into fully colonial units which emerge anew from syzygy or clone each other to birth approximately the same colonies in which their offspring play congruent roles.

Cells reproduce themselves as their tubes and fibrils line up and pull apart. Each fabric and fibril in the cell is replicated in a watery mirror splitting while the whole image cleaves into two identical icons as smoothly as a velvet glove comes off a royal hand. Strings pull apart and re-form inside the seals of new orbs.

Mitosis is like duplicating a statue not only three-dimensionally but in terms of the full three-dimensionality of each of hundreds of semi-autonomous blobs inside a dynamic structure.

A single germ cell coming from our body can fuse with a similar such cell from another body to assemble an entirely new man or woman—a person with a unique mind, separate essential organs (heart and kidneys and eyes), an identity, and a will. Our own body was cultivated that way, built up in its functional entirety from two consanguineous but separately arising gametes. In chambers of ourselves and each other, we make other people—younglings resembling the beings from which their foundation cells come. Possums, magpies, and termites do likewise.

Not only do we stick together, but we replace every atom, every proton, every molecule, and thus the substance of every cell in our body from minute to minute, month to month, decade to decade without any deterioration of the field holding all the fresh monads and their molecules together in pretty much the same slots.

Whatever put our molecules and cells in place holds them in place and, without reminding, puts brand new ones in the same places like clockwork. How do you keep sticking stuff together, not only agglu-

tinatively and viscerally, but phenomenologically, until you get "us"?
What kind of heritage is this for cabbages and kings?

• • • • •

Cells have been impressed into floating spongeous factories; jellied,
tentacled bells; aquatic pulsators; stop-and-go amphibious tubes;
preening lionesses; even Audrey Hepburn in *Moon River.* While
shaping whole sophisticated anatomies and destinies, they remain
existentially covies of cells.

What is the relationship between plant and animal forms and the
cells comprising them? What role do cells have in the logic of complex behavior? Are gigantic, ornamental entities that stalk and chirp
still basically and existentially cells or have their emergent properties made them wholly different things?

• • • • •

Something at the kernel of each multicellular creature obviously
takes over from the lives of its subunits and expresses a unique identity. System limits evaporate; their molecular shackles do not flow
upward—so scorpions and falcons who are only cells are not cellular at all in their behavior. Their primitive selfhoods claim full and
final dominion over their minds and bodies despite other, primige-
nial inhabitants. Their hearts will sacrifice limbs to survive.

We do not think of ourselves as communites of cells. Our minds
assert the irrevocable consolidation of tissues, extending an ego that
pays no attention to the discrete life forms underlying its existence.
We look at another creature and see a body—a ferocity and a will—
not a mobile pond packed with proxy paramecia and amoebas.

When we are sick, whole cell complexes (organs) are what we fix.
Even when we return to single cells in molecular/genetic medicine
(with either stunning or negligible results), the outcomes are expressed
only in terms of tissue knittings. Even in cancer, where cell cohesion
dissipates regionally, then metastasizes globally, pathology is identified as tissuelike tumors.

We act like normal, full-fledged entities, entitled to strut. In fact, we write the book of life; we describe our situation unsentimentally, even while we live it as some other thing.

And though we feel safe in our transcendence of cellhood and will never lose our emergent properties as long as we are alive, we have a kind of wariness because we do not know why our levels cohere, why our identity emerges through other life forms' conjoining, why our bloody channels and pulsing organs have a single imperative.

At a very real level, we are integrally and inseparably cellular and are no more stable or safe than a pile of snakes or tumulus of worms. One rock in our midst and apart we come, scattering to the eight winds. It is a legitimate miracle that we adhere at all. Meanwhile, every morning, every breath, every moment, we charge on oblivious. We have big plans, plans that seem grandiose and optimistic for something that depends on maintaining a moment-to-moment cellular, electrochemical unity that will all come apart, as if it never even existed, from the intrusion of one well-aimed projectile. And will come apart anyway of its own inertia someday.

Does any of this make sense?

A cell is more deeply structured than a star. . . . .

When nineteenth-century biologists recognized the primacy of protoplasm in plants and animals, they annointed the cell as the elemental subcomponent of life. "All our experience," wrote one, "indicates that life can manifest itself only in a concrete form and that it is bound to certain substantial loci. These loci are cells and cell formations.... Life activity is cell activity; its uniqueness is the uniqueness of the cell.... The cell is really the ultimate morphological unit in which there is any manifestation of life, and ... we must not transfer the locus of intrinsic action to any point above the cell."[3]

Despite the afore-mentioned superiority of tissues, cells are us. The intelligence of a single cell quintessentializes our condition.

For arising at all among the universe's stardust, for conscribing whole lineages of tinier creatures, cells are irreducible embers, diamonds of elemental phlegm, and oscillating pumps of primal intelligence. Protoganglia with far more memory and capacity (each) than the best silicon chip, they are self-sufficient, self-organizing, self-generative, and originless. With their plasma membranes, secretory vesicles, centrioles, vacuoles, and nuclei, they are the precise rendering of the antiquity, opacity, and infoldedness of the universe itself. They are each, in fact, little naked universes.

A cell expresses the innate radiance of matter—the dawn into which the pure mind of creation suffuses.

. . . . .

The cell as first and only principle of Renaissance biology came to mean the cell as robotic organic machine, as tissue-amalgamating mote, but it should also be the cell as vortex, as transdimensional bubble in the extensibility and boundary condition of cosmic ash.

For being small and primitive, cells and their organelles are neither slave particles nor mere building blocks, nor are they witless fractals of barely sensate tissue. They are transdimensional rents in the fabric of space-time, gap junctions in the cordage of matter. Meta-things releasing meta-energies, they foreshadow and potentiate the full rainbow of species.

Our consciousness is not just a synergized sum of cellular charges and sensations; it is the attunement and harmony of all their energetic junctions and chimes, of their stops and pipes in the howling nullity of matter—bottomless eddies that appeared unannounced and unpredicated in the waters off Babylon.

. . . . .

The singularity and explosion that astrophysicists propose at the beginning of time and space—the nuclear cauldron of the Big Bang—are finally an irrelevant and materialistic metaphor for the inviolable unity of existence. On the outward plane we are dwarfed by the astronomy of creation, but inwardly we comprise it.

Largeness and atomicity, beginning and end, thickness and extensivity, all things crushed together in density (before time) as against the porridgy vacuum of stuff in space (manufacturing time) are identical circumstances. This is the esoteric meaning of the Big Bang—a singularity followed by ignition, boundaryless dispersal, and reorganization.

If you glimpse deeply enough, you will see not "cells," but totally mysterious, unknown processes—bubble-packed crystals, crystal-packed bubbles, refracting hyperdimensionally through vibrations of atoms and molecules, shimmers that only seem to be "things."

Microcosm and macrocosm are whirlpools in each other's oceans. Inside-time and timelessness wage no dichotomy—in fact, vanish here. Cells percolate like moments, up through elemental debris.

The Appearance of DNA . . . .

No matter what opinion one may have of genetic determinism as an ultimate ideology, deoxyribonucleic acid is a majestic and incalculable object. Thin almost to the point of nonexistence, packed with text in a miniaturized format far beyond anything that an electrical engineer could devise, capable of writing whole life forms which then arise and morph seamlessly from each other, bearing its own utility programs, DNA behaves like the crowning achievement of an advanced technology on a planet circling a sun much older than ours.

One reason for such hyperbole is that DNA has the meticulous, intelligent design of an artificially digitalized and semanticized seed. Its "chip" was (hypothetically) produced in a lab on an unimaginable distant world, then packed in a capsule and launched out of its star system. Along with tens of thousands of other such vessels, it crossed light years of space and plunged happenstance through a young atmosphere into a primeval sea. Bursting its friction-scorched cannister, DNA came to life and, subject (of course) to local weather, demonstrated its full portfolio.

Most of its sister spacecraft are still hurtling aimlessly across the void that makes up most of space. A few of them probably fried to crisps in stars or crashed onto barren, frozen worlds. One or two of them seeded other planets, far from here. But then they were fictions to begin with, right?

. . . . .

DNA betrays the molecular fingerprints that a visitor from the stars might. For one, it debuted completely designed, surprisingly soon after the Earth's fiery inception. It invaded a terrestrial clime resembling present-day Venus with toxic metallic rivers and radioactive air more than Chamber of Commerce waters around Hawaii. It "terraformed" this land into a place habitable by its own alien offspring. Within a mere billion years, DNA-coordinated bubbles (fossils now show) flourished in heated pools and springs, releasing oxygen, initiating the Gaia effect.

But then who made the extraterrestrial creatures who fashioned it?

None of DNA's nanotechnology and archaeology, of course, proves that it is machine-sculpted or alien to the Earth; it merely shows how an unprecedentedly complex and refined object appeared suddenly in an otherwise primitive situation. If we dismiss science-fiction plots,* life seems either a triumph of algorithmic nihilism or the descent of a hyperdimensional kingdom—or the outcome of intelligence that is intrinsic.

Whether DNA was invented by super-engineers or arose from indeterminate networks in the body of nature or truly is (as science claims) a chance artifact of Big Bang debris, it is certainly a perfect solution to an unknown problem.

The Changeling Molecule . . . .

Once written, DNA proved stable and resilient both, projecting myriad fabrications while remaining unchanged in structure and

---

*Interestingly, for all their interstellar drives, time machines, and flying automobiles, no science-fiction writers of the 1940s, '50s, or '60s foresaw the microchip.

configuration at its core. Its hyper-intricate mechanism presumably provides the plan to construct any plant, animal, microbe, virus, or mushroom on Earth, as well as an infinite number of possible other bio-entities. Original DNA has conceived every life form from the dawn of the Earth's biosphere. Yet—despite enormous discrepancies of scale, design, somatic patterning, habit, habitat, and behavior—in each biont it remains identical. Despite millions of remarkably varying species and motifs (from toadstools to baboons to sea urchins), there is a profound accordance of all living things at a cell-molecular threshold. Shapes as diverse as turnips and bears are organized by synonymous globules. What was used to design primeval lichens and cyanobacteria needed no modification to become lamprey and mastodon building-blocks. In fact, four hundred modern human genes retain basic codons that are still found in baker's antiquarian yeast.

Over hundreds of millions of years, while nothing has amended the basic genes and their structures, the entire landscape has transmogrified around them—every river, every mountain, every sea, every cranny. And genes have played a critical role in that upheaval, while maintaining their original design.

Now a random handful of soil contains some ten billion bacteria and over a million fungi, comprising "more order, and information ... than there is [by present reckoning] on the surfaces of all the other known planets combined."[4] Every meadow and droplet is inhabited by thousands of mostly tiny creatures, both winged and wormlike, or both; some bearing long rows of glasslike locomotory bristles oscillating in frantic synchrony; some with plates of armor, condemned to battle; some with muffs, some with horns, some with palps, some with antennae, some with spiny fur. At the bottom of the ocean, muscular disks and fans tiff with polyps and medusae, crawling shells, many-tentacled balls and bells, stinging filaments, gelatinous lobes, multicolored tunics, and glowing worms and fish, chubby and flat.

The unity/diversity koan represents a paradox at the heart of nature. How can so much feracity emerge from one configuration?

Biologists extrapolate incremental pathways from microbes and protozoa to macaws and ferrets, but the living reality in a tarn or jungle, wild and self-possessed, caring not a whit to be endowed or named, disdainful certainly of cellular consanguinity, still startles and baffles them.

• • • • •

Crystal-like and hieroglyphic, DNA is borne within the nuclei of cells in discrete, encrypted, inheritable cubicles (packets of biological information called genes), organized themselves in long, thread-like agglutinations (chromosomes). All the genes in all the chromosomes of an organism comprise its so-called genome (short for "gene/chromosome").

While storing and replicating gargantuan amounts of compacted data in its genomes, DNA has no access to a world outside of its nuclear membrane. It provides only templates for amino acids which themselves construct enzymes and protein fields. It never sees the Promised Land, which is the bequest of proteins in an entirely different nonlinear field of activity.

Proteins make up more than half the dry weight of cells while conferring their actual shapes and structures and providing mechanisms for them to recognize and catalyze one another. Proteins (not genes) are thus the purveyors of life's organized topologies. Yet, unable to mint themselves from scratch or to reproduce, they are exclusively requisitioned by DNA.

The two realms rest in unstable equilibrium subject to each other's algorithms and dynamic phase transitions. This is a bizarre legacy and blueprint for biological systems. Somehow distinct duchies— one linear and metalinguistic (DNA), the other multidimensional and topological (proteins)—came together through a glass darkly to architect form and meaning.

The Design of DNA . . . .

DNA is an extremely long and thin filament—more than a billion times longer than it is wide. Two yards of it in a human cell measures only ten atoms across.[5] Imagine a limb attenuating from New York to Tokyo to grab a gyoza. DNA is in fact so close to the limits of materiality (more than a hundred times narrower than the sheerest wavelength of visible light) that it cannot be seen through an optical microscope. Yet, at such total remove from optics and blinder than a bat on a sunless moon, DNA's proteins "know" how to jell into multiple, very different light- and hue-sensitive eyes— for fishes, insects, and humans—each commandeering photons and relaying data to a central ganglion that interprets their binary patterns by image and context.

The hundred thousand billion cells of the human body store something like 125 billion miles of DNA—which one writer compares to seventy round-trips between Saturn and the Sun.[6] The collective deoxyribonucleic string inside any one of us can circle the Earth about five million times. The only way that two yards of it can compress into the capsule of the cellular nucleus is by their twining around themselves millions upon millions of times, forming the lightest, most condensed software conceivable. That is not only a lot of tangle but a whole lot of string.

DNA is the only archive (we know of) that the physical universe has of itself, of its actual body. It is absolute memory, caching far more information than conceivable along strings only ten atoms wide. In fact, it stores, mythically, the instructions necessary to create all technology, through (first) forging microbes and cells; then organizing these in colonies and intra-extraverting their symbioses into jellyfish, worms, sea hares, and amphibians; and (finally) reorganizing the composite crinklings and gap junctions within those (via generations of ova) into layers and membranes of bird and mammalian organs and brains. All this is somehow accomplished through juggling, reformulating and reinterpreting the same basic syntaxes in shifting protein fields.

The Language of Life . . . .

Genes contain self-generating series of glyphs resembling the morphophonemic sets that underlie language. That is, they comprise random words bearing no fixed relationship to their letters. Without submolecular poets to "hiss" and "babble," there is apparently even less onomatopoeia and punning in the genetic language than in any human one.

DNA's short alphabet of four letters is made of nitrogen-containing ring-compounds each linked to a five-carbon ribose (or deoxyribose) sugar, itself attached to a phosphate cluster. The base letter-rings, tatted of cosmic clabber, are named adenine, guanine, cytosine, and thymine (or uracil when occurring in a different form in RNA, one of DNA's guises). Adenine and guanine manifest five-unit rings, cytosine and uracil comparable six-unit ones.

There is no inherent relationship between codons themselves and their expressions; adenine, guanine, cytosine, thymine, and uracil are classically and inherently morpheme combinations making up *ex post facto* words. Some units inefficiently correspond to more than one amino acid or punctuation; for instance, in RNA, stop-start is represented nonprejudicially by uracil-adenine-adenine, uracil-guanine-adenine, and uracil-adenine-guanine. There are four identical forms of the amino acid alanine: GCA, GCC, GCG, and GCU. Serine is spelled alternately AGC, AGU, UCA, UCC, UCG, and UCU; lysine AAA and AAG. Scientists regard this self-created code as though it were written by our kind of intelligence. But DNA is actually the first language discovered by humans with no personified intention behind it. It doesn't behave like any cipher we know.

Genetic coding is a cipher only because we say it is. It is actually the consummate noncipher underlying all ciphers. No one programmed it; we do not know the class of design it entails. We make it look like language; we represent its activities by our models of signification—but that overlooks its fundamental nonlinguistic qualities. It is "speech" of an entirely different order, and its intelligence is of a different order too.

* * * * *

DNA's alphabet is much more defective and synonym-ridden than ours. Though its characters are mathematically capable of inscribing sixty-four different (4 x 4 x 4) words, each representing a distinct amino acid, in fact they confabulate only twenty-two: twenty amino acids plus two "stop-start" punctuations, all following the chemical formula $NH^2-CH(R)-COOH$ while differing in their side-chains. These semantic units—in groups usually of a hundred, several hundred, or considerably more—transcribe the almost limitless repertoire of proteins (and possible proteins) that amalgamate life forms.

Melanges of homonyms, acronymns, heteronyms, palindromes, puns, spoonerisms, and gaps are par for a message woven randomly in stormy seas under unmanaged conditions. Repetition and overlaps were mechanically and statistically inevitable, and they were preferable insofar as they provided tiers of insurance for loss or damage. Ideograms came to identical results along different amino acid/protein pathways and then were imbedded by millions of years of system development and interpolation. Once the genetic foliage was dense enough to bear fruit, lineages of life forms could proceed with some promise of stability, even as they were assaulted and discombobulated.

Transalphabetic potentiality holds the key to the relationship between layers of linear information and innate complexity, i.e., between biochemistry and biology. Fallback redundancy is invaluable. Duplicate representations are not wasted; they are permutable. Destroy or disrupt one pathway, and another identical or equivalent one, with a different developmental history, stands a fair chance of supplanting it. Genes and proteins, remember, are in complex disequilibrium with each other. The specific genes, while only adventitiously linked to their amino acids, are also unrelated molecularly to metabolisms and biological themes and meanings they initiate. Alphabets may vary, even catastrophically, over time, but then bionts get made by their aliases and subalphabets.

Biological syntax is pliant enough to generate manifold structures and functions almost regardless of intervening semantics. DNA is in fact so thick and versatile it is as though any aspect of it could code and stabilize virtually any morphology by one route or another. Crisscrossing pathways not only redundantly lead to the same results, identical life forms could probably be generated upon totally different girders and from nonparallel strings.

The reverse of Murphy's Law is in play: If something can make an organism, it will. Earth's botany and zoology—maybe not precisely sunflowers and roses, quaggas and hyraxes, but browsing quadrupeds and spirals of aromatic, sheathed petals—are lexically inevitable.

This signage remains just as crucial today because intricate creatures quickened by delicate, multidependent messages are also subject to continual environmental degradation. Heat, light, radioactivity, and thermodynamic entropy all work against life. DNA must continuously repair itself to keep its message from falling back into gibberish.*

DNA is an actual object.

Physically, the DNA ribbon is a continuous double helix: two chains wound quasi-symmetrically around each other and the molecule's central axis of symmetry and connected by their four base letters which clasp only adenine to thymine, and guanine to cytosine (the most effective hydrogen bonds being forged in these kinds of pairs).

Even so, DNA is hardly a fixed or concrete configuration. Helices multiply and degrade as the molecule interacts with replicative enzymes and is transcribed, recombined, cut, permutated, and bound to its transcription complexes.

A DNA helix is almost always right-handed but sometimes left-handed. Other configurations of DNA resonate in dynamic equi-

---

*As life forms age, the frequency of environmental mutations increases, though at a very slow rate—one more in 10^5 cells from ages twenty to sixty.

librium with one another, including three helices coiled about one another (triplexes).

As noted, there are variant dialects of DNA and RNA. For instance, a cell's intracellular mitochondrial code originates in quite a different province from its nuclear DNA.

. . . . .

DNA's double spiral is such a snugly and elegantly quasi-symmetric crystal that perhaps only it and other sorts of two-stranded aperiodic solids can collate life anywhere in the universe.

Physicist Erwin Schrödinger conjectured that some sort of aperiodic solid must lie at the basis of all complex organic entities. Irregularly vibrating crystals in quantum states have inherent advantages. They do not squander form in entropy; they do not succumb readily to error momentum, nor do they (like onyx or ivory) freeze information in dull molecular motifs; they juggle and reorder chance mutations in discrete sets—i.e., they reinvent themselves.[7] For a complex crystal, DNA is extremely stable and resilient, but then its sole "job" is to store and maintain series of data for very long periods of time.

Quite apart from DNA, or even life, a double helix and other helical structures may represent the primary form in which information is stored and transmitted across the universe at large. We will explore this possibility in Chapter Ten.

Writing the Protein Text . . . .

The structure of its helices reveals how DNA topologically encodes and replicates its content. Spiralling about each other in deep microspace, each molecule's chains are opposites—one a scrupulous text for transcription enzymes making up amino acids and the other, running in reverse, its backup. The backup cannot be sensibly read because, being inverted, it is nonsense (much as this sentence would be if retranscribed with its letters from right to left).

Although the accessory chain cannot be used to make amino acids and proteins, it is hardly superfluous. Mirror imaging allows DNA to imprint its cipher in a second, emerging strand being etched by enzymes—like a photographic negative in a positive. Amino-acid tristitches are thus cached in a format that can be accessed and copied with moderate ease. Their DOS is archived across time, as text continues to be inscribed, back to front, negative to positive to negative, reincorporated from organism into successor organism.

There is another use for a handy negative. If one ribbon's message is damaged and distorted, the other ribbon, serving as a kind of spell-check, is able to restore the hereditary sequence. This would have been an indispensable utility function when DNA was evolving in the Earth's corrosive, radiation-rich habitat, for serious damage to its fragile proto-codes would have halted life's whispering at a formative stage.

· · · · ·

As organisms grew more complicated, there were increased opportunities for random copying errors in their reproductive cells prior to mating. Thus, design protocols in genomes were permanently altered. Two hundred distortions per germ cell are a rough consensus for human transmission. Fewer mistakes likely occur in less complex, shorter-lived mammals and even fewer in invertebrates, but heredity is never flawless.

While proof-reading and repair work by enzymes naturally correct many copying mistakes, the best insurance for defect-free offspring is sexual reproduction—the luxury of four copies of genetic instructions, a double helix from each parent. Though obviously not identical to each other—or we would all be twins—they originate from the same common ancestor and each carries the species plan. Within the branches of a lineage, variation in traits (such as color, size, shape, musical or navigational ability, etc.) is natural; the traits themselves are more or less functionally interchangeable. They can be sorted in virtually any combination and still fuse healthily in offspring as long as they all express the core template.

New couplings of sperms and eggs meld in such a way that their chromosomes in homologous pairs (from each parent) exchange DNA sectors between themselves. Genes are also exchanged—and, with them, potentials of amino acids and proteins. As these sets of alphabeticized segments are shuffled and recombined during syzygy, lethal mutations are usually elided, a "correct" version generally overriding a garbled, misrendered one in the same position.

Most circumstantial, uncorrected errors are outcome-neutral because of the densely redundant network along which their information travels and is ultimately expressed. Though such miscues do not typically either enhance or undermine the absolute survival potential of offspring in a uniform way, they improvise within the domain of species perpetuation and, when not lethal, provide the basis for genotypic diversification and new design potential. Some will hit upon unexpected positive results, underwriting novel tissues, varieties, and even whole species. These then teem into available niches in seas and on continents.

A certain number (perhaps five to ten percent in humans' six to seven billion units of DNA) range from causing birth defects to being fatal.

Researchers who artificially intrude upon this deep-seated equipoise are meddling in a design process of which they may know the visible performance sequences but not the deep operational principles. They are tampering with the Old Speech, rattling the cages of demons they never meet face to face. They never meet them because they exist at a different level from the *fait accompli* of genetic transcription. Biotechnology, without some remedial gene therapy (not yet devised) to identify and replace disfigured bits of messages, dangerously boosts the number of mutant codes in human or other pools.

* * * * *

From the temple of inscriptions, the scroll is read. Enzymes reword DNA sentences, transmitting them in messenger RNAs (mRNAs). While eliminating noncoding DNA sections, these couriers splice information differentially and selectively in a manner (subject to constant microenvironmental control) that ribosomes can interpret and translate into polypeptide macromolecules. While DNA is tame, RNA is mutable and unstable, garbing itself first in standard hereditary code but then shifting chemistry and deriving renditions and prototypes of itself.

* *

Enzymes: a category of proteins that serve as biochemical catalysts in organisms, changing the rate of change in reactions while surviving the effects intact. **RNA (ribonucleic acid):** *a single-stranded molecule involved in protein synthesis, whose structure is specified by DNA. Its variants include transfer RNA (tRNA), which transcribes nucleic acid in protein languages; messenger RNA (mRNA), which is synthesized directly from DNA in the genetic material and attaches to the ribosome; and ribosomal RNA (rRNA), which, together with proteins, form the structure of the ribosome.* **Ribosome:** *a spherical, cytoplasmic cell organelle manufactured in nucleolus of the nucleus; it uses RNA to accomplish protein synthesis.* **Polypeptide chain:** *from approximately ten to a hundred amino acids joined by peptide bonds in a polymer.*

* *

Apparently far more ancient than its own present father, RNA (with its uracil base and ribose, not deoxyribose, sugar) is a descendant of an ancient self-catalyzing, heredity-bearing microbe that carried out some unique, originary aspect of the thermodynamic transition from inanimate chemistry to life. If any one naked object could be called "life," it is probably RNA. However, as we shall see, RNA may not be "life" so much as its immediate signature-vibration in a three-dimensional squiggle.

Atavistic, degrading even as it forms, almost impossible to purify absolutely, RNA molecules are the electrons and quantum particles of biology. Before they can vanish, their codes turn into complexly folded proteins catalyzing and metabolizing reactions in organic space.

Via emissary RNA strands with brief, fluctuant half-lives of up to an hour, the static kingdom of DNA designates, particularizes, catalyzes, and collocates proteins in moebius congeries of amino acids. Twenty-two amino-acid words are, in effect, propelled from DNA in the cell's nucleus in self-assembling but multivariant strings of hundreds each without the core molecule itself being compromised in any irreversible way. If not literal input, then system potential is actualized situationally in gradients of trans-DNA modules that come into synergistic interactions with one another. Bits of information activate vast arcane morphologies that organize themselves in three dimensions. This provides the ontological, structural, and physiological basis of cells.

The variety of cells reflects the chemicodynamic potential of DNA. Differentiation of cell types is the root of the divergent tissues culminating in organisms. For humans, there are in the range of 260 functional cell types, including kidney, brain, heart, liver, bone, muscle, nerve, etc., each constellated by a semi-unique set of genes and amino acids leading to discrete protein glossaries but oscillating between multipotential states until fate seals them into a metabolic vortex with its own specialization. Other terrestrial species display phosphorescent cells, stinging cells, photosynthesizing cells, etc.

There is nothing about the biosphere that "feels" intelligent in an extrinsic sense. . . . .

Germs (seeds, gametes, eggs, root-forming twigs) are produced by all life forms. In benign circumstances these living relics assemble layer by layer into full-blown bionts: giant trees and flowering bushes from solitary spores; huge feathered birds from eggs the size of bee-bees or golf balls; thirty-foot-long scaly fish and lizards from pee-wee clusters of jellied milt and roe; mice, bears, pigs, and camels from microscopic fusions of petite pellets; delicately wired ants, flies, mosquitoes, beetles, and spiders through multiple transformations of cylinders with yolky cores; and bacteria, viruses, and other ancient bionts from nucleic dust.

The material of life has been thoroughly dispersed throughout this planet and is as common now (if not as downright plentiful) as water and stone. From its initial agglutination in Precambrian tide-pools and surf, germ plasm has morphed and travelled from scale to scale, shape to shape, vehicle to vehicle, carrying its protean formula in weatherproof containers to almost all realms of sea and land. These are discovered now in submarine volcanic springs and other boiling hydrothermal vents; inside glaciers; in toxic wastes and acids; at the bottoms of caves; thriving in Antarctic lakes; eating metals in mines; under megatons of rock; in arid, frigid deserts; in pure salt; in closets, attics, and the ears and guts of other creatures; even clinging to the apparatuses of Chernobyl and various other radioactive citadels. The Earth has been overrun by magical embryogenic beans. It would not be all that surprising to find their correlates on Europa, Titan, Mars, or even swirling in Saturn's clouds.

• • • • •

There is no explanation—no interpretive hermeneutics—for how seeds are made, how and whence the infectious letters constituting DNA supervened long ago on Earth to sponsor fecundity. These wound themselves in alphabetic banks of codons, each germ comprising a few hundred atoms at best, to make a second, far more complex syllabary of proteins, both of them (apparently) prior to the imbedding of the mRNA microbe in a cell.

There is also no simple way to explain the ingenious manifestation and interplay of so many mammoth, dense organic molecules capable of folding into stable, economical shapes and collaborating with one another to synthesize tissue layers and bodies.

How DNA and RNA regulate these architectures *ex nihilo* without outside direction and design is a mystery far older than that lodged in any Sphinx. It is perhaps the mystery that Egyptian and Martian Sphinxes share behind the symmetry of their meta-human gazes.

To creationists, this suggests designed specificity—God or aliens—but that is paternalistic thinking. There is nothing about the biosphere that "feels" intelligent in an extrinsic sense. It is rough, turbulent, inexorably algorithmic, and quite bloody as well. It has a pagan, innate, metastasizing feel to it, as though unscripted surfaces are being peeled raw to admit other surfaces, infiltrating through themselves with placental and other debris. Maelstroms are resolving in patterns. Stuff is cloning and heaping under gravitational, informational spells, hemorrhaging over its own boundaries, yet compelling its surplus back into funnels of design. Everything is young and new at the same time that it is ancient, wrinkled, and hieroglyphic. Yes, a God—a divine intelligence—might work this way, but it is the God of Rooster Priests and Gnostics, the God of Matter, not the God of the Discovery Institute or the Southern Baptists. Embryogenesis is molecular transubstantiation, the work of Thoth and Hermes Trismegistus.

How do molecules become alive? . . . .

It is a wondrous, mysterious universe in which energy can be transferred from positrons and electrons to molecules to cells to systems of cells to creatures to herds and tribes; in which that energy can be conserved indefinitely in xylems and membranes and stirred to activity and thought. It is a miraculous planet that lures solar particles to the ground where they torque, sublimate, and flower in petals and eggs.

. . . .

Xylem: in a plant, the woody nonliving aspect of the vascular system, which conducts water and minerals from the roots to other cells.

. . . .

There is no fiat that life must coalesce at all, however sticky the primeval pool, however packed with putative amino seed, however dynamically charged. Why didn't proto-life formations here dissipate at the level of precellular bubbles or lesser effervescences? Why not rocks and mud forever instead of membranes and trembling scarves? Why a whole planet packed with foliage, animalcula, vibrating eggs, and hatchlings? Why organize them one by one by one, keeping some intact for a hundred solar revolutions or more? Why cultivate desire and knowledge? Why go through all that peril and labor just to return them one by one to entropy?

It would be far simpler to let coarse celestial mechanics dispose of every promising organismal start, leaving the universe as empty as it should be after the Big Gang rode through, levelling space-time and setting fires throughout a sterile abyss. That would have been the easiest path for matter to have taken, the one of least resistance, the obvious course. It is what matter should have done according to our laws of record—which explains our schizophrenia: life is a glaring anomaly.

Look at slippery seaweed on rocks, tetrahedronal and quincuncial bionts swarming in tidepools and ponds, a menagerie per cubic centimeter. Through a convex glass, dirt under a conifer crawls alive. Consider a dewdrop, a dead log, a morsel of dung or carcass.

How did original thermodynamics get trapped and its turbulence sealed and tamed in biosynthesized bundles? How do chloroplasts convert the daily charges of photons, while modulating, decocting, and dispatching them into the quite different energy tiers of sugar molecules? How are the essences of thermochemical bonds and gravity stepped down into metabolism of cells? "How does the vast web of constraint construction and constrained energy release used to construct yet more constraints happen into existence in the biosphere?"[8] How did sod become segregated, infrastructured, and assimilative such that bionts of exquisite delicacy and paradox continue to differentiate, sort their own texture-masses, and transfer quanta of them? How did aquatic microstructures consolidate bodies and then regain independence in macrostructures, identify nutrients, and dispatch their tendrils into offspring whelps, ovum after ovum down through generations? What principle makes life an *almost* perpetual-motion machine spreading as lichen, cestids, worms, mites, weevils, and wrens on both behemoth and Lilliputian scales to every cranny of this planet?

• • • • •

Despite the preponderance of evidence against us, nonlinear, ordered arrangements like biosystems are apparently as ordinary in nature as higgledy-piggledy—and for reasons likely having precious little to do with natural selection as such. What has been happening on Earth the last 4.8 billion years is not discursive molecular activity hitting upon order and escaping degradation by random agglutination and wild luck. It is everyday business, routine activity.

Regulative, self-organizing structure comes flying out of chaos here as if it owned the universe. Design principles seem, wherever there is matter and clime, to match chaos blow by blow—to personify it as well. Complexity increases spontaneously such that chaotic dissipative structures—structures that consume entropic flow to establish and maintain their forms—are ubiquitous. They start out at large as stellar furnaces and atmospheric vortices on worlds like Saturn and Neptune and then evolve on bodies like Gaia (perhaps across gaseous worlds too) into things as complex as cells, organisms, ecosystems, and polities. Morphogenesis is as thermodynamically meted as entropy.

Complexity is not the stepchild of trial and error. Complexity *is* matter. How else to explain bionts toying so recklessly and indifferently with ballets of perpetual motion. Why even go there?

• • • • •

Maxwell's demon tries to figure out which piece of dust to start with, which gas molecule to measure. He is flummoxed. When he tries to measure them all at once, he must put in a gargantuan amount of work only to find that there is no displacement from equilibrium. There is nothing from which he can extract or compensate his work. He is ultimately so frustrated that he dissolves into gas particles; in fact, he could not have existed in the first place.[9]

But there he is! Eddies form in entropic streams and sport in places very far from equilibrium. In such unbalanced regions, amazingly complex systems—specialized, low-entropy structures—form and dissipate spontaneously. They spring into being without regret. A cell is one such structure; the collection of cells we call an embryo, or an adult, is another. Energy pours into these networks that have deviated far from equilibrium, albeit (remember) at the expense of increased entropy in their surroundings.

Creatures, once alive, go through appearances of operating like man-made appliances. Biophysically, energy seeps into cells; for instance, chemical energy in glucose. In the same reaction carbon dioxide and water are released, as the energy is used constructively but not always efficiently. Some is lost at every step as waste; this is entropy's cut of the pie. In the end, there is less "free energy." Life is born a-dyin'.

In sum, the entropy equation must always balance. The Second Law cannot be broken in nature as a whole. Energy flows through labyrinthine structures and is dissipated, while an area of lower entropy forms from the use of the dissipated energy pumped out of each network as a by-product.

Indeterminate, Immanent Possibility . . . .

It is amazing that even a billion years constitute enough time for DNA and its ensigns, disciples, and suckers to propagate ten trillion or so mobile proteins to assemble a biosphere here.

If we take "two hundred" for a rough amino-acid calculation of protein size, the chances of any one protein being written is about one in 10^{260} or trillions of trillions of times greater than the number of atoms presumed to occupy the known universe. Thus, most "proteins" never get made at all, nor should they—and they *don't* in the cynical world of *Looking for Mr. Godot,* a hybrid title submitted to a newspaper contest (in which readers were asked to combine works of two authors): "A young woman waits for Mr. Right to enter her life. She has a loooonnnnng wait."[10]

We should still be waiting for Mr. Right; yet he has been here all along. In fact, by the count of cosmic time, we hardly had to dawdle at all. If this seems too good to be true, it is. But then what is it if not "true"? *Some enchanted evening* ... for sure.

• • • • •

The obvious is so opaque that it needs constant restating. The Earth's biosphere defies any other known thermochemical system. It bucks the fluxions that make slaves out of stars and asteroids. Transcending mere linear sums of entropic, enthalpic activities—turbulent waves of pressure and temperature—it weirs ordinary sunlight, while transforming it through vibrating lattices into pond mulm and mitochondrial epiphenomena.

Regions of RNA aggregate and dissolve and proteins catalyze one another's alchemy. While mechanically conserving energy, the biosphere turns trillions of separate adiabatic and nonadiabatic interactions into higher-order molecular and cellular functions. It breaks out of even its own deeply closed original protozoan and poriferan configurations to sponsor radically new motifs and higher-level designs (crocodiles, ants, and the like). It seems smarter than its own algorithms and less belabored and grim than its house thermodynamics.

If you explored the hot, acidic surface of Venus, you would likely find not one molecule with agency—no algae, no crab, not a bug or sprout or virus. Every stone would be barren, sterile.

• •

Enthalpy: a thermodynamic function defined as the internal energy of a system plus the effects of pressure and volume. **Adiabatic:** *describing a reversible thermodynamic function at constant entropy without loss or gain of heat.*

• •

Stuart Kauffman distrusts the facile excuse of a ghost in the machine but equally skirts any aseptic reductionism that condemns crabs and dandelions to a status of "nothing but the atoms [of which they are comprised] and their locations and motions in three-dimensional space."[11] Yet they are no more than that, right?

He keeps returning to a paradox that can never be negotiated: the contradiction of simple atoms moving innocently and more or less predictably and yet comprising massive, unpredictable, philosophical structures nowhere accounted for in their indifferent movements. What can't exist exists, and the riddle (that there is anyone to pose it) is also the disparity that makes it a riddle: "The concepts of atoms in motion in three-dimensional space do not appear to entail the concepts of an autonomous agent, self-consistent constraint construction, release of energy, [propagation of] work..., and the closure of catalysis, tasks, and other features that constitute ... an autonomous agent or a coevolving ecology of autonomous agents. In one sense, of course, there is nothing but atoms in motion in three-dimensional space.... But the historical coming into existence in the universe, of autonomous agents, and of ... [their] bioworld is nowhere accounted for by Newton's laws."[12]

This is, in a nutshell, the gap between thermodynamics and morphogenesis—the whole enchilada.

The game was already being waged by trilobites and dinosaurs, untold birds and fish, moths and worms, hunters of the bush and dancers of the Dreamtime, Persian battalions and crusading knights, long before Newton, Darwin, Kauffman *et al.* came along to warn us that we were in grave danger of either trying to win a ruleless game or playing without rules. That didn't stop Caesar or Napoleon. It didn't stop Joan of Arc, Johannes Kepler, Daniel Boone, Emily Dickinson, or Babe Ruth. We are made of cells, cells of molecules, molecules of atoms and energy. Yet, distinct from the most clever, syncopated bubbles and gases, life forms have personae. They are more than just the incurable flow of time.

* * * * *

One way Kauffman arrives at living systems is from a continuously expanding, inevitable "next set" of thermochemical events, the "adjacent possible"; i.e., phenomena are always in states of dynamic transformation into new novel states, with complexification merely one consequence. Thermodynamic inevitability allows him to make a presumptuous, though clever, rationalization in favor of a very complex but still purely mechanical cosmos:

The adjacent possible consists of all those molecular species that are not members of the actual, but are *one reaction step away from the actual.* That is, the adjacent possible comprises just those molecular species that are not present on the vicinity of the Earth out to twice the radius to the moon, but can be synthesized from the actual molecular species in a single reaction step from substrates in the actual to products in the adjacent possible.

... [T]he adjacent possible is indefinitely expandable. Once members have been realized in the current adjacent possible, a new adjacent possible, accessible from the enlarged actual that includes the novel molecules from the former adjacent possible, becomes available.

Note that the biosphere has been expanding, on average, into the adjacent possible for 4.8 billion years.[13]

RNA enacts this invisible augury from lotteries of paired DNA stalks. Messenger units infinitely spread, as they derive alternate phase states of themselves and derive novel self-organizations from prior emergent complexities, all of them at best semi-stable, hence intrinsically, infinitely changing. This happens not because of dumb luck or wild chance and not because enough time has passed for anything at all to happen and now "anything" is us—it happened because it had to happen. This is the only reasonable, half likely explanation. The "adjacent possible" is inevitable, and it prefers zebus and magpies to hurricanes and volcanic ash. It metes that most precious commodity of all, the one we call life, the thing we only take away, forever (from chickens in factories as well as prisoners before firing squads) and never give back.

Not that it doesn't throw out a lot of lava, rocks, flares, and methane typhoons too—but it refuses to let them make the rules.

Somehow complexity increases *spontaneously* in nonequilibrium situations, with entropy dropping inside the emerging anomaly while increasing outside. Where life congeals, like clockwork entropy decreases inside its membrane. This proceeds not just in living sheets, but also, against all intuition, in inorganic and adamantine systems as well—BZ substrates, crystals, and Jovian cyclones parading shamelessly across their planet's mask. The mathematics behind such spectacles was documented in 1984 by Isabelle Stengers and Ilya Prigogine in their book *The End of Certainty: Time, Chaos, and New Laws of Nature,* as they explored how something as snarled and dynamic and far from equilibrium as a developing embryo is *almost dictated* by our planet and by the cosmos itself.

Their genius was to show that structures (like cells) that do this sort of thing, i.e., maintain areas of low entropy by consuming energy, are inevitable throughout the universe, creating habitats far from equilibrium, much as they have with aplomb and modesty on this planet—and then escaping into them on hooves, wings, and millipede legs.

This is not quite classical mechanics; it is not even quantum mechanics. It is a hybrid default mechanics that allows us to indulge in an "ideal" universe without the hope of life—for, as long as we live in that universe, as long as we ourselves contaminate the equation, what we say about it, or anything else, is as much fair game as it is moot. After all, sticks and stones may break my bones, but words are just words. And Br'er Rabbit, like Maxwell's demon, says, "Don't throw me in dat der briar patch," because dat der briar patch is exactly where he wants to go.

Life is an expression of irreversibility and organization at the core of thermodynamics, at the heart of the universe.

• • • • •

It had to happen—and we no more know why than Saharan hunters with their oracle bones or members of the Emu guild casting ostrich charms and turtle tokens. In viruses and protozoa up to the most sophisticated life forms, sequential designs are not only inevasible but a phase of a much greater complexity potential (such that even what is dormant now is profound and vast enough for revolutionary new species and body-plans).

If it were all just sterile, carefree atoms dancing up a storm, 4.8

billion revolutions of the Sun would not be long enough to wind this clock. If a few billion years (or less) is ludicrously too brief to log the total atoms of the perceptible universe (by billions of monkeys pounding away somewhere on billions of virtual keyboards), then it is certainly too short for statistical fluctuations of molecules to produce the variety of heritable organic forms from blackberries to irises to porcupine quills. It may even be too short to write a single viable two-hundred-letter protein without first crumbling into one or another pit of disorder.

Something turned rivers of heedless dust into peristaltic streams of sensate protoplasm inside a self-luminous skin.

Is it not remarkable that the most powerful engines and subtle effects are supposedly propounded and maintained in an idiocy comparable to billions of monkeys pounding mindlessly on billions of computer keys? These machines then make other machines.

But how can a mechanism made by nothing make itself?

Either the Earth got a ridiculously happy draw or a DNA-bearing craft from another world plunged into the dawn-time ocean.

Or nature is different from what we think it is.

• • • • •

Buddhist scholar Charles Stein (see p. xxxiii), in a rejoinder to Kauffman, suggests that, in addition to the adjacent possible, there may be "indeterminate, immanent possibility"; a universe may unfold through its unabashed nontemporal complexity.

Stein writes:

> For Kauffman, physical systems "advance" upon the "adjacent possible." The adjacent possible is the determinate set of species that, though not yet actual, are reachable in a single "step" from species currently actual. (The detailed example Kauffman provides is the evolution of species of proteins.) Now, Kauffman demonstrates that the complete "configuration space"—the field of determinate possibilities—for a biosphere, an economy, or the universe as a whole is not con-

structible, and yet he holds that the direction of time is precisely an advance on this space. Thus, we catch a glimpse ... of the emergence of a conception of indeterminate possibility—a realm of being that, though not yet capable of being discriminated in detail, nonetheless is the context for all that does manage to come into determinate being. Possibility as such is revealed as a fundamental ontological modality immanent in nature, yet irreducible to a configuration space of determinate possibilities. This immanence may be the ontological condition for the emergence of mathematical structures themselves—at once eternally given yet entirely requiring human construction—as well as (perhaps) the reservoir where consciousness awaits its own coming to birth.[14]

* * * * *

Whew! Saved again by the kid from Krypton...!

Farmers can return to their fields, financiers to their computer screens. Rappers can continue to rap; even love is still possible. Our affair here on Earth was ingenerate and implicit from before the debut of elementary particles, from before the creationary event for which the "Big Bang" stands in our ontology.

Matter is either intrinsically alive or possessed, as Loren Eiseley divined, of amazing, dreadful powers. The most intricate and meticulous formats could not assemble ego-bearing entities out of molecules if those molecules did not tend toward animate status anyway, regardless of genes, regardless of specific proteins, for reasons of spontaneous organization we cannot begin to fathom or excavate. The chromosome game is a cover story for the Indo-European migration across the surface of the world, through precepts of first causes and phoneme etymologies.

The universe is already familiar to every creature that finds itself alive on a planet.

No wonder we are calm and at home in our bodies and not in eternal fugue.

4

The Principles of Biological Design

• •

Physical Forces in Nature

Man is coagulated smoke.—Paracelsus, 1493–1541

How did soft pebbles get synthesized while enlisting catalysts to accelerate their designs? . . . .

Biologists presume that primeval plants and animals were generated on Earth as burrs of small hungering chrysalides, fissioning geodes, and ripely budding garnets, much as snowflakes are forged in clouds, gravel in riverbeds, stalactite clusters from mass pulling on damp lime. Everything, whether inanimate or alive, is deemed mineral at its origin.

A biont is more complicated than gneiss or a zircon, but that is ostensibly from molecules mingling and specializing over epochs, hoarding energy, herding stray particles into their cobwebby fields. Living things and stones cannot come from different orders. Physico-chemistry is their mode of design. Since creatures are in the universe, they must obey its canon of laws, even as they mature from one-cell replicas, swim about, deviate to snap and swallow, hiss, swell again with gametes. They are essentially dust—albeit pretty dust.

• • • • •

I stand watching wind blow the Moon through clouds, though that is not what is happening. Heat effects are rushing to equilibrium, raising congeries of vapor. As fast molecules outstrip slow ones, stirred particles of air pull the vapors along in their irregular stream, casting the appearance of a travelling moon, which is really a massive rock off which solar splash is radiating.

The clouds are what is moving. The Moon is moving too, impelled by masses of larger stellar objects—by comparison much more slowly and in actuality much faster than the dross in the sky.

All of this is textbook physics: rules of engagement. Molecules, culled in stars, go where they must, forming landscapes. Their gatherings recruit fog and currents, craters and tundras, cyclones and solar flares—the whole solar-planetary domain. They explain the Moon mirrored in watery ripples.

But there is also me, apple blossoms, a raccoon, moths, two bats. What is drawing molecules into us, into our elemental cycles? What entices matter, while still obeying the precepts of thermodynamics, into biochemical bonds and metabolic pathways? What could possibly convince a particle, seemingly against entropy, to do my body's tangled bidding rather than follow the breeze? What is so persuasive about the call of tissues that molecules flock to assemble in myriad beasts prowling landscapes, each one glimpsing the world by electrons flowing into their ganglia? Where is life's pied piper? How is his summons drawn and issued? How can a system as simple as "Moon rolls through clouds" contain an event as complex and outrageous as "parrot dive-bombs marmoset"?

This is not just a case of a few eccentric molecules either; it is trillions upon trillions of them per biont per lifetime, fleeing the stratosphere, soil, and seas across the planet, mingling in rigorously enough hewn grids to maintain whole schools of fish, teeming ant colonies, litters of hares and lorises, redwoods, old tortoises, and human patriarchs—bark, brain, shell, and guts—to say nothing of gophers, sheep, dandelions, mosquitoes, mushrooms, crayfish et al. What turns loam plus salt plus oxygen into protoplasm; fens and puddles into gardens and zoos? What is so attractive about life that it holds attentionless atoms in thrall? What is the grid which, on the spot, remorselessly routes glassine threads and membranous fields into visceral designs, conducts nucleic acids through keels of beads, and separates swallows from sharks?

What compels the stubborn, entropic denizens of chaos into bios?

The proximal cause is gravity. . . . .

Imagine what resists and propels a canoe in its glide through tidal chop—not the oars but the ether through which the oars act and of which their woodenness is honed. Gravity is everywhere, pressing relentlessly at every seam. It suspends a deer differentially in its leap over branches, a deer whose particles are already positioned by gravity, branches that have ascended while propping their interstices against the planet's mass, bifurcating irregularly in its stream millions of times, finally hunched and brought down by its weight. As gravity brazes stars and worlds in their orbits, it lassoes individual molecules too, gluing matter together without stickiness.

Gravity not only propels the canoe, it is the skin around which the canoe's domain has congealed.

What is the relationship between the laws of thermodynamics and gravity? It surprised me that I didn't have a clue. Does gravity work by thermodynamics? If not, what is it doing with its mitts on everything? I scanned physics books, presuming that gravity had to have a thermodynamic side. After all, these are the twin locomotives, the fundamental agencies and forces of the universe. When I found nothing on this point, I asked my buddy Dusty Dowse.

It turns out no one knows. "That's the big question," he informed me. "Einstein asked it too. They meet 'out there, somewhere....' It is generally assumed that all forces meet in a grand unification theory that has eluded everyone since its concept was first voiced."

• • • • •

Some believe that gravity is actually the fourth dimension, which is why it leans on three so hard and from such an indiscernible angle, giving our domain a vaster, thicker, deeper depth than depth.

Some say that scalar waves ride gravity into electromagnetism along the dimension where all trolley lines converge, where they rejoin the curvature itself of space-time—"out there, somewhere," where time flows both ways, where gravitational and electromagnetic fluxes become telekinetic.

Whether gravity has a transdimensional or scalar aspect, it behaves much as its discoverer, Sir Isaac Newton, defined it—the hand of God resting on every particle in the universe; every particle extending an influence over every other, no matter the distance between them.

• • • • •

The same astronomy and meteorology that requisitioned all mass-designs continues to reconstruct and modify them under variations of the same conditions. Structure engages gravitational constraints along every trajectory and at every seam. Life designs are pre- as well as post-genetically gravitational. They share the planet equitably with solar winds, intersecting streams, rain in puddles, cloaks of fog. Fluctuating among dynamic and inertial vectors and integrating them in its modest toposphere, a living entity arrives at its transitional and mature forms. Its shape-mass is spun at the hub of vortices tapered by gravity and heat in electromagnetic, thermodynamic, and strong and weak nuclear contexts.

In complex landscapes, these forces are represented by their local minions: density, shear force, intermittent flow, and vagrant stirs of relative temperature and molecular heterogeneity. Other regional influences include: convection (heat transfer), compression, adhesion, elasticity, substance boundaries, disequilibrium of adjoining materials, surface tension, molecular weight and charge, chemical phase separation, diffusion in solution, electrochemistry, and material buoyancy. All of these gravitational proxies impinge on every membrane and and translate into their emergent shapes and functions.

A life form represents distributions of space, mass, energy (electron chains), and assimilation and elimination pathways. It becomes compacted in pelts linked by scalar and magnetic charges, ligatures, and transition gaps, all overridden by gravity.

On Jupiter, if anything piscine lives, it too must be a child of gravity—giant, flexile, and submarine, capable of negotiating stiff megacurrents, radiation, and tremendous pressures. Nothing else could evolve there. On Earth likewise, organ shapes emerge out of gravitational fields, though with less compression and turbulence to hone their mass-designs.

Inertial Structure of the World . . . .

In a universe like this you apparently cannot have pockets of chaos without order (see pp. 113–115). Given how matter is constructed, how space is organized, and how energy is distributed, multilayered, semi-symmetrical shapes and intricate, spiralling designs will ravel virtually everywhere. They will be delicate, rounded beads and luminescent orbs; methane bands and whirlpools; hieroglyphic signets on stones; flows meandering into fan-shaped deltas; jellied deposits with speckled ridges, all foreshadowing gregarious things like cells and eggs.

Fractal striations, strange attractors, connectivity mutations, and percolation sets provide hints that tissue-like lattices arise unbidden in Mandelbrot and Julia series, BZ reactions, and the like—long prior to photosynthesis, genetic regulators, and chemically hierarchical signals.

Segmentation and replication are matter-inherent properties. Packets of galactic soot, ahitch on asteroids, are crudely cell-like. Wormlike fossils imprint themselves in meteors; iterative incisions coagulate on small moons. Molecular laminations and badges on rocks floating in outer space foreshadow microbial inklings in the Earth's primordial seas. Their prototypes later guide Golgi folds, vertebral segments, worm metameres, and the quickening seams of larval tentacles. This is gravity as we know him, but it is also more

Mandelbrot set: a collection of points mathematically generated in a complex plane, originally by Benoit Mandelbrot: "its disks studded with prickly thorns, its spirals and filaments curling outward and around . . . the Mandelbrot set seems more fractal than fractals, so rich is its complication across scales. . . . [Yet] to send a full description of the set over a transmission line requires just a few dozen characters of code."[1] Julia set: a class of shapes invented by Gaston Julia and Pierre Fatou during World War I before computers: "Some Julia sets are like circles that have been pinched or deformed in many places to give them a fractal structure. Others are broken into regions, and still other are disconnected dusts. . . . '[S]ome are a fatty cloud, others are a skinny bush of brambles, some look like sparks which float in the air after a firework has gone off. One has the shape of a rabbit. Lots of them have sea-horse tails.'"[2]

than gravity, as Jack was surprised to find coming down from the beanstalk.

An embryo is a chaos-based forcefield radiating through other natural chaos fields and giving rise to morphogenetic strings that then reconstitute according to inherent configurative potentials. Scalar forces then transfer motifs across exponents almost limitlessly.

Scale in fact grasps the entire universe in its seemingly spindly, inadequate arms, for scale embodies a principle equal to thermodynamics. In its exempt domain a thing is specified by its magnitude alone—pure mass, pure speed, pure depth, width, or length. Those scalar arms reach effortlessly like Alleghany vines, from under atoms to beyond stars. They regenerate wherever matter needs them. Instantaneity and absolute knowledge spread, irrelevant of direction, beyond constraints.

The shaman mask of the Earth espied from space reveals a bird's deep cosmic design. Jupiter's belts and red spot dynamically match color bands on fishes, flower petals, and insects.

A Crab nebula long ago was a crab in the making. A dream is a complexity pattern too, an oneirosphere filled with weightless life forms.

The poet Charles Olson credited Herman Melville with intuiting the shared inertial basis of bivalves and sonnets, of phenomena and phenomenology, i.e., the harpoon thrown accurately from repose, "the whale itself's swiftness, Ahab's inordinate will, the harpooneer's ability to strike to kill from calm only. *The inertial structure of the world is a real thing which not only exerts effects upon matter but in turn suffers such effects.*"[3] (italics his). After the giant mammalian head was hoisted, dripping blood, against the whaling vessel's side: "Silence reigned over the before tumultuous but now deserted deck. An intense copper calm, like a universal yellow lotus, was more and more unfolding its noiseless measureless leaves upon the sea."[4]

Where a Cetacean gravity-sinkhole had been, eternity was now flowing back to its source, the hint of an ineluctable force as omniscient and mysterious as the waves of its departure.

Olson was inspired by *Moby-Dick* to grasp at an abstruse, almost inarticulatable bond between poetry and physics:

> [T]he metrical structure of the world is so intimately connected to the inertial structure that the metrical field ... will of necessity become flexible ... the moment the inertial field itself is flexible.
>
> Which it is, Einstein established, by the phenomenon of gravitation, and the dependence of the field of inertia on matter. ... [M]atter offers perils wider than man if he doesn't do what still today seems the hardest thing for him to do, outside of some art and science: to believe that things, and present ones, are absolute conditions; but that they are so because the structures of the real are flexible, quanta do dissolve into vibrations, all does flow, and yet is there, to be made permanent, if the means are equal.[5]

There can be no gap finally between the reflection of a moon in a pond and DNA.

As Melville wrote to Hawthorne in 1851:

> By visible truth we mean the apprehension of the absolute condition of present things.[6]

A cell is a bubble with a skin. ....

The two major vectors in the assembly of life can be characterized as topokinetic and morphogenetic—that is, thermodynamic fields comprising digitalized subsets.

Mid-twentieth century German anatomist Erich Blechschmidt regarded embryogenesis as biodynamic self-organization, relegating any DNA controls, genomic organizers, and other chemical gradients to subsidiary status. He placed generic force first, genetic activity second. The material in a cell's nucleus cannot by itself, pronounced Blechschmidt, trigger development, put proteins in place, blueprint tissue, or decree organization because genes provide only responses to differentiations, "replies to instructions."[7] They come into play after gravitational-thermodynamic patterns choreograph matter—and then only in terms of those patterns. "Growth is not an endogenous process, but one induced from outside."[8]

• • • • •

Gross organization begins in pure space, in equilibria of stellar dust distributing themselves at sonic speed—not too hot, not too cold; not too dense, not too dilute—congealing, igniting stars. On sun-orbiting rocks like Earth, molecules cool to express new properties.

Waterfalls send cascades of matter swirling through tidal funnels, converting stasis, peaking and dipping to generate and transmit energy. Mini-spouts and currents boost and impede kinesis, shaping themselves inertially along a multiplicity of detours. Many-skinned bubbles stream from stilled rapids.

Life debuted in trillions of such primitive bubbles floating in the Earth's waters. Nature had to seal a container for anything else to happen, to prevent dissipation of its gathering recipe. This was the role of the early nuclear membrane, the pellicle of the cell.

Eddies, globules, moirés, and other foreshadowings of jellyfish and seaweeds semi-stabilize and disperse across fluids. On a rainy day on a modern city street, oil floating rainbowlike in slicks provides a haunting cinema of precellular activity.

In Blechschmidt's biology, tissue networks relied first on their own intrinsic organized stuff. They were expressions of "forces in the physical sense," not "chemical properties of special substances."[9]

As more time passed, through millionfold days and nights, trinkets began to emerge like loose change in the Earth's primeval spume, bogs, and puddles (what genes? what ancestors?). Swirling and crusting with soft boundaries, some of them turned ribonucleic and eventually composed blebs and coils (early RNA?). These gradually attenuated and became spindly, developed cybernetic qualities, and replicated themselves until they grew as long and thin as transoceanic kelp, while wrapping their spiralling span in cables. Other semi-configured clumps oozed out of rills, bubblings, eddyings, tagmata, undulations, splashes, etc., to braid around those cables—to be encoded, dilated, fused, and rearranged. Torrents of gravity folded semi-liquid integuments into sheets, many sheets into layers, layers into membranes.

By the time tissues and genes were locked inside a system, the biodynamics and homeostasis of their networks carved the basic organ and creature molds by a combination of gravity-mass (cell density) and energetic electron transfer and discharge (metabolism). Movements "in the metabolic fields of growing and proliferating cells and cell aggregations are always movements against resistance and therefore are true work in the sense of vital functions.... In these aggregations, which we call 'spatial metabolic fields,' cells attract each other by uptake of matter and repulse each other by output of matter and are in this way held together by locally different forces."[10]

Cosmic fields gradually became more and more abstruse in relation to the complicated living spirals at whose heart they were buried and over whose range of potential states the environment plays like a musician at a keyboard.

• • • • •

Chromosomes are dense yet diaphanous, deeply ciphered strips of molecules that count, catalogue, tabulate, code, and store information, and turn one another's expressions off and on by the enzymes and microstructures their messages induce. Without some earlier agglomeration of inanimate substances *prior* (both phylogenetically and ontogenetically) to genes, there would have been no genes ever and no biological characteristics for genes to digitalize.

Nucleic cables contaminated organic patterns and began to cipher and install their designs in microspace because the nature of RNA wire is to feel, mirror, index, and replicate. They melded already-synopsized motifs with one another in long whorling chains of themselves.

Ultimately, spools of biological systemization became imbedded in semi-closed microstructures at every level at which membranousness could emerge: not only nuclear but subnuclear, not only cellular but visceral. Tissue itself is a macrostate of thin sheets, a kind of metabolizing shale. Sables and minks carry the exotic accumulation of such macrostates on their backs, fronts, and sides. Collagen is stringy mica. Neurons are synapsing waterfalls in jackets.

The metabolic requirements of tissues and organs must modify purely gravitational considerations. . . . .

As thermodynamic vectors shaped diadems, structures within them drew emoluments from solar radiation and the dark chemical reactions of sulfide, methane, iron, nickel, and other metals. Prototype chloroplasts and kindred organelles absorbed, degraded, regenerated, and consumed heat quanta they received as either photons or photon-degraded molecules from other bionts. Different tiny shape constructs—the forerunners of mitochondria, Golgi bodies, etc.— folded into complex thermochemical circuits that used remittent pathways such as glycolysis and the citric-acid (or Krebs) cycle to run basic ATP-generating systems. The chemistry of the ocean became interpolated within thick films.

In cells with a membrane-bounded nucleus and separate membranous organelles (i.e., all modern cells, or eukaryotes, which make up fungi, plants, and animals), glycolysis occurs in the cytoplasm outside the multimembranous, enzyme-packed mazes of their mitochondria. Each glucose molecule is shattered into two molecules of pyruvic acid, which then cross the double membranes of mitochondria and invade their matrices. There the Krebs cycle decomposes them into carbon dioxide. This reaction generates some ATP; however, most of life's fuel is produced within the tangled inner membranes (cristae) of each mitochondrion, where chemical energy in the form of reduced coenzymes from the Krebs cycle streams into electron-transport chains.

. . . .

*Krebs (or citric acid) cycle: a series of enzymatic reactions following a cyclic metabolic pathway within the matrix of the mitochondrion, this reaction occurs in most aerobic organisms, mainly during respiration, as NADH, FADH², and phosphate-rich ATP are generated from organic substances and converted for cellular energy. **Glycolysis:** the one metabolic pathway in all cells, glycolysis is the splitting of glucose sugar into pyruvic acid (CH3COCOOH). **ATP (adenosine triphosphate):** a molecule that releases free energy as its phosphate bonds are hydrolyzed, driving metabolic reactions.*

. . . .

Biochemical activities commute thermodynamics into autopoietic systems. They tie the inanimate physics of suns to the movements and metaphysics of organisms. Enzymes catalyze and combine kinetically—independent of nucleic regulation—to form metabolic reaction networks, feasible routes from nutrients to outputs.

You do not need genes to run the Krebs cycle, hydrolysis, or photosynthesis any more than you need genes to evaporate water, store it in clouds, and send it back down as rain, to be vaporized again.

Biochemically transparent, metabolic control cycles pretty much go on their own forever, changing one molecule into another in ways

that can be tracked by various equations and computational algorithms with control coefficients. "[T]he phenotype (the output of
energy and matter) is predictable from known laws of chemistry:
laws of kinetics and thermodynamics. Metabolic networks such as
glycolysis and the mitochondrial tricarboxylic acid (TCA) cycle and
electron transport system are common to all cells."[11]

• • • • •

While life forms follow the same thermokinetic rules as preanimate
systems, they add survival-energetic requirements of tissues and
organs to pure gravitation and enthalpy. Beginning as semi-porous
basins, cell colonies become more and more fettered and differentiated by their mutual symbiotic functions. They accrete, twist, and
bind into layers. The individual ecologies of the body function like
sectors of a coral reef. As tissues twine dynamically around axes,
their thickening precincts have to be fueled, oxygenated, and irrigated. This contributes to their gaining functional shapes and infrastructures.

Gobbling nutrients and discharging their wastes while competing
for the same microspace, cells attract and repulse one another in fields
of increasing intimacy, generating topologies and designs much as
macro-regions on Earth marshal isobars of weather and landscape.

• • • • •

The early Cambrian saw the formations of complex spheroids, cells,
membranous layers, and simple invertebrate tissues. These translated energy independent of the means of transformation—so that,
ultimately, anything (any mass and energy combination) was able
to become anything else. "[A]utonomous agents coconstruct[ed]
and propagate[d] organizations...."[12]

As biodynamic phases established themselves in successive, interdependent layers from the Cambrian through the Triassic, simpler
creatures morphed into denser, more compressed creatures, then
into organisms spirally transposing their mass outward. Phylogenetic events became ontogenetic templates. Ontogenetic motifs

accrued in phylogenetic fields, rolling along the topokinetic high-way, irreversibly following time's arrow. Heat was still disintegrat-ing and becoming inaccessible all over the place, but near mimicries of perpetual-motion machines now held its bucking bronco longer and longer within their tangled pellicles. They danced with ther-modynamics while twitting and challenging it to degrade them, which it always did, preserving *both their* secrets.

Creatures are outlaws obeying laws.

The fetus must be a complete animal at every stage. . . .

Life for Blechschmidt is a series of outside-in differentiations — suc-cessive repercussions of universal physicomechanical (i.e., gravita-tional-entropic) forces that incite a condensed adaptation inside a pliant membrane. His thesis is that "positional, morphological, and structural changes of the minute parts of the body in the nascent and developing embryo"[13] are instigated anew each time by emerging tissue complexes and are pressed into the immediate service of topo-kinetic designs from the nutritional and excretory needs of layers of cell-life. Tissues are made up of integrative cellular "collisions" — mechanical and energetic solutions to metabolic crises. Topokinesis is thus the cell-inertial, tissue-specific magistrate of thermodynam-ics and gravity.

Position, shape, and function are ever interdependent. "[B]efore any new structure can begin," Blechschmidt notes, "there must first be adequate space and immediate kinetic occasion."[14] For instance, shifting ratios between an embryo's body surface and the changing volumes of its internal subfields lead to unique types of builds and sizes and shapes of organs. Blechschmidt and his co-researcher R. F. Gasser emphasize that "the shape of every organ is not an inher-ent independent property of each structure . . . [but] dependent on surrounding structures and . . . continuously adaptable in a biodynamic way."[15] Even though gross entropy is regionally overridden inside membranes, strict conservation of energy among tissues is preserved, for metabolism remains an outgrowth of thermodynamics.

Cell masses develop activities at each stage that alone define their "meaning," as tissues dilating to fill space must also contribute to useful chemistry. These two requirements overlap to the point of meshing their trajectories indistinguishably. There is no space to squander and, as fractal designs pack protoplasm with fascicles, cavities, and tunnels, their throbbing belly must eat, breathe, and eliminate in uninterrupted waves.

Any embryo is an evolving organism enacting at any particular moment the characteristics its cells and organs have acquired to that juncture—a continuity of plastic dynamic events, receiving and dispersing energy, each basing its existence in its predecessor and transferring its metabolic potential to its successor. Not only are successive embryogenic states linked; they are precise derivatives of each other, each next phase and body plan resolving both its own entropy and the tissues arising in its immediately prior phase while addressing the exigencies of its new condition and introducing an endemic behavioral style and means of survival. Every stage is a novel, perfected animal. Organisms have no other dynamic or functional rationale. No organ is ever static. "Nonfunctional organs do not exist."[16]

All tissue structures, from the beginning, perform work, and contribute to nutrition, molecular transport, excretion, immunity, and the like—thus, they derive their shapes, positions, and purposes. A combination of shearing mechanical forces and metabolic emergencies consecrates a heart or kidney or lung.

Functions emerge, Blechschmidt deduces, "gradually by adaptive processes during growth...."[17] There are no inherited organ templates, only accumulated dispositions. "All organs therefore possess, as constituents of the organism, formative functions that are elementary functions of the organism. Each organ is functional within the limits of its formative functions according to the properties which have been developed up to the prevailing developmental phase."[18] This design chronology cobbles tissues into activity centers and confers their final shapes and roles.

• • • • •

When given more space, cells divide to fill it. When crowded, they die. When pressed into uncomfortable arrangements or suffocated, they struggle to maintain continuity of nourishment. When contaminated, their colonies anastomose and cut or squeeze out corridors with exit points. No matter how spheroid, spongelike, or coelenterate the original zooids in Precambrian time were, the relentless dynamics of pressure fields and shear patterns imbibing and dissipating molecular energy and then eliminating toxic by-products continued to reorganize them, losing design principles at one level, recovering them or their complements at another. Of course, the urgency to find food and escape being consumed shaped the swift and sessile body plans of whole orders of sea creatures.

Organisms not only deliquesce totally each generation (literally "refiguring" as germ cells); they inflate again as irregular spheroids (embryos). They become hieratic (genetic) anew, generating blueprints of their own shapes out of the characters comprising them. Life was transferred from pulpy masses to germ cells back to masses; gap junctions were carved in them; they squeezed together in high-density fractal fields; their cavities filled with liquids. These shapes were regurgitated from ovum to ovum such that the figments of whole organisms became composite organs in creatures evolving out of them. Deconstruction into signs led to reconstruction as squirts and whelps.

As energy and information flowed back and forth between membranes and helices, each always-transitional creature was reallocated and reorganized inside a subsequent design, both phylogenetically and ontogenetically. The forerunners—encrypted, condensed, compressed, depersonalized, quantized—became the inertial basis of their successors, their genomes concomitantly regenerated, reciphered, and then synopsized anew.

Throughout this process genes must have provided contexts for development: "[G]enes," affirmed Blechschmidt, "are simply a specially stable (reaction) substrate for all developmental stimuli and may thus be likened to the origin point in a series of coordinates at which the developmental movements take place. This means that genes can only act in the interior of metabolic fields."[19]

From the radial axis of the jellyfish, a new tangent crawled along the bottom—flattened, bilaterally symmetrical, segmented tapeworms. From these grids came more intricate stability—thicker, fancier vermiforms, eel-like swimmers thrashing their way out of larval milt. The first propulsive, wave-penetrating topoid was a roundworm followed by cadres of threadworms, ragworms, clamworms, etc. Fishes and other eely designs are little more than muscular, cartilaginous worms with ectodermal fluting and deluxe accessories. Millennia later, more multidimensional complexity coalesced atop those grids—scalar reptile hulks exploding outward and upward seemingly without restraint of massiveness except for an apical grind of gravity. Their snapping jaws caught deeper, tighter fractals (insects pullulating inside-out yolk) and cadres of fleshy, warm-blooded, quadrupedally stable plexuses (moles). More fractal complexity followed (monkeys). And so on ... into the executive realm.

Ontogeny does and does not recapitulate phylogeny. . . .

An infamous quasi-biological parable tells us that ontogeny recapitulates phylogeny, which is paradoxically both completely true and utterly false. What is true is that creatures have no conceivable way to formulate themselves except by the additive histories of their ancestors' fabrications, a lineage of embryonic designs and layers all the way back to the first cells. This itinerary repeats more or less in each embryo.

Biology never has to truly reinvent itself from scratch. It rests upon living prototypes imbedded in its own body, which become the body of its liverworts, ferns, porgies, crows, frogs, wombats, coconuts, etc., as well as their seeds and eggs. In fact, it could not invent itself or any aspect of itself again if it had to establish its phylogenetic sequence on a purely ontogenetic basis. The only thing that makes it embryogenic as well as thermodynamic is the entangled history of membrane-enclosed morphogenetic shapes underlying it—however these are potentiated and powered.

By gradual nonlinear aggregation of integrated microstructures through billions of generations a modern animal fetally assembles its full mechanism. Without this self-activated precedent of design, not even the simplest biont could form.

What is false about recapitulation is that the ontogenetic stages of modern embryos do not resurrect anything like the exact phenotypes or phenotypic ontologies of their forebears. Complexification, i.e., morphogenesis, is not in itself anything close to a faithful replaying of the pedigree of prior structures and organization patterns. Instead it reenacts principles of organization imbedded in these by a combination of nonequilibrium states, daily episodes in the world's fray, and the ciphering and deciphering of mostly recessive templates.

An embryo is a continuous, reenergized assemblage of dynamic relationships out of the partial residues of ancient and proximal ones. While keeping its integrity, it is under constant attack and degradation. It adheres to a basic axis of species congruence so that organisms more or less reconstruct the fields from which their germ cells emerge—and also so that they are more or less viable.

In short, the stages of an embryo must recapitulate enough ancestral form and design to hold their tissues together but enough immediate function in the context of that history to feed, generate energy, and expel shit. These two exigencies—one historical, the other dynamic—coincide in most structures because that is the course of both greatest safety and least resistance. Yet they will inevitably diverge incidentally in some respects, if not in the present generation then in the next, or the next, because any adventure, big or small, ultimately imposes its rupture of precedent.

An embryo thus oscillates between its long-forgotten ancestors and its immediate kinetic and ecological requirements, recapitulating something that represents neither, in order to weather the storm of its own development. No one could possibly track this oscillation because it comprises a series of already-made but also newly-made split-second decisions that look like commited paths, a range of uncertainty states generated with swift, seamless precision and resolving cumulatively in the miracle of biological identity.

An egg turns into a salamander, a salamander takes on the Halloween countenance of a frog. A mosquito rises like a hyaline prince from the mud, unwrapping sylphy, tissue-thin wings. This chain of being proceeds generation by generation, larva to larva, such that, although all ancestors are lost, their underlying uncertainty states keep returning (grubs within butterflies). The extinct equivalents of all quasi-caterpillars and their moths share one dormant genome, one plan.

As old genetic messages are rearranged, the structures they formerly expressed are replaced semi-congruently and semi-redundantly but also semi-noncongruently. This fluctuation unfolds in newly harmonized fields during the maturation of the egg. Almost all actual ancestral traits—both their morphological and genetic aspects—have been distorted, elided, mutated, or replaced by successor designs. They bear now only the hollow, correlative imprints of recapitulated former presences, their erasure and/or revision. Even where design criteria and prototype have been totally eroded, the new pattern lists along the empty pathways of deletion because *something* must shape it, even hollowly, must keep it alive and metabolizing. What is recapitulated is a deep design, a fluid homeostasis, not its taxonomic pathway.

Put another way, what is recapitulated is the *absence* of ancestors—not their absence in the sense of nonexistence but in the sense of the creative refiguration of their forms.

• • • • •

Ontogeny cannot precisely resurrect its phylogeny. It has no such map, no such directive, no innate phylogenetic energy or engine, no injunction. The thermodynamics at the core of each embryo lacks any path to a literal predecessor or facsimile of its lineage. No ancestral comb jelly, acorn worm, fish, newt, tree shrew, or lemur kinetically impels our cells to aggregate at any phase of their ontogenetic development.

Tissue energy cannot be inherited in codons, so new bionts must assemble themselves, and self-activate chemicodynamically each time as if from scratch. They are impelled by gravity, heat, and their own hungers, by immediate topokinetic requisites. Their drive to feed, grow, attract, repel, congregate in clumps, and extend their domain inward and outward is their principle of order and organization. Despite the genetic weight of their phylogeny, they must rediscover the stages of their development in terms of one crisis after another which, because of their phylogeny, leads paradoxically and deceptively to the same outcome, almost as if they were following a diagram.

They must happen solely because the molecules in their membranes *have no other choice.* Their design has only one mechanical option: *to unfold.*

• • • • •

According to Blechschmidt, ontogeny is itself an independent, synopsized, miniaturized phylogeny. Jumpstarted by blastulation, an organism continues to differentiate and fuse in fields and subfields from energy transferred among rapidly germinating blastomeres and then throughout the amassing tissue layers they become. This is of course the same thing that dynasties of cells and organisms do while evolving over generations.

The fetal procession of coelenteratelike and fishlike states in mammals is a deceptive masque because modern embryos enact their own ontogenetic "phylogeny" based on the adaptive, thermodynamic requirements of each of their tissue stages in the watery pressurized field of the womb. Each embryo must be a realized functional animal at every phase of development, or at least enough of a complete animal to metabolize its tissues and survive its perilous situation. It cannot embody a series of semi-functional or nonfunctional museum phases merely to conserve archaic designs; otherwise, it would perish before it became.

Fetal adaptations reenact phylogenesis on Earth so compellingly because the pondlike womb involutes and concentrates many key aspects of the saline marine environment in which cells and multicellular creatures evolved. In fact, the uterine bladder is biochemically just another body of water, as each cell is, preserved from antiquity within its own replicating lenses of membranes.

Embryogenesis reenacts the proto-dynamics if not the precise phyletic history of microbes allying in tidepools to make up the first membranous creatures and lineages of their descendants. By the time two modern cells—a sperm and an egg—share their contents, they immediately begin resynthesizing equivalent lamina, lumens, tissues, and organs. Somehow the force of their lineage is preserved in the functional cellular machinery, the topokinetics of their accruing molecular structures.

Adaptation to a mineralized brine, internal or oceanic, is the historic fate of the early stages of both the vertebrate embryo and primordial life itself on Earth. Embryos and ancestors converge, even as they diverge. They are propelled both to reinvent and return to the sea. They *are* the sea.

The ontogenetic "fish" is in all likelihood a different "fish" from the phylogenetic one. They are not totally unrelated, but they are mostly so. In fact, an archetypal fishling has probably been lost and regained fetally many times through the arduous development of amphibious, reptile, and mammalian orders. The "fish" that is now recapitulated may have some paradigmatic features that go all the way back to the ocean, but these are unlikely to be even the attributes that make it look fishlike to us. We see a fish mainly because of later reptilian and mammalian reconciliation to a fetal sea.

* * * * *

"Ontogeny recapitulates phylogeny" describes a cosmic system in which the difference between a womb and a tidepool, between eonic trends and local developmental sequences, is temporary and contingent. From this perspective, ontogeny recapitulates phylogeny (as the uterus recapitulates the dawn-time ocean), and together they recapitulate cosmogony—the overall evolution and destiny of chaos systems in the universe.

Plant-like and animal-like entities similar to those on Earth must bud, crystallize, congeal, layer, and derive ontogenies out of phylogenies everywhere that they are not scorched, freeze-dried, poisoned, or asphyxiated out of existence.

Hominization . . . .

The subscience of physical anthropology rests on the widely accepted premise that, for the last several million years or so, primate embryos (the blastulas of lemurs and apes) have served as pliant molds for selective feedback leading to mobile phalanges, reduced muzzles, upright vertebrae, and cerebrally enhanced brains. Evidence for the phases of this process is littered beneath the plains of Africa and Eurasia in isolate fossils of skulls, vertebrae, and limbs, hunting sites and middens, and are extant in another form in genomes of living primates. By successive transformations over generations, ancestors of tree shrews turned into forerunners of lemurs and primates; they spawned creatures that were equally pre-ape, pre-monkey, and pre-hominid (as well as ancestors of lemurs and other primates); a few of those creatures became *Homo africanus* and his descendants.

Yet without an embryonic wheel providing a fluid, nonlinear repertoire of traits, there is no way that bodies—or their fossils—and DNA could converge. If ontogeny is the molten spool, phylogeny *must be* its bronze thread. Anthropologists don't actually demonstrate this connection; they presume it.

Fossils and genes otherwise record two mute, quarantined lineages—the former (like yeti footprints) a forbidden archaeology of us; the latter, like the Vegas line on college basketball, a pseudo-algebra of probability. Genes themselves leave no detritus, no abacus, and petrified bones do not contain the forge that molded them. The primate embryo is their sole melting pot—the alembic between extinct hominid lineages and the denizens of a modern world. It is a black-hole butter-churn into which academics cast all of Darwin's and Mendel's propositions, plus everything else that bears on mortality, fertility, and probability.

Without an equilibrating, molecule-binding embryo to fuse phylogeny dynamically to ontogeny, there would be no case for evolution at all.

. . . . .

The intimation that ontogeny recapitulates phylogeny as well as the more esoteric one that together they recapitulate cosmogony are ultimately intuitions of a vast cosmic localization on Earth. They are different ways of saying that the microcosm, helix by helix, *is* the macrocosm, galaxy by galaxy.

Thermodynamically, ontogeny and phylogeny are the same distributed chaotic event: the relationship between a creature and its germinal cell is a form of a logarithm or square-root function between a morphogenetic sequence and a lineage of prior embryos potentiating it.

As multicellular life gets its mazes going, a serial scale is generated such that ontogeny continues to be a repeating, self-resetting, reorganizing theme in which phylogeny is its melody. Even as creatures are incubated, they project variant invariant offspring. There is no other road, either algebraically or existentially, down which ontogeny can go.

Embryonic flux keeps transforming Ur templates. Ultimately it elongates, vertebratizes, mammalizes, and humanizes once-coelenterate, once-reptilian tissue masses. The fossil of a gibbon is (through a glass darkly) the pictograph of a gibbon carved on an Egyptian tomb is also the inmost tablet of amino-acid codes inside an ape cell.

5

The Dynamics of the Biosphere

Deep Time and Space

> Yes, these eyes are windows, and this body of mine is
> the house. What a pity they didn't stop up the chinks
> and the crannies though, and thrust in a little lint here
> and there. But it's too late to make any improvements
> now. The universe is finished; the copestone is on, and
> the chips were carted off a million years ago.
> —Herman Melville, *Moby-Dick.*

Biomachines . . . .

Life was invented long ago in a manner we can no longer interro-
gate. Decoding chemical letters on helices (much as we trap fogprints
of electrons and neutrinos), we cannot discern either the direction
or dimension of the sources casting form and motion into the web.

According to today's nomenclature, all life forms are mechanisms
of molecules: extremely tiny biomachines made of and by even tinier
machines. Machines by definition are power-deploying, task-per-
forming devices, some of which also acquire and store information.
They are, however, presumably inanimate and artificial.

Yet the biosphere originated from nanomachined assemblages of molecules (polymerizations), energized by ultraviolet rays binding emergent polymers long before ozone buffered the Earth from their hot rain. These macromolecules were somehow potentiated such that they spontaneously twist and fold into three-dimensional architectures that carry out interdependent biological activities. Self-organizing molecular devices that invented life now maintain it.

Designed in quantum environments at nano-scale (which signifies parts with size tolerances of less than a micron), meta-devices operate by moving particles around and aggregating loose atoms and molecules into functioning structures. Themselves alive (or at least organismic), these little machines then manufacture more complex entities—bionts—by snagging molecules from inside their own plexuses or on the fly. They amalgamate themselves into larger nano-structures, which are aggregated into giant (by comparison) micro-structures, and finally into cells, tissues, organs, and multicellular units.

Every nanosecond, trillions of nano and micro apparatuses continue to build and repair cells, screen and conduct molecular effects, remodel and mend one another, archive and retrieve biological information, catalyze one another's organic processes, and reconstruct themselves from one another. Or so we are told.

Some nanomachines manufacture lipids that deploy in flexible sheets as membranes. Others, in the bloodstream, tumble as submarines through Brownian typhoons, tracking and destroying invaders and cannibalizing diseased cells. Some function as pygmy thermostats, rheostats, and magnetostats, stabilizing body-interior parameters.

The factory to end all factories is preindustrial and meta-mechanical.

Events at the Physical Limits of the Real World . . . .

At a nano-scale, events pertaining to life must run along the frontier of classical and quantum physics, operating in both mechanical and uncertainty states and flirting with Cheshire-catlike jumps that defy simple measurement. Quantum mechanics cannot routinely be applied to affairs involving organelles or cells, but it is also not just an artifact of abstract mathematics and string theory that applies solely to artificial environments (e.g. suspended isolations of atoms in magnetic traps in vacuums); it does have a physical reality. Quantum phenomena, after all, are inherent in the complex behavior of the universe and have emergent properties in the classical material world which includes the biosphere.

That quanta are hard articles is revealed by such laboratory phenomena as high-temperature superconductivity or interactive particle orbital impacts on electron-system plateaus in magnetically susceptible two-dimensional fields (my lay attempt at describing the fractional quantum Hall effect). Such highly specialized performances may not admit easily into our humdrum universe yet, but they suggest that high-energy relativistic photons as well as electrons and quasiparticles with fractional charge may influence atomic, molecular, hence enzymatic systems, and thus contribute to the origination of life.

In Young's so-called double-slit experiment, the reality of quantum effects enters the everyday physical world, as a coherent beam, shined on two slits, yields an interference pattern generated by wave-constructive and -destructive events. The light is emitted as photons, or individual packets, so, once the apparatus is set up and the ray fired, they interfere. The photons land where they would if the beam had multiple photons in it; they are impeding one another across time.

This effect may be part of the deep syntax of photosynthesis and biomutation of electron orbits.

• • • • •

Qubits are quantum-level packets of information in "coherent superposition"[1]—linked possible quantum states. They fall somewhere between actual binary bits in the real world and hypothetical constructions which, if real, lie at the physical limits of microelectronics. A computer using classical bits only needs to "know" whether you have a 1 or a 0. A qubit, by contrast, is both—and, as Werner Heisenberg aphorized so memorably at the birth of the quantum era, if we observe a state to see if it is a 1 or a 0, we interfere with it. So we can't count or "pixel" qubits.

By the same reasoning, if we set up a macrophysical quantal system to monitor qubit states, we can only glean information from it in a statistical manner. Young's double slit already confirms quantum behavior at a local level, so there is a hardware basis for qubit memory. With a quantum bit in a cybernetic network, instead of 1 or 0, we get either or both (depending on your interpretation)—not just uncertainties but precisely uncertain states that are also data-rich.

When several such qubits are made to interact, they develop entangled quantum states that defy classical physics at a fundamental level that is also impossible to describe—

> impossible even if one could change the laws of physics to try to emulate the quantum predictions within a classical framework of any sort . . . [for] the members of an entangled collection of objects do not have their own individual quantum states. Only the group as a whole has a well-defined state. This phenomenon is much more peculiar than a superposition state of a single particle [which] does have a well-defined quantum state even though that state may superpose different classical states.
>
> Entangled objects behave as if they were connected with one another no matter how far apart they are—distance does not attenuate entanglement in the slightest. If something is entangled with other objects, a measurement of it simultaneously provides information about its partners. It is easy to be misled into thinking that one could use entanglement to send signals faster than the speed of light, in violation of Einstein's special relativity, but the probabilistic nature of quantum mechanics stymies such efforts. . . .
>
> [E]ntanglement can assist the sending of classical information from one location to another (a process called superdense coding, in which two bits are transferred on a particle that seems to have room to carry only one). . . .

As with individual qubits, which can be represented by many different physical objects, entanglement also has properties independent of its physical representation. ... For example, one could perform quantum cryptography with an entangled photon pair of atomic nuclei or even a photon and a nucleus entangled together.[2]

My understanding of any of this is quite provisional. With little background in physics, I am intuiting through many glasses opaquely. Even physicists who live with qubits daily have difficulty comprehending or explaining them. Like the Cheshire cat, a qubit is there and not there. But as long as it is there *and* not there rather than there *or* not there, it can be forced to store not only information but our kind of information. Once we attach baggage to it, a quantum bit can't slip in and out of classical states without carrying our baggage along with it. Thus, information in the baggage may derive the complications and benefits of entanglement, though we have to tether it to a real rather than a conceptual event.

If we could build a device that used this software—even a row of just fourteen atoms holding currents flowing simultaneously clockwise and counterclockwise—it could carry out more computations in tandem than the fastest supercomputer in the world (at Los Alamos). Quantum computers of "normal" pocket size could find factors and solve problems that would take their Los Alamos counterpart trillions of years.

The operative point seems to be that entanglement, with its unique relationship to information and entropy, may be subject to its own set of thermodynamic-like laws that operate on chemical, nuclear, or metabolic energy regardless of the energy's state. They are laws of a totally different nature and domain from ordinary thermodynamics. In this situation information and entanglement are transferred between remote objects in a manner more reminiscent of teleportation than television. We are in a different physical universe, but that universe is also *in* us:

When two qubits are entangled, they no longer have individual quantum states. Instead a relation between the qubits is defined.... If dice could be "entangled" in the manner of quantum particles, each entangled pair would give the same outcome, even if they were rolled light-years apart or at very different times.[3]

This kind of stuff doesn't occur on the street, but it does happen somewhere, and it is most likely to dominate nano and subnano worlds. Particles arising from quantum states and "structures" might find ways to access entangled qubits, self-correct, and regenerate at deep levels while seeming simultaneously to destruct. "One can encode an infinite amount of classical information in a single qubit ... but can never retrieve that information from the qubit."[4]

We don't have to retrieve it. All the individual consequences of its expression don't have to be "real" or accessible at our scale for them to transmit *real* information into nanosystems and nascent biological fields—into us, into our cities.

• • • • •

The genetic molecule is an outgrowth of biology's innate complexity, a dynamic translation of its quantum effects onto a thermokinetic, metalinguistic level. Perhaps subnanomachines—teeming through realms beyond the imagination of contemporary physics—manipulate atomic nuclei and transmute chemical elements into one another. Perhaps they set entangled objects oscillating across conventional elemental states, giving rise to emergent properties. A primal cyclotron would go a long way toward explaining the alchemy of life and its synthesis of that ultimate uncertainty effect—the mind. Where else could Schrödinger's meta-Cheshire cat* find itself in a such mysterious state that, neither dead nor alive, it pounces on a

---

*In this gedanexperiment a cat in a box with a vial of poison and a quantum triggering device is both dead and alive as long as the electron in the trigger hangs between two possible positions while occupying both.

qubug somewhere in between the two? Where else would a protean creature like a pre-amoeba, neither alive nor inanimate, emerge from a mineral to meta-stable state?

It would seem that, in thinking and experimenting concretely upon these things, we approach a phase of either mind or matter (or, more likely, both) that can entertain uncertainty states (quantum leaps) in its own neuron fields while locating them exquisitely at the boundary where its own imminent nonexistence is entangled with the emergent property of its existential awareness. In other words, we are paradoxically both creating and reducing entropy by our imagination and investigation of relationships between classical and quantum states in the universe. We are approaching etiological characteristics of a reality we embody *because* we embody them.

The deepest cellular effects come from nano-constructions operating on the borderline of classical and quantum realms. . . . .

The conceptualization of cell activity as nanomachinery re-casts organelles as purposeful devices engaged in microfabrication and transport:

• **Ribosomes.** Like robotic welders in an automobile plant, ribosomes string amino acids from mRNA templates into polypeptide necklaces that self-assemble into proteins. More profound and transmutational than any digital machinery, ribosomes represent the fundamental interface between information (nucleic acids) and structure (proteins). They convert genetic molecules into tissues.

• **Flagella.** Constructed of spiralling aggregates of proteins, microfilament and microtubule springs of about ten or twelve nanometers conduct rotary motion, churning the flagella of one-celled organisms in such a way that subsequent whiplike activity shoots them through water. Flagellar motors have nano-spindles and casing structures like electric-motor shafts and armatures. They do not, however, run on electrical currents siring magnetic fields; they draw their fuel from ATP, using its decomposition to re-form molecular shapes around a ratchet that sets a protein shaft racing in swift revolutions.

• **Chloroplasts.** Chloroplasts construct optical antennae—light-harvesting wafers that, ingesting photons from the sun, mutate them as chemical fuels. In the process they transubstantiate light (photons and electrons) into quanta of energy, action, information, and design.

Long ago and on a planetary scale these primitive factories released massive amounts of oxygen, a by-product of their collective activity, out of oceanic waters into an atmosphere, thereby filling an ozone layer and shielding nascently oxidizing bionts, while providing nutrients for them. This opened a biochemical pathway for the entire animal kingdom.

Every plant—every weed, seaweed, alga, leaf, and moss—contains, as a matter of course, millions of these sophisticated machines. They blow to the ground, obsolescent each autumn, by the uncountable trillions, any one of which—any discarded oak or yellow maple leaf—would be a priceless marvel if produced from scratch by a biotech factory.

• **Mitochondria.** On the model of engines that drive electrons through wires into electrical motors, mitochondria synthesize molecules of adenosine triphosphate and diffuse them through each cell to initiate biological reactions. Generators of chemical energy in life forms, these little power stations conduct the controlled combustion of glucose and other organic molecules, thereby providing fuel for metabolism.

• **Topoisomerase.** Molecular tightening in closed double-DNA loops increases tension. Helices resist deformation by springlike bending into supercoils, as tiny enzymatic devices unwind molecules when they become too tightly kinked in reproductive separation, sorting their ribbons and loosening their stresses and snarls by eliding filaments and resealing the gaps. In the process they nick and splice DNA so that its molecules rotate unimpeded around their own strands. Using the energy of ATP hydrolysis, DNA gyrase keeps pumping out positive supercoils that relieve and replace negative ones.

No artificial machine conducts this activity. We still await its analogue in human terms.

• **Cell nuclei.** Colloids of ancient solvents in proto-genetic machines were able to crystallize and compress raw data in configurations thousands of times denser than any artificial cyber-device. Nucleic-acid machines now hold nanoparticles in suspension to store ultra-high-density data. Patterns are mirrored through water molecules in such a way as to transmit instantaneous mirages of magnetic atoms between foci of elliptical rings of atoms, while crystals composed of a few hundred atoms each (quantum dots) discharge fluorescences and electrochemical thresholds across nervelike synapses. Other nanomachines then store the accumulating data by microelectronic effects similar to the way film negatives are photolithographed. Genes thus arose from states of information rather than the other way around (see below).

An embryo is a hologram projected from a microchip by a primitive technology resembling a laser but far more penetrating and exact. It inscribes textures of monolayers on more topologically complex, involuting layers.

• **Chromosomes:** A chromosome may be an antique, but it is cybernetic enough to run itself and clone progeny after progeny, until expunged. More compact, efficient, economical, and superdense than transistors and microchips (which are back-engineered from them), these genetic devices invent themselves. Wireless, they manufacture units operational and self-sufficient in every respect.

⁕ ⁕ ⁕ ⁕ ⁕

Where did all this design specification, from the origin of terrestrial biology, go? Where are the source nanomachines now? How can we unravel the labyrinths they have ravelled about themselves?

The full miracle is not appreciated. . . . .

Biomachines surpass in subtlety and inventiveness any twenty-first- and likely twenty-second-century machine. It takes at most a decade or two to manufacture a functioning appliance (a refrigerator or washing machine) from flow charts and raw materials (metals, hydro-carbons, et al.), a century or two to construct a radically new design (a light bulb, jet, or computer). It takes eons and galactic equinoxes to make a creature, breathing and alive, out of scud: a tentacled device that scuttles off through sand on one of the universe's planets; a replica giraffe that drops from a membranous canal into the dirt and struggles to hoist its top-heavy, dappled figure onto gawky limbs while its similarly dappled mother licks away the placenta. We can fabricate nothing close to snake and eagle dolls, wasp and crab motors, tiger motherboards—we who have mere minds.

Plus, bionts are not aggregated or programmed like other machines. They are autonomous, self-assembling clusters. They do not need diagrams or parts. They arise from nanomachineries pro-cessing mud, water, and sun. They plug not into electrical sockets but nature-direct. They may not brazenly violate the first law of thermodynamics as they transmute heat and energy and dispatch kineses through future tissues and generations but, as noted, they operate right on its edge.

The evolutionary embryogenic process, spinning out ergonomic microbes, dodecahedral and icosohedral radiolarians, geodesic pump-filtering sponge masses, taut mimetic insects, and blubber-bearing, diving mammalian factories, is performing at a level of grace, refine-ment, and subtlety that makes all human technology early Stone Age by comparison.

• • • • •

These days, when we call conception a miracle, the full miracle is not appreciated. Something inviolable is getting puppets alive out of wholesale gunk so that they cohere seamlessly, move by their own will, and notice the universe.

There is no feat of stage magic here for skeptics to deconstruct, technocrats to plagiarize.

This is no mere sum of additive effects. "This, is no bare incoming/of novel abstract form, this/is no welter or the forms/of those events, this,/Greeks, is the stopping/of the battle."[5]

This is where the sun taking "off his clothes/wherever he is found,/on a hill,/in front of his troops,/in the face of the men of the other side, at the command/of any woman who goes by,/and sees him there, and sends her maid, to ask,/if he will show himself,/if the beauty, of which he is reported to have,/is true...."[6]

This is where inertial structure—the gravitational field—meets waves of quanta arising from and dissolving into vibrations.

• • • • •

The essential act of creation eludes us. Scientists using a lugubrious technology insert linear programs into banks of already-existing biomachines and compel them into partial embryogenic responses. That is the extent of their skill. Not only do they not synthesize life itself out of molecules, they have not the remotest clue of how this might be done. Tinkering with prearranged sets and recontextualizing their messages, inserting exogenous snips into germinal cells, they observe the outcome of their acts only after hatching the eggs. Ignoring any originary and epistemological problems of creation or morphogenesis, they attempt mainly to intercept inheritable disease potentials and breed better crops and herds. Allocating bits of machinery fabricated elsewhere, appropriating biosynthetic aspects implicit in that machinery's own deep organization, they do not break into actual gears of the clock. Cells embody "hidden information that we can manipulate but not access directly."[7]

Little of the ultimate ordering of life forms or mind and behavior is codable; these seem to arise simultaneously from a higher order and a deeper, more discrete apparatus.

Creatures are shiftier and more serpentine than their substratum. With multiple levels of visceral and metabolic cohesion, the capacity to keep themselves alive and heal their own tissues, transmission through time of their own template, tiers of alternate strategies of behavior, as well as layers upon layers of conscious and unconscious mentation, they are too nonlinear to be the proximal offspring of genes.

Jump-starting a frog or paramecium, imparting its independent motion and desire—that would be the true art. It would require getting self-awareness to surge into a wreath of artificially melded membranes and filaments and having them rouse to attention, wriggle, and try to flee—"all [Herman Melville] obedient to one volition, as the smallest insect...."[8]

Life is a *Gestaltungskraft*—a configurative agency transcending its components. It is not a metalinguistic cipher that a DNA sequencer or computer might chaw on. It is not those pretentiously digitalized strips that make up the twenty-four volumes of the Human Book of Life, published in *Nature* (October 2000 edition; Volume 409, 15 February 2001, pages 860–921).

Pest-free soybeans, frost-resistant strawberries, rabbits with human ears, luminescent monkeys, and other biotech stuff are mere inventory modification in the Precambrian warehouse.

Experimental Flies and Sonic Hedgehogs . . . .

Before we attempt to gauge the limits and contingencies of genetic control, we should make clear that inheritance by DNA plan is fundamental, irrefutable, and independent of species. If a gene can do nothing without cells, cells are correspondingly limited to the simplest movements and collocations without genes. But add the right clips of DNA to a protoplasmic medium, and a dramatically acrobatic awakening follows. When legs can be grown in place of antennae on the heads of termless generations of fruit flies and extra pairs of wings on their sisters' thoraxes, compassion regarding sentient beings may be in doubt, but cause and effect are not.

Discovered in the 1980s, a sequence of homeotic selector genes (called the homeobox) function almost as molecular address labels.

Since each parasegment of a fly has its own unique set of labels, when the labels are changed, the expression of a spliced gene behaves as if it were located somewhere else—and other genes follow in step. The activation process works in exact register with the boundaries of insect parasegments and in perfect accordance with pair rules and segment polarities along the animal's torso. It is a closed case.

• •

Homeotic selector genes: these genes regulate the fates of other genes and thus influence the overall body plans of animals. **Homeobox genes:** *these discrete sequences of DNA participate in the formation of segmented body patterns of certain animals (like insects) and thus help control the developmental fate of each segment relative to the others.* **Segmentation genes:** *the cues for these relatively late-acting insect genes (named hunchback, Krüppel, hair, runt, ftz, gooseberry, even-skipped, engrailed, zerknüllt, etc.) are provided by earlier-forming egg-polarity genes. Controlling systems of discrete bodily parasegments, they include gap genes, pair rule genes, and segment-polarity genes, groups that govern one another's expression. Mutations in these genes can alter the number and basic internal organization of segments without disturbing the global polarity and internal positioning of the developing egg.*

• •

Additionally, traits that have been inculcated in the wrong place by genetic engineering (like legs in place of antennae) have been rescued in subsequent generations by transgenic implants. Equivalent chromosomal strands grafted from elsewhere, as long as they are the right ones, restore lost anatomy and eliminate defects in future generations. Even more amazingly, lines of flies have been repaired by genes from humans—although this should be no surprise given the universality of DNA and the interchangeability of its alleles. Homeomorphic elements shared by flies and primates activate nondiscriminately in either's blastodermal tissue because they likely represent an ancestral state preceding both insects and humans and from which both evolved. That is, they are older than Methuselah. Their persistence indicates the depth, durability, and modular integrity of primeval genetic products.

• •

Blastoderm: the layer of cells surrounding the blastocoel (blastula cavity) that gives rise to the basic germinal disc from which the embryo develops in most placental vertebrates.

• •
• • • • •

It is this kind of systemic reliability that drives modern genetic research. For instance, the homeobox line leads into subsequent Tinman and Hedgehog editions.

> [O]ne of the three Hedgehog genes—Sonic Hedgehog, named in honor of the cartoon and video-game character—has been shown to play a role in making at least half a dozen types of spinal-cord neurons. As it happens, cells in different places in the neural tube are exposed to different levels of protein encoded by this gene; cells drenched in significant quantities of protein mature into one type of protein, and those that receive the barest sprinkling into another. Indeed, it was by using a particular concentration of Sonic Hedgehog that neurobiologist [Thomas] Jessell and his research team at Columbia recently coaxed stem cells from a mouse embryo to mature into seemingly functional motor neurons.
>
> At the University of California, San Francisco, a team led by biologist Didier Stainier is working on genes important in cardiovascular formation. Removing one of them, called Miles Apart, from zebra-fish embryos results in a mutant with two nonviable hearts. Why? In all vertebrate embryos, including humans, the heart forms as twin buds. In order to function, these buds must join. The way Miles Apart gene appears to work, says Stainier, is by detecting a chemical attractant that, like the smell of dinner cooking in the kitchen, entices the pieces to move toward each other.[9]

By this logic, life is a Deep Blue computer game played inside our bodies.

Real genetics doesn't exist yet. . . . .

Though there are many cause-and-effect links between specific genes and identifiable morphogenetic outcomes (e.g. nonviable hearts), there is no explicit chemicomechanical link between genes and morphogenesis as a whole.

Mapping a genome doesn't tell what protein a gene produces, why it produces that protein in one creature and another in a different species, how the protein behaves in the emerging organism, how it folds pretzel-like into myriad intricate shapes, or how its interaction with other proteins leads to specific tissues, or how those tissues stick together, develop cohesive organismic metabolism, and convey reproductive energy. It doesn't even distinguish in its timeless equation where the vestigial instructions for one extinct animal cease and another creature begins ... because they don't cease and begin, because they are not even species—the planetary genome is a whole. Relics of worms and apes, wings and eyes, phosphorescence and mimesis are all mixed up together (to the point where their limitless connotations in the expression of single creatures change meanings and hierarchies routinely).

Despite the regency of a brahman class able to unravel source codes and engineer and rewrite parts of them, along with rampant pontification about biotechnology and our engineered future, genetic reductionism ignores the fact that, read as text, genes represent linear information without a plan, divorced from not only a functional consequence but a basis for existing at all (either initially or inertially).

· · · · ·

Genes were never alone, were never the first trump or evening stars; from the beginning they were avid collaborators, mirrors of constellations long vanished, devoid of meaning and functionless when extricated from a context which they now determine and which (even before that) predetermined them. Far from being originators of biomorphology, they are products of a separate semeiology—their regional alphabets culling secondarily from life's thickenings, coilings, and microcrystallizations. They precipitated within amalgams that became the first biological entities on Earth, then were interpolated in all subsequent motifs descending from those. By-products of primitive life, they took on a critical role in the development of biology. Initially geochemical and generic, they became genetic only after getting trapped in the soup of the emergent cell.

Barry Commoner writes:

> Because of their commitment to an obsolete theory, most molecular biologists operate under the assumption that DNA is the secret of life, whereas the careful observation of the hierarchy of living processes strongly suggests that it is the other way around: DNA did not create life; life created DNA. When life was first formed on the earth, proteins must have appeared before DNA because, unlike DNA, proteins have the catalytic ability to generate the chemical energy needed to assemble small ambient molecules into larger ones such as DNA....[10]

Yes, the reason it is naive to presume that genes make embryos is that embryos, in fact, made genes—very primitive embryos and very long ago. The embryogenicizing process of organizing and layering uncertainty states into cells and then tissues is what established life. Genes arose as an artifact of this event, as a result of the process they now seem to invent and regulate by their codes.

Chromosomes are microforms that copy-copy-copy ... index and copy, index and copy what is already there. They gather incipient forms around them in order to function as information rather than gibberish—to sift design through molecular interactions. Brittle and vulnerable, self-obliterating as they shimmer into existence, they are dependents of life's cocoons. Minute by minute they borrow context and meaning from a transcendent event they ride splat in the saddle of.

Genes are real (they wouldn't be here if they weren't "real"), but they are not what we think they are.

The twinned helix is (from the onset) an abacus that counts the dancer in the dance—wire precipitated from matter as it thickens into life, the gruel of a living centrifuge, finding itself in space and holding form and energy together with memory-imbedded line. Once fractal patterings of microbial effects, its strands mutated into cellular memory. The deeper their activities installed themselves in life, the more morphologically profound and dynamically critical they became.

. . . . .

There is no bridge between genetic reality and ecological reality (other than the obvious one, which is not a bridge). Biology is not a genetic phenomenon, not even an emergent property of genes. The order of life forms originates from myriad interdependent epigenetic factors without a central regulator, without (in fact) unilateral regulation as such.

> Early life survived because it grew, building up its characteristic array of complex molecules. It must have been a sloppy kind of growth; what was newly made did not exactly replicate what was already there.... DNA is a mechanism created by the cell to store information produced by the cell.[11]

Genes are a record, an after-effect of protoplasm's confluence of linear and nonlinear properties. They are not the cause but the by-product of life—its rudder, not its provenance.

DNA codes are no more than a stray thread or two hanging off a quasar-deep orb of yarn. . . . .

Something existed once and was alive, some sort of proto-plant or -animal. Its wriggling, coarse body germ was read by helical filaments, then coded, subminiaturized, supercoiled, buried; tens of millions of years later a descendant of it lived (a tiny jellyfishlike thing, an mini-octopoid or rotiferian ancestor . . .). It too was coded, subminiaturized, and fractally packed many layers atop the original biont. Then something else . . . and something else. . . . Every layer of organization and complexity sank into and was swallowed by a subsequent layer which itself became integrated into further emergent, uncountable layers that were not only collectively melded but protracted in relation to one another. These came to include the invention and distribution of metabolic pathways with their multi-tiered regulation by activators and inhibitors, cascading molecular networks, optimized output of energy from photons and other nutrients along integrated, dependent pathways, ". . . the precise molecular and functional characterization of environmental sensors; of ligand-specific receptors; of membrane-spanning ion channels, ion and metabolite transporters, ion and metabolite pumps; of membrane-located signaling proteins; of immune systems in mammalian and submammalian vertebrates; of acclimation-specific isoforms of contractile proteins; and of protein kinase/phosphate-based phosphorylation control systems and intracellular signal transduction pathways, all of which interface the organism with its environment."[12]

Through eons of births, deaths, and offspring, these were coupled with:

> . . . overexpression, underexpression, or knockout of specific
> genes . . . in tissue-, organ-, and cell-specific physiologies. . . ;
> the roles of exons, introns, promoters, enhancers, transcription factors, and regulator signals (inducers, represssors,
> growth factors) in the genetic regulation of classical physio-

logical and biological processes...; molecular dissection and reassembly of physiological cascade systems including clotting and the immune system in lower vertebrates...; specific brain localizations, brain activations; and the molecular basis of neuronal growth, differentiation, and regeneration.[13]

The biosphere was sealed, a little more deeply each generation, into its own background where the primordia of sea lipids tucked into their progeny, membrane absorbing membrane, not only condensed and ellipsized but strewn diffusely among levels of its own emerging gradients and control functions. Nano-architectures and their feedback loops are now both unfathomably consolidated and irreversibly opaque. No one, of course, saw it happen, or how it happened; it left behind no forensics and, in any case, it is not the kind of thing that one "sees." After all, where would you look?

• • • • •

The primeval regulators, the source codes, design specifiers, circuit triggers, and quantum switches—while traceless and inaccessible—are still present, just as crucial as ever to the assemblage and structures of organisms. They are the lattices, frames, and base symmetries upon which life cultivated its once and future designs. Their submechanisms run under the mechanical surface, camouflaged by their own legacies of condensation, sublimation, transitorization, abbreviation, concision, tabefaction, ellipsis, syncope, meta-coding, deeper condensation, further abbreviation—and by the sheer layers and deviations of molecular motion itself.

Initial developmental principles, whatever they were, have been camouflaged asymptotically within their own deep embryogenic structures—nanotechnology within successive nanotechnology, strings synopsized by other strings in such a way that all that is latterly exposed is the leftover scaffolding, the fringe, some debris, and a scant surface of frayed taffeta from a project that packed up shop long ago.

The phylogenetic grid is underwater—the bulk of its iceberg archived there. Its principle of organization lies in a space far denser and less penetrable than even a black hole or Zeno's Paradox, for it is made of absence, thus is pure, dumb regulation—infinitely, infinitesimally thick. It is the weight of the wind, the sum of all waters. Its trails and paths are so many and packed so fractally they both fuse and diffuse, disperse and meld into one another, nary a clue.

Every deceptive layer of tissue is packed with the design-information of its prior construction and history at levels that are both genetic and nongenetic. Nucleic footprints become faint and indistinct until they vanish altogether into the noise. The nanomachinery underlying and specifying them has been sheathed, condensed, simplified, made redundant, and compounded many times over in substrata, as new sheets of information and structure unimaginably long ago superseded prior ones that superseded others even longer ago—now it is utterly intangible. It is so nano and its speed of assembly so much greater than even our most sophisticated tracking devices that it effectively doesn't exist. Its subtlety, its entangled coding, its nonchalant mastery of relativity and space-time overwhelm our senses and our reason.

• • • • •

Because of the constant miniaturization and condensation of matter through its own valences into nucleoplasmic and then cellular layers during the long (hypothetical) annals of Darwin's evolution, the true information conducting organization and biopoeisis dwells now (if anywhere in what passes as the physical universe) cryptically beneath the level of genes, beneath atomic and subatomic cauls; thus, is subnuclear to both cells and atoms, to protons and electrons as well. Job spoke for centuries to come:

> The deep says, "It is not in me";/And the sea says, "It is not with me...."/It is hidden from the eyes of all living,/And concealed from the birds of the air.... God understands its way,/And He knows its place./For He looks to the ends of the earth,/And sees under the whole heavens,/To establish a weight for the wind,/And mete out the waters by measure.[14]

The source of life is buried beyond genes, beyond molecular structure and calculabilities. Its last stray threads are all biotechnicians can presently untwine. The rest has been interred so far inside the cell nucleus and the cytoplasm, the best microscopes merely bounce off its outer skin. Only it is not skin. It is more like the perpendicular distance between a line and the moving points on a curve that approach zero as they travel an infinite distance from their origin.

Modern sequenced chromosomes and their amino-acid offspring are relics from the pre-protein environments of five billion years ago— rebuilt ciphers with Precambrian tags. Even 32,000 decipherable strands from the system's origination, though cardinal and inextricable, represent less than a dimer in a haystack; they are as tenuous in relation to the axes and core body plans of undulating life forms as mist rising from an ocean is to the mass of intricate, swirling waters beneath.

They are like fog blown by winds rooted far deeper in the atmosphere. They are not responsible for enforcing order or guaranteeing safe passage for bionts.

RNA-catalyzed mechanisms may be inherited intact in modern times, but the specific and precise rules of their "controllers" are fragmented and scattered discontinuously and intermittently throughout cellular space, though they still work, in fact work better, sleeker, and more efficiently than ever because there is hardly anything left of them (only what is absolutely necessary—single morphogen trails fading as they arise). They have been simplified to states of barest drag on their quanta. Time and turbulence have eroded away everything else. Since the planet didn't kill them while it was scraping them down to almost nothing, they are the sleekest cats in any parade. Their aliveness alone runs them. And everything else has fallen by one wayside or another.

What remains couldn't possibly work; how can an engine run missing more than ninety-nine percent of its parts? But it *does run* because it has survived. It runs simply because it runs. Everything works only because everything *else* works. That is why it is alive and why being alive is unique and special. That is why we can't find its premise, its operating circuits.

Life has a synergy and momentum that can dispense with tiny levers and axles, the absence of which would halt any artificial machine in its tracks. A fertilized egg is so simple because it is so complex. Its pellucid layers, deprived of most of their puppet strings, still proceed through a remarkable metamorphosis without a hitch. They can assemble so much more because their weight of mechanism is so much less. This is a science we don't yet comprehend.

All synapses and triggers were heaped into the morphogenetic mechanism and buried beneath the null of its vestigial chemistry; they reside there now (like the weight of moonlight) without a scratch or chirp.

"God made everything out of nothing," remarked Paul Valéry. "But the nothing shows through."[15]

· · · · ·

The key to life is either too small or too discursive, too tautological or too piecemeal. It is concealed by chemical packaging which has come to look exactly like packaging, not mechanism. Even if we found it, we wouldn't recognize it as different from anything else. We couldn't detect why it succeeds, how it shoots unsigned commands across now-coalesced gaps, commands that no longer speak, that communicate essence by units of silence. Eons ago its context spilled down the sinkhole of the universe. Its mark was thrown away, or splattered into nature with the letters of God's alphabet.

We see it, but we don't believe it, for this kind of precybernetic information is faster than either light or time, quicker than white on rice. The impalpable originators and triggers of distributed networks of genes, enzymes, and proteins travel independently like the tentacles of an invisible octopus*, trailing acausal chains of effects in another dimension. They are operating an alien factory by an unknown principle of organization.

---

*I am reminded of my friend Wynn Free's joke: "What's an octopus's favorite gospel song?" Answer: "Put your hand in the hand in the hand in the hand in the hand in the hand in the hand in the hand of the man who crossed the water."

Junk DNA and Introns . . . .

Only about three percent of DNA's codes and rebuses appears to be readable genes prescribing functional proteins and enzymes. These codons congregate in gangs involved in similar processes and strategies of activity, an indication that cohesive embryogenic zones and basic tissue processes may have been organized independently in nature and then consolidated to make novel life forms.

Modern sequences that designate functional proteins or "start/stop" are scattered far and wide among DNA gruel, single sparks amidst darkness, "as rare as city lights seen from an airplane at night."[16] At least two-thirds and perhaps as much as ninety-seven percent of the human genome (and equivalent amounts of other genomes) is debris.

Over millennia of system development, sequences of active genes have been intruded upon and separated by copious material composed of long non-protein-coding strands. These are not read by transcription enzymes because (according to us) they have nothing cogent to say about how to make proteins or, for that matter, anything. This would suggest that the process that composed DNA did a lot of wasted scribbling too in order to arrive at useful sequences— a deceptively literal confirmation of the randomizing framework of natural selection. Over epochal time certain sections of nucleic code may also have had their integrity shattered from haphazard interference patterns and the inevitable error component of repeated copying. It is generally believed that the more garbled debris is, the more ancient its origin.

It doesn't seem to matter that the arrangement of nucleic blueprints lies in a discontinuous three-percent because their activation is nonlinear too. At the deepest level of system function we cannot actually tell what is (or was) essential. It may also be that we are looking at something quite different from waste or debris. We may be looking at raw omniscience.

• • • • •

Human genes cluster in zones favoring the guanine-cytosine regions by a ratio of about three to two. Some such precincts are as much as thirty times more gene-dense than outlying areas. These true "intelligent" genes are solitary gems amid residues, anthills on a plain of piffle. The role, if any, of the vast amount of DNA is unknown.

Almost a third of "junk DNA" consists of tens of thousands of protracted sequences that repeat again and again like the decimals of certain fractions—sometimes the same two letters hundreds of times, thousands of times—UAUAUAUAUAUA....

One chorus of three hundred letters iterates half a million times. Other sections of "junk DNA" consist of palindromes, formed as if by mirror images of the same sequences running into each other from opposite directions—and different epochs—providing another clue as to how they were formed in the dawn time.

Some junk language may represent reverse transcriptions of edited, mutated, and amplified codes that have become, at least temporarily, superfluous and noncoding. These may contain large banks of unused biological and environmental information, perhaps maintained in semi-Lamarckian modes (see pp. 247–249).

Junk DNA is mostly restricted to adenine-thymine zones. Exceptions are the relatively recent, short repetitive strings among primates known as the Alu sequence. These may be migrant transposons or mobile pseudogenes from outside the genome (see pp. 236–241). Congregating in guanine-cytosine zones around genes, according to some biologists they consist solely of cell-repair mechanisms. Like computer viruses, they bear indigenous instructions for their own transcription. Their RNA promoters pirated local reverse transcriptase in apes and hominids in order to transpose themselves. Occurring roughly once every 5000 nucleotide pairs, Alu now comprises about five percent of human DNA. During the Ice Ages its abrupt arrival possibly helped finish off the hominization process.

• • • • •

Tangled within primate genes, yet other major noncoding sequences embody as much as ninety-eight percent of the sum DNA (though their relative quantity and size vary from genome to genome). Known as introns, these encroaching segments lie between and segregate ones that write identifiable proteins (exons). Introns make up much of the supposedly gainful DNA clustering in already sparse pockets amid the junk. Because they can be altered without affecting gene expression, they are considered functionless. However, this is a partial truth.

During the 1990s intron sequences were observed catalyzing their own ligating. Oxbow-shaped RNA loops automatically spliced in new pathways during ribosomal processing (see "topoisomerase," pp. 146–147). Thus, cycles of transcription and elision are strategically and intrinsically self-organizing, using (in novel ways) even the material they discard, so nothing is discarded. In a hair-sensitive, self-referential system, waste is simply reinterpreted; it is hardly ever negated or trashed. This is partly because the process must remain precise enough not to change the reading of the exons by even one nucleotide, which would reframe them nonsensically. In morphogenetic landscapes, compasses come in mysterious forms.

When DNA is transcribed into pre-messenger RNA, exons and introns are both imprinted. In the fabrication of an interrupted gene (i.e., the working definition of "gene"), the introns must be edited out of intron-exon sequences by a molecular machine made up of RNA and its intron-removing, exon-cinching spliceosome. This part of the assembly line jettisons the introns and knits together the exons to make up mature messenger RNA embodying a legible formula for a discrete protein. During the operation, latent regulatory elements of the genome may instruct a spliceosome (see p. 211), while removing introns, to recognize some introns as though they were exons or vice versa, thereby shuffling the recipe.

Introns and other deceptively null sequences thermochemically come to favor looping in lariats and snipping themselves out of certain RNA strands in order that overall meaning gets generated, yet continue to resurrect themselves equally dynamically in order to be elided.

If introns clearly contain regulatory DNA sequences that, while not operating as genes, help genes to sort, transcribe, and improvise proteins, they play a key role in the multipotentiality of heredity. Innate RNA splicing to remove needless and jeopardous introns yields multiple and unpredictable tissues from the same genes.

When biotechnicians manipulate transcription artificially backward from RNA to DNA, they produce so-called complementary DNA (cDNA)—an intronless genetic strand that becomes the exception that proves the rule. In the natural world the very inscription of introns includes their excision, but in such a way that their temporary presence provides potential recombination sites for exchanges between sister DNA molecules, a process leading to evolution of new and alternate proteins. But cDNA can have only only one version or state, one expression, because, from the way it was copied, the gene *is already spliced.* Its transcripts form no lariats for splicing by intron elision (because, of course, there are no introns); it cannot be spliced again. Whereas normal genomic DNA responds to multiple signals with different expressions, cDNA is a one-trick pony.

If DNA is mostly slag, where are the remaining instructions for sea urchins, daffodils, and swine? . . . . . .

The possible meanings of introns and other noncoding elements need to be reevaluated. The putative meaningless if hauntingly repetitive letters of junk DNA are imbedded by historical and phylogenetic ineluctability and could bear quite profound messages beneath their illegible surfaces. They might contain attractors for alternative splicing and heterochronic, heterotopic, and other transcription and expression shifts in gene pools and assembly templates (see pp. 219–221).

Perhaps junk DNA stores a whole meta-language in which quite different orders of information are preserved and transmitted—a diction of uncertainty states.

In some circumstances, of course, repeats lead to organismic instability and diseases or birth defects. DNA testing shows about one percent of all children carry junk bands among tandem repeats not present in either of their parents, apparently indicating mutations of base pairs. These can express their mistakes phenotypically. In the case of Huntington's chorea, forty or more repeats of a triplet means nervous degenerative disorder, the more replications the earlier the age of occurrence. Thirty-four or fewer produce no symptoms at all. Between thirty-four and forty yield uncertain symptomologies.

. . . . .

Could iteration be information-packed at such an impalpable level that its long entrainments of nonsense serve to deepen and specify it while giving its variants broad, laserlike precision? One temporary discrepant sequence among hundreds of thousands of repetitions of short sequences of single molecules in genetic environments on a scale of millennia, if strategically placed and timed, could designate and particularize a chaos shift of information.

By the discreteness of its microdose (probably at a level of electrons rather than even atoms) a homeopathic remedy, when amplified by succussion, is presumed to act with global therapeutic effect. An evolving genome could similarly trigger genes between basins of attraction and alternate expressions by tiny oscillations and toggles within deceptively meaningless reduplications, reordering their sequencing to enlist different enzymes. Eventually a tipping point is reached, a complexity barrier crossed.

Where dynamics seem to lead nowhere in a system of recursive meaning, they paradoxically lead everywhere because a hidden principle of order retrieves every sparrow and caryopsis of grain.

Perhaps subcodes control master codes, and subsubcodes regulate *them.* That would make the notion of "nonsense" or "junk," especially as applied to the bulk of genetic material, presumptuous and provincial.

If the instructions for life lie among the whole of nature in a metagravitational field, then the terms of their specification are so large they are intangible and, for being intangible, immeasurable. In a field that broad, even a desert of junk partakes of and conducts emergent aspects.

That's what chaos in the universe at large is all about. Keep adding more and more disorder and gibberish to what is already total bedlam, and suddenly ordered patterns begin marching out. At the moment when all is lost, at least for creatures who honor patterns, meaning of such profundity it takes one's breath away comes flowing through the maelstrom like clowns and dervishes in shameless promenade.

The genetic code is not a code. . . . .

Nucleoplasm and cytoplasm are now maintained in codicils subject to other disjunct, defective codicils, themselves subject to even more impenetrable metatheses: phantom semes and alphabets in vanishing, tagless series. The tissues they record are turned into information in germ cells, then entangled, reindexed, and condensed—not only through being read by an aperiodic solid, then mirrored, sorted, and commuted back into life; not only by being shrunk fractally into micro-designs while turned inside-out and then outside-in as both amino-acid signs and actual membranes; but also by being "savagely" ground again and again from squirming three-dimensional objects into Braille and back into organisms again, all in molecular-electron fields generating each other.

From gastrulation to gastrulation, rebus to rebus, fluid-filled layers make vessels, vacuoles; afterwards (because of their histories) they unravel, open out, reconstitute, redeploy, invert in new layers, virgin pockets, fresh meanings, all the while storing disjunctive transcriptions of this molecular activity in coded bits entrained amidst junk; then they start over and fashion a new body from the nucleic text. Function and motif are drawn out of distension, transposition, resupination—Latinate words laid on a Pliny-Aristotle universe to reflect bottomless corners and gyres. Every process is partially and imperfectly imposed on and through itself trillions of times, trillions of deepening leagues down, always in and out, in and out of text, in and out of design, in and out of bodies, mutating as it goes—"since," Richard Burton sings in *Camelot*, "the whole rigmarole began."

• • • • •

Deciphering the genome is not like using Hebrew to read Aramaic, or any other *ex post facto* translation by Rosetta stone. Biological encryption is not only written in a totally foreign and unknown language; it is not "written." It is an originless meta-code—one set of partially determinable hieroglyphs linked to another, and another, in a series without a denominator or accessible index (even though— the ultimate mummery—it is meticulously and profoundly indexed with damnable specificity at every level). These series have formed around one another in tendrils and runners branching through invisible geographies and textures across all of cellular space and all of time, most of which we read as void. Cosmic wrinkles underlying syllabaries, they presently look about as deceptively alphabetic as crisscrossing webs of spiders of different species (and spiderlike runes from other galaxies). This is the mirage of genetic literacy, its mask— the illusion of an eroded Atlantean coin, the hieroglyph we cannot penetrate or copy because it doesn't exist in our space-time.

Creature development does not follow a plan like an instructional manual for a model plane or a series of blueprints for a house. Genes are epiphenomena of their own nonterminating sets even as organisms are epiphenomena of cells. At this level, there is no distinction between text and morphology, object and process; the script arises from the form it gives rise to, and the form translates, juggles, and redesigns, reembodies the script. Transcription and implementation of an extant DNA message are inseparable from the message itself.

Embryogenesis has no starting point, no engine, no machinery, no source code, no datemarks, no demarcations. Its software is its hardware; its hardware is its software. The computer that executes its program *is* the execution.

• • • • •

Everything critical about being alive, everything that matters to us, is here—so blatantly that we can do nothing about it except live: life is our experiencing of life.

And when we take it apart, poof! it vanishes. Of course something is still there in its place—chemicals to dispose of. Short of a vacuum, something is always there: valence, hyper-charge, strangeness, transformative obliteration.... Open a nut; you see its kernel; within that kernel, another kernel; kernel after tinier kernel, until we pass beyond the capacity to discern at all. What is ultimately inside a thing is not what a thing *is*. Life is not atoms, molecules, cells, or genes. It doesn't have a point of origin or first cause inside it. It doesn't submit to any of our descriptions, even when scientists claim to nail "the mothafucka" to the wall.

It is actually nothing. It was perhaps once some *thing*—qubits spinning off nanomachines, nonlinear attractors ordering turbulence, molecular couplings of Boolean reactions, equilibrations shattering open the adjacent possible. These spawned autocatalytic sets in phase transitions. But, then again, these are just words. They come *from* life.

• • • • •

We have made the biological riddle and molecular biology into what we think we are, not what *it* is. In fact, we *are* what it is, all by ourselves, without elaboration. Prehistoric men and women lived this; they knew it silently, from down under.

Out of the eyes of a gorilla or jaguar, it is not the polite knowledge of science staring back but the universe itself. It is not the information of genes but the pure intelligence of unknowing, unconscious nature. Lordly, flouting, even insolent—it does not defer to us or back down when faced with our machineries.

What are those shaggy animals pigmented onto the stone of Lascaux or the lofty visage-and-head poles in Alaska except (ultimately) chromosome maps? Long before there were pixels of cells, tribespeople left originary signatures of bison and kangaroos.

Totemic glyphs reflect cosmic glyphs through the jellyfish mindmass of the ocean. This is the resistance of matter itself against the abyss.

Whole classes of creatures simply appear. . . . .

Bacteria, one-celled plankton, and multicellular algae comprised the whole biosphere until about 543 million years ago. Then, within a mere ten million years of the early Cambrian, "creatures with teeth and tentacles and claws and jaws materialized with the suddenness of apparitions"[17]—the predecessors of sharks, squids, crabs, terns, iguanas, etc. The matrices for virtually the complete animal kingdom were invented in a flurry of protein activity, as though DNA, once created *in situ* or delivered to Earth by meteorite or artificial craft, spent a preliminary few hundred million years settling and organizing itself in the great sea, after which it burst into intense template experimentation for the epochs it took to fill the basic oceanic and terrestrial habitats and subniches, then (over the next several hundred million years) improvised from there with moles, newts, jack rabbits, dolphins, doves, foxes, foxgloves, weevils, armadillos, and billions of other, now-extinct designs.

That is a lot of system creation and invention from the same genetic repertoire.

* * * * *

The fossil record of the first biological renaissance does not seem to depict bevies of random selection or myriad willy-nilly activities. It is more as though an implicit complexity lay dormant and, once aroused, exponentialized.

The prevailing Darwinian creed that each lineage represents creatures that gradually evolved into each other, genotype by genotype, is belied by the sudden arrivals and sharp breaks, their body parts inscribed indisputably in the record. Genomes apparently get juggled in their entirety. Whole classes of creatures simply appear like Cherokee totem-beasts through the creation portal.

Undoubtedly many, if not most, of the novel protein configurations and rudiments of organs came into being in the Earth's biosphere long before they were needed or had any adaptive value or explicit purpose. Palaeontologist Stephen Jay Gould has given these preexistent complexes the name "exaptations": cartilage that one day will propel proto-musculature through water; submarine pharyngeal bladders through which future lungfish, amphibians, and their descendants will distill air; calcified membranes with which to munch other protein and later crawl upright on stilts; flaps of feathered skin between limbs with which to achieve flight; slow-twitch muscle fibers which change their chemistry as they are transposed to regulate body temperature. Exaptations all *preceded* their later niches. They filled different functions until they were displaced to meet more pressing environmental needs elsewhere with absurdly fortuitous results.

> ... [A]utonomous agents coevolve to be as capable as possible of making the most diverse discriminations and actions, take advantage of the most unexpected exaptations ... as readily as possible to coconstruct the blossoming diversity that is, and remains, Darwin's "tangled bank."[18]

A mode of transformation imposed itself seemingly beyond any limits except those of gravity and fractalization. It is a system that responds to environments, aquatic and forested, geological and anatomical (i.e., caves and intestines), though in a nonlinear fashion, by a raffle that oblivious to its own end-products, though all we see alive and among fossils are end-products. It is undaunted along a gradient from giant lizards to the smallest insects, and back. For instance, over 55 million years, a hippo-like member of the artiodactyls—an order of even-toed hoofed mammals that includes cattle, deer, camels, and hippopotami—metamorphed into ancestors of Moby-Dick and his kind. Current molecular and fossil analysis linking the ancestors of whales to the ancestors of hippopotami shows that a fat African river horse itself emerged midway through its cousins' progress toward full oceanhood.

The notion that genomes just roll the dice and happen upon coincidental adaptive morphologies makes no sense. A Lamarckian model of immediate, explicit flow from the environment into somatically responsive genes makes no sense either. Either pure natural selection, mutation, and gene reassortment have cultivated remarkable self-search methods, or they are contaminated with inherent design—or some combination thereof.

Local order arises from tumults of cosmic disorder. Playing the game itself—the natural sorting of elemental factors on median planets—in fact seems to change the galaxies' rules *in medias res,* to design (for no extant reason) biodynamic systems.

How else did ancestors of worms and newts also become monkeys and men? . . . .

Whales, elephants, maples, and mayflies all brought signature templates into the fossil record. Yet they must have evolved once upon a time from templates hatched by other species. Does a new creature reorganize itself spontaneously first from nearly the same genetic information, after which the genes comply enough to perpetuate the innovation? Or do genes rearrange themselves first to stabilize alternate states? How does an original message "loosen" enough to tolerate and even sponsor a different organism from almost the same genome set? How does the logic of nucleotide sequences hold together while vibrating between successive life forms? What supplies the energy or meaning to get from one activity-shape to another? What imbeds new information in a manner that cell regulators can interpret and incorporate? Is there an in-between state, or do genomes quantum-leap into fresh motifs?

· · · · ·

Mice and humans have not only near identical genomes but similar gene clumpings; the two species' X chromosomes carry the same five areas of rich gene associations. A map of the human genome set atop one of a mouse matches its central shape with only minor and peripheral variations.

Some cell-nuclear prestidigitation must have turned a few precursors of mice into primates, for a primogenial creed still binds us to such remote foragers. We share cells, philosophies, and phobias. Yet we are (at least by the standards of exterminators and lab technicians) totally different.

Diverging pathways to mice and men were traversed from a common ancestor close to a hundred million years ago, using the same core families of DNA organization. Morphological and phenomenological gaps were bridged without appreciable changes in genetic material. Homeotic and deep-systemic genes and proteins recalibrated to achieve new hierarchies and motifs.

Can a group of codons decohere such that single creatures have offspring not of their own species? What then keeps changelings from "rolling back" down the basins of their genomes? Must geographical barriers to interbreeding arise for molecular variation to become speciation? What, otherwise, prolongs an outwardly healthy, "normal" creature that is genotypically shuffling the deck?

What single codon or series of codons makes something else into a spider or a hedgehog? What was a wasp before it was a wasp, a whale before it was a whale? How does the transitional creature pass through its provisional not-mouse status?

* * * * *

Environments change; then genes change. Except, generations before that, identical messages generated heterogeneous microstates that continued to diverge along lineages in semi-closed ecosystems, permuting their codons and reinterpreting their own texts. The expression of a genome likely changes at a faster rate than its codons; yet both must account for and interpolate environmental dynamics—weather, population density, carrying capacities, asteroidal dust.

Of course, if biological systems originate before chromosomes, permute faster than their codes, and use text merely to stabilize protein configurations, then DNA is not the place to look for grand answers. Mice and men could share at their roots a more slowly changing ancestral archive, with most of the radical differences between them representing cascading networks of alternate protein states and other epigenetic landscape transitions and phase shifts.

[E]volutionary theory does not require the slow accumulation of small changes to produce body plan differences. Relatively early-acting, small genetic changes in genes that affect features of body plans such as axis orientation, segmentation, and appendage formation ... have substantial and immediate phenotypic results.[19]

In fact, the metamorphosis of somatic designs can come from very few genes if those that change affect the most primary aspects of development and network design.

What about the origin of life itself? ....

The first animate form was likely organized in a tidepool or around hydrothermal vents. Were its tubules hammered out *ex nihilo,* codon by codon, or did they emerge as full-blown protozoan scrolls—wise "worms" exploring the deep around them? Did pockets of inanimate goo amend themselves unstably until they became a state-of-the-art bacterium or a prebacterial Archaeum, or were they contaminated early on by hegemonous filaments capable of indexing a new basin of attraction? What was such goo in the moment before it was a life form? What states of organization did its genes pass through in going from serial rivulets to tight helices?

If a nucleus-less Archaeum is daunting to write, letter by letter, translate, mutate, and then recode, how difficult is the electrochemical clockwork of a wasp? How was *it* originally inscribed?

It takes 580,000 DNA characters to tabulate the smallest bacterial genome, about ten million to stabilize one ordinary bacterium, about a billion for a tiny fungus. The nematode genome is a modest hundred million letters. What about the 3.2 billion letters necessary to store panther or orangutan essentials in germ cells? How could DNA compose this record, codon by codon, on a time scale appropriate to living equations, experimentally and by random mutation?

• • • • •

Stuart Kauffman would rather assign self-organizing "machines" to an atomic principle and Boolean vertices than pilot them by a ghost. He offers another interpretation of a "fourth law of thermodynamics for self-constructing systems of autonomous agents. Biospheres enlarge their workspace, the diversity of what can happen next, the actual and adjacent possible, on average, as fast as they can.... [F]or the adjacent possible of a 6N-dimensional phase space to increase as the biosphere's trajectory travels among microstates, a secular increase in the symmetry splittings of microstate volumes must occur, such that different subvolumes go to different adjacent microstates. Eventually, such volumes hit the Heisenberg uncertainty limit.... [O]rganisms do touch that limit all over the place in the subtle distinctions that we make, turning genes on and off, smell sensors on and off, and eyes on and off, this way and that."[20]

Are we dealing with "unit[s] of quantum information rather than an infinite number of inaccessible quantum bits,"[21] or both. Does either nano-universe explain the advent of large animals, their synapses, thoughts, and paws connected to a trancelike flux of qualia?

Once again, the landscape is Cheshire cats to the horizon— Melville's single volition behind a wall of flesh, the "most buoyant thing within."[22]

• • • • •

Human hearing provides a model for probabilistic gradients of selective resonant genetic activation. Sound waves striking the tympanic membrane of the ear cause it to shudder, setting the oval window vibrating. The vibrations pass into the perilymph and, from there, the endolymph in the bony, snail-shaped labyrinth of the cochlear duct. Embryogenetically, cells differentiating from the tectorial wall of this duct assemble the sensitive organ of Corti with tiny hairs. The movement of the endolymph stimulates the hairs to generate nerve impulses which travel over the cochlear into the eighth cranial nerve and the auditory cortex where they are interpreted as sound.

Frequencies of vibrations transmitted from wherever into lymph are discriminated at a cellular level as each one synapses ultimately to a sensory neuron, its generator potential turning it on. The electrical resonance of this vibratory chain is altered by an ion channel encoded by the gene Slowpoke. The frequency depends on which of over five hundred potential splice variants are produced by the gene. The specific splice variant present in a given hair cell depends critically on where the cell is in the long gradient.[23]

The refinement of the sequencing stands out. It not only allows us to hear, sound in the mind is the basis of name, tribe, and mammalian calls.

If gene-regulatory programs and molecular switches are subject to uncertainty states and alternative events straddling the quantum/classical boundary, and even crossing into the quantum realm, genes and their mutation, transcription, and enzymatic phases behave more like photons than billiard balls. They code expressions like Slowpoke, as tissue plexuses rather than molecular discriminations.

If this is the case, then the relationship between genes and electrons, like the one between organisms and information, is subject to factors buried in kinetic energy quanta and fractional magnetic fields. It *might as well be ghosts.*

Long before science we were demons born of nightmares, strangers at the gates of a city yet to be built.

Biology is more like the work of a great artist than the product of methodical assembly lines. . . . .

Macromolecular music plays; portraits in three dimensions are fashioned by cells and their proteins speaking the Old Tongue. Tadpoles ravel from frog ova, geese from goose eggs, hedgehogs from hedgehogs. That is not only the rigor of genes and proteins sewing their layers together and unifying their pulsating mesh but the inimitable signature of a Bernini/Rembrandt "thinking" them, her mark on every canvas. The sculpting of a swan from an ovum, through all its miraculously seamless phases, resembles Vermeer rendering a scene in oils, not a unionized shop turning out plastic molds and televisions. A snake has ineffable style and elegance; so does a tortoise, an orchid; so do snow leopards, cranes, and raccoons. These are price-less objects, what immemorial Greek sculptors and Tlingit carvers aspired to—forms beyond imitation, beyond equation, beyond prior imaginability. Club moss, a magnolia tree, a cobra, a flounder, a drag-onfly, a porcupine, a crow. A mushroom, a peacock, a walrus, a sycamore—nothing in a museum rivals their raw novelty.

Life is an incomprehensibly deep set of responses to the existence of a cosmos, an implausibly dense, meta-stable configuration, a cal-cination, conjugation, sublimation, and (as the alchemists intuited) projection of matter through layers of its own ceaselessly differen-tiating and materializing mass. It is simultaneously an icicle, a prism, a plumb line. Morphogenesis is the twisting, radiating, and shim-mering of the subtlest aspects of molecularity through the deepest, densest precincts of planetarization. It is the capturing of a sun-star's matin glow in membranous throbbing, a gill of a wind-spreading pond sparkle in sense nodes, a forest of post-volcanic rivulets in pho-tosynthesizing trees.

• • • • •

Biology provides mechanisms for the bottomless within, and it explains each cohesive facsimile as a synoptic function of DNA-con-ducted molecules organizing in minute structures, coalescing and transmitting energy. But biology cannot explain what makes a func-tional shape-construct emerge from its embryonic layers and fly away like an osprey, hatch and crawl toward water like a turtle.

The native genius that composed Bach's chords and Mozart's melodies, Pawnee medicine wheels, Norman cathedrals, and Chi-nese vases must lie at the foundation of cellular life. Each species and template is a thematic thread, an inherent style, a latency exploding spontaneously through matter, drawing on the characteristics of not only genes but molecules, atoms, subatomic particles, gravitational fields, cosmic and psychic energies, as they vibrate and orbit about one another, radiating and enacting algebraic and geometric patterns while subjected to the flair of a master artist conducting them.

Ontogenesis is rooted in the background of the entire universe.

Stimuli are the tinder of creation. . . . .

The simple sexual organs of fish evolved into the genitals of amphibians, then the penises and vaginas of semi-reptilian lizards. These quadrupedal, scampering designs, as they explored land, used marine-adapted flaps and vessels to envelop underwater egg fertilization and reproduction. Aquatic creatures were condensed and synopsized in the proto-organs of air-breathing mongrels, their genitals fashioned from fusions of excretory tissues with swarms of germ cells, ultimately bifurcating into scrotum and labia (male and female) from the same multipotential set.

There was no one parturition at which a primeval fish became a newt, or even a coelecanth, nor was there a moment at which one ape sported a penis, his companion a penis-enticing vagina and clitoris. Sexual organs coalesced, like all organs, concretely and functionally, yet at no one time nor in any one place. There was no apotheosis, either physically or mentally, when sex became gender. These things happened incrementally, which means that they happened unconsciously and subliminally, likely through hermaphroditic phases before becoming separate conscious identities.

Sexual organs were also formed complete and whole. The creatures who bore them knew implicitly, perhaps archetypally, what they were, and they used them to express primordial states; to invent dances and other ceremonies; to recover bodies and worlds that were, even then, almost irrevocably lost. Their longings suffused cells with deeper morphologies via heat cycles and other activations and deviations of desire. Instinctual acts and courtship rituals through millions of generations of foreplay and orgasms continued to transform and specify organs themselves, while the organs provided the unconscious patterning behind the acts.

The force of cells inside us culminates in desire, attraction, tension, and release—submission to originless flux. Tissues surge to the termless ringing and shapely corridors of life, soften, and transfuse. Internal liquidity is what cell life feels like, what germ cells are. The baffling proprioception of molecular stuff flowing through us is our original nature more than even our thoughts.

. . . . .

Whether there is a fourth law of thermodynamics (or not), things apparently intend to get complex; they want to. They want to live on planets, pick berries, chase caribou, lie under darkening skies, damp their fur in glacial lakes; they would go through hell and high water to sniff immortal pollen in the air, to embrace another of their species, to romp in the fields of eternity, to awake to the invitation of another white dawn They want to become the adjacent possible, to plumb the indeterminate immanent. And that must be because what is happening in the "universe" at large is not entropy but antientropy—or even counter-entropy.

We don't just have bodies; we have the bodies embryos give us. We don't just grow useful organs; we inherit organs that can be crafted and shirred by morphogenetic events. An embryo not only hatches tissues; it organizes acts for them out of its layers. It maintains those acts through organs that come to embody them.

We have the style of life and consciousness that exudes from matter organizing itself in introverting nodules, morphogenetic grids, silken sheets, fleshy caverns. What we feel is the sum of dynamic semi-liquid processes colliding and wringing into shapes. Inside us their synapse is thousands of times louder than thunder and brighter than a trillion suns. This, the sacred texts tell us, is not only how we are braided of elemental cable but into what we are blown asunder when the cables snap. We are atoms that embryos have strung in thermokinetic lattices and set (in a fusion of subconsciousness, self-consciousness, and superconsciousness) to dance the Great Dance.

. . . . .

From the beginning of the biosphere, predation drove molecular entities into cores that, as they became impossibly dense, erupted, fissioned, and differentiated, pulling out shapes that hardly seem possible to have entered or agglutinated there. Tissues were initially protist colonies maintaining the fusion/dispersion that underlies all life. Reproductive zones inside membranes converted feeding and engorgement into desire, cannibalism into fertility. Billions of changeling spat, fetal salamanders, and other biological entities now teem from fain assignations of these seeds. So the universe gets to feel the wonder of its own coil, to participate in the tremor of its procreation.

At a tissue level we are a mosaic of convergent histories and membranes, each with its own arcane meanings. Our shapes have been assembled through the creatively unstable matrices of billions of jumbled creatures, conferring not only their algebras and biokinetics but smells, tastes, vestigial excitements, primal reckonings of space. We can never escape the lineages of plants, oceanic beasts, amphibians, and other entities that incubated our catacombs—fleshy and pelagic—in their ova. We spend our entire incarnation, while transforming biological density into cultural acts, becoming what those creatures were, what they portended.

＊ ＊ ＊ ＊ ＊

Where does eros come from in the adolescent boy and girl?

At least three different beginnings fuse in young lovers—desire, organs, and the sex act. Each of these radiates from a separate source through phylogeny, genetic synopsis, ontogeny, and erotic latency to arrive at an ineffable moment in which the universe itself seems to tremble and hang as deep as sky and delicate as a sparrow's egg. The feeling is one of ancientness—time, primitivity, and immersion in something vast, bottomless, and unknown.

Among the components of any act, conscious or unconscious, there is no way any longer to distingush the erotic from the non-erotic—what would the latter be? Whether originating from the urgency of germ cells to sunder, or taken from creature volition and later grounded in tissues, or both, desire roosts in every organ and every cell. "Existence permeates sexuality and vice versa, so that it is impossible to determine, in a given decision or action, the proportion of sexual to other motivations, impossible to label a decision or act 'sexual' or 'non-sexual.' There is no outstripping of sexuality any more than there is sexuality enclosed within itself. No one is saved and no one is totally lost."[24]

Life curls upon itself, releasing new forms; organs are made and desire finds and inhabits them, or desire is piqued, and tissues congeal around it—or germ cells continue to migrate; organs bloat and fold to receive them. They were once sea cucumbers and urchins, parts of sea pansies and feather stars. Now they are terrestrial viscera, residues of clams, eels, and sponges, bearing fetal cells in bladders that swarm with the craving of matter to be alive, ready to bridge eternity's gap and return to their pent source in spasms of waters.

There are no set genders; there is only desire itself. Genders are nature's transient defense mechanisms, keeping the molecular basis of life congruent with the social lives of cells in communities. Sexuality is what breaks the inchoate trance of membrane stasis. One sees it in every singles bar—whether urban, oceanic, or amphibious. Creatures everywhere want to shatter the mirror even if it reveals only another perfectly smooth reflection inside. Obscure in its origin, indeterminate in its objective, desire is the engine behind the coalition of rudimentary creatures once upon a time (again) into thicker, fuller-synapsing life forms.

The meaning missing from molecules emerges from concordances of proteins. What is not disclosed by solitary cells arises in the collaboration of cellular designs. What lone genes cannot program is synergized by macromolecular configurations. Cells not only get amassed in tissues; they invent tissues even as thoughts contrive symbols and events. Thus, paradoxically, it takes contingent (extrinsic) meaning to arrive at meaning's inherent design.

Life forms develop solely by interaction with and attachment to one another—not only cells but brains, nervous systems, cultures, and cosmologies ... not only living forms but waters, minerals, brines, muds, electrical currents, stars.

Stimuli are the sole tinder of creation.

Without the creative engagement of self with other, complexity must forever lie dormant, even in systems of its greatest potential.

• • • • •

What drives metabolism, strategy, and meaning at the scale of tissues and organs charged by neuron bursts and ganglionic corporation are polities of cells raising one another's unconscious mitochondrial and Golgi-body mechanisms to the threshold of conscious will. Cytoplasm confederates as a tentacle or impulse. Cells have *only other cells* with whom to collaborate in this universe-transforming enterprise. Sentience is a group act, a social event. Moon reflected in mind is a community of particles, phonemes, and suns. Otherwise, all that exists is matter strafed by stellar currents across oblivion.

Metaphysical considerations aside, biological agency and order are submerged and distributed among billions of molecular and submolecular decisions made over millions, even billions, of orbits, transferred back and forth among one another's intricate nonlinear networks, dimensions, and chronologies. Every new organism is lodged in every prior ancestral one in a chain of ontogenies perpetuated by the first life form and every succeeding one, yet without retaining most of their actual circuits or links. This is not a machine. This is not even an industrial metaphor. It runs without maintenance or context. It is called life.

• • • • •

An embryo is an "Oedipal" object in the most fundamental sense; it contains information that is both irresistible and unbearable. It entangles us in bodies and genders. It tells us exactly what we don't know—and what we can never know—about who we are and where we come from.

Nature is a Medusa masquerading as a Sphinx.

6

The Limits of Genetic Determinism

Dimensionless Epigenetic Landscapes

> Most people do not take heed of the things they
> encounter, nor do they grasp them even when they have
> learned about them, although they think they do.
> —Heraclitus, 540–480 B.C.

The Reign of Quantity . . . .

The current attempt to decode the human genome is the culmination of a materialism that has been unfolding inexorably for a century and a half. Its indoctrination has been passed from generation to generation and incorporated into everyday thought in such a way that people now promulgate it without realizing it is only an ideology.

With Prussian thoroughness, materialism has underwritten an alliance of politics, economics, and technology that frames the modern world. Everyone who transports his "being" in a car or plane, phones a friend, or undergoes an MRI is subtly and profoundly reindoctrinating himself every time.

Materialist propaganda takes on many forms (both obvious and latent), but mechanization lies at their heart. The multigenerational, transnational campaign to extend the laws of physics and chemistry to biology, psychology, and society has continued to gain momentum for centuries, leading to a culture running a technology in which the only reality is either molecular or derived from molecularity.

As a spreading, self-replicating environment of machines and commodities coopts both natural and social landscapes, the human race moves out of a fluid and mysterious engagement with life and phenomena into a fixed and hyper-tangible one, not in every thought and deed but as the umbrella under which our civilization organizes itself and sets policies and goals. Through science's reduction of everything in the universe to atoms and molecular properties (industrial metallurgy), labor, time, experience, and meaning have been commoditized, quantified, and vitiated. Beyond biological chauvinism and the reign of quantity, the super-realism of a shallowly concretized universe permeates literature, art, war, business, crime, religion, rock 'n' roll, daily manners and morals; it determines where people live, how they get themselves about, how they eat, seek pleasure, and, in general, inhabit lives. It is the bottom line, the offer that can't be refused, but it is also the all-black canvas, the sadomasochistic orgy, the serial killer, Hiroshima. Paradoxically this obsession with concrete acts, solids, and their valuations and ownership leads to a sense of emptiness and nothing seeming real.

The path from life as a mystical and sacred phenomenon transcending quantities and "things" to its subjugation by remorseless physicochemical vectors has been traversed in stages. As in all transitions, billions of separate, often-unconnected events have collaborated for a cumulative result. At the beginning of this sequence, a human soul or animal spirit was said to confer individuality and vital essence on its bearer; midway in the transition the life force may have been unknown, but it was assumed to operate along a psychophysical vector; now the genome has totally replaced the soul as the unit of self and identity.

. . . . .

Western philosophy has always sought the physical basis of things. In the mid nineteenth century Charles Darwin codified a rummage of prior unresolved scientific concepts regarding the origin of life forms and species into a theory of biogenesis by random physical processes. That bionts evolve and complexify so elegantly by mere disjunction is a mystery Darwinians routinely assign to millennia of random selection with survival of only the best plans. By this argument a far greater number of less fortunate designs quickly became extinct, so we do not see the full range of trial and error traversed to arrive at today's habitants. Presented with only blue ribbons, we falsely conjecture leaps of systemic intelligence. What we have may be only monkeys working on typewriters, though they ultimately wrote *Hamlet* and *Paradise Lost*.

Darwin made life mechanical by making it both inertial and accidental. Creatures arise through heat effects, gravity, and sorting. An organism possesses any "characteristic" (and exists at all) solely because its requisites have come into being through an interaction of vectors. From properties inculcated by trial and error, a life form has necessarily survived multiple onslaughts on its integrity from the environment and subsumed them in its very nature. It is what it has to be, and other, more vulnerable versions of itself either (mostly) never materialized or are already extinct. Each of its traits, segregated in a substratum, has been given a sole identity through the accumulation of contingencies.

The neo-Darwinian, Newtonian model uses only entropy organized by gravity, heat, and their minions. Because it is spare, it is elegant. It has become the credo of the modern world and, though most people either do not believe it or do not think about it, it is the basis for progressive technology, transnational capitalism, and contingency ethics, comprising the eidos that rules everyone's life.

• • • • •

Once the molecular structure of DNA became the exclusive province for all inherited traits in living things, no one had any need to delve further or consider other influences. Evolution, despite its dissimilar fruits, was a straightforward mechanism. The Darwinian cable car had been sealed, with us inside. An extraordinary epistemological leap had been made—assigning everything alive to ball-bearings in grids—without most people realizing its revolutionary implications or long-range impact on human destiny.

By the mid 1960s biologists were mapping DNA's helices, identifying codon sequences in amino acids, deciphering the order of molecules on chromosomes, and thus rendering at least the extant aspect of life forms in runes resembling (prophetically) bar codes. Genetic engineering originated in the 1970s, based on eliciting the sequences of genes in specific regions of DNA, then contriving recombinant strands from enzymes of microorganisms and patching new codons together (or molecularly synthesizing them from scratch). Transgenes were implanted in parasitic forms of DNA—viruses, plasmids, mobile genetic elements—which infected cells and spread their traits within organisms as the cells did.

The objective inheritance of biological traits suggested to ambitious technologists the possibility of creative interference with the process. That is what scientists do every day at the office: they cut life up, look inside, move around parts, educe new results.* They push the envelope wherever it looks as though nature might give way and yield further secrets and means. In friendly (and not so friendly) competition with one another, they try to invent new technologies from the cutting edges of prior ones. Where one crypt has been opened, they scour its booty for other vaults to pry:

> Man puts an end to darkness,/And searches every recess/For ore in the darkness and the shadow of death.... Its stones are the source of sapphires,/And it contains gold dust./That path no bird knows,/Nor has the fierce lion passed over

---

*An ornithologically oriented friend was recently told by a biologist colleague that his work on the language of birds wasn't scientific. "I know," he replied, "it's not science until you kill something."

it..../He overturns the mountains at the roots./He cuts out channels in the rocks,/And his eye sees every precious thing..../What is hidden he brings forth to light.[1]

The Genetic Paradigm . . . .

A mostly triumphant run—from Darwin and Mendel to Crick and Watson to human-genome decoder Craig Venter and Ian Wilmut, the cloner of Dolly the sheep—has come to imply that biologists have succeeded in deciphering life. They have certainly deciphered something. They are running machines and models, and investors are coming away with big bucks.

Yet without a way to address embryogenesis at a level of unity above genetics, biologists have come to view creaturehood itself as a subset and secondary effect of genes. Embryos are thought to develop as outputs of logical switches. They exist because genes "make" them. Genes have ontological priority as their sole efficient cause. Philosophers of science not only ignore genes' own uncertain and incomplete molecular etiology but act as though embryos have no formal components apart from the coded instructions that are present in extant genomes.

* * * * *

Having declared that what we see (and can see) is all there is or could be, we simply look harder if some effect is not covered by the reigning paradigm—and do more cumbersome experiments of the same kind. If these don't fill the breach, something somewhere else is deemed deficient or broken. It will be fixed. Defect-ridden clones in labs represent, to their proctors, unsuccessful technology rather than misapplied metaphors or oversimplified theory.

The paradigm looks cleaner from the outside, for it rests on a biased assumption that, since any action must have a linear, physical solution, the sparest digitalized vectors will always be the truest agents. Genes are the software and proximal cause because they are exactly the kinds of control mechanisms that today's world expects and wants inside their bionts—hardcore programs storing source codes through which they program effects. These programs minted us, so of course—now that our tools are perspicacious enough—we find them *inside* us. Genes are given sole priority for "inventing" animate nature because there are no other candidates and they neatly fulfill the prediction of their own existence.

No matter that the application of the "genetic determinism" paradigm is full of gaps, failed predictions, and rigged assumptions, biologists continue to believe that systematized, thermodynamic molecularization is precisely and quintessentially the way matter *should* assemble life, how we would run a factory. We have copied our metaphors not directly from nature but through the opacity of our projections onto nature: a model is real because it must be real. Most educated people swallow it—hook, line, and sinker. Faced with printouts of chromosomal parchment, they dutifully consider themselves cyborgs manufactured by source codes comprised of additive expressions of fixed traits—the playback of a program deposited algorithmically in each genome.

Any other imaginable candidates incur far greater dilemmas. They portend—God forbid—noncybernetic (or meta-cybernetic) channels of information.

Genetic determinism is a rigged metaphor to serve a totem.

Genes as Commodities . . . .

The "superscience" of trait biosynthesis is a game fixed in advance by its winners. Post-modern genomes have been designed expressly for cultures which count beads, levy taxes, keep censuses. Look closely at the digital printouts of chromosomes. These are—guess what?—business ledgers. It is evident why such semblances are so precious to both scientists and executives: commodities can be quantified, patented, stored, inventoried, and exported. Even if the qualities that are being concretized are illusions, the sleight of hand of laying claim to them serves corporate myth almost as usefully as molecular assembly lines. Along the way, some properties are lifted by their "gene-straps" and transposed into products. They are the ones that yield more or less identical traits from species to species, phylum to phylum; thus they are patentable.

The ancient wisdom, both environmental and tribal, underlying native crops and medicinals is discounted. All that suddenly matters are the chimerical genes inside cells by which properties of flora and fauna can be acquired and controlled. A few months of hi-tech manipulation in a laboratory is valorized above thousands of years of heuristic, indigenous inquiry.

The characteristics of plants and animals are privatized, taken from a public and egalitarian realm, enfranchised by multinational companies, and then auctioned in the futures marketplace—often back to their previous custodians as if they were true new products invented by progressive free enterprise.

Fake property perhaps, theft nonetheless.

• • • • •

The militarized mathematics supporting both money and genes is a deviant nineteenth-century mutation. It is not a paranoid fantasy to suspect that the unconscious goal of genetic determinism has been, all along, to serve emerging capital agendas.

According to rules established by the Union for the Protection of New Varieties of Plants under international convention, the "common heritage of mankind" (i.e., the wisdom of tribes and peasants) is *not* subject to either public domain or ownership by the people who have been its stewards for generations. Thus, traditional knowledge cannot be safeguarded from the unholy alliance of business and science—modern bandoleros—any more, in fact, than endangered species can be defended in every cove and at every waterhole and coral reef from poachers, pirates, profiteers, and recreational killers. Corporations have self-stamped license to rake Amazonian rainforests, ponds of Asia and Africa, and coastlines of the Pacific Rim—in short, the ecosphere—for all available booty. When necessary, they kill or buy off enough potentially meddlesome tribespeople with the same inflated currency that already has impoverished them.

Indigent natives can even be enlisted to plunder and decimate their own habitats.

"Northern countries are allowed to take freely from the south, as 'common heritage,' genetic resources which are then returned to them as priced commodities."[2] Medicinal neem, decocted since the dawn of time in India as an antibiotic and insecticide, was supposedly "discovered" and then patented by W. R. Grace Company. The privatization of the plant by its genetic code caused its market value to jump a hundredfold in two years.[3]

Colonial exploitation has finally reached its *reductio ad absurdum*.

. . . . .

DNA codes are what corporate biotech is now primed to merchandize as its first order of business at the dawn of a new century—the future of not only philosophy and medicine but the world's markets. Premiums of nature, genes are poised to dominate the twenty-first and twenty-second centuries, twilling fabrics, foods, and potions undreamed of. They have already been annointed the supercommodities and future rulers of civilization. Genetic science seeks to turn the basic characteristics of life into acquisitions to be repackaged and sold. It thereby conflates the two central currencies of modernity—shareholder value and biological identity. Companies are rushing to beat one another to the patents for everything with a pulse, every circumscribable trait, in order to control the global economy of the future, forever.

Seeds with properties snipped from jellyfish, insects, bacteria, and viruses are meant to replace conventional crops, as domesticized grains, vegetables, and fruits once replaced wild plants.

Using implanted glass "gene" chips, doctors expect to monitor patients' biological software codes and states of health and to administer to their pathologies with targeted tags of DNA.

Cloning is proposed as a top-end reproductive option for the elite (along with their domestic pets). Thus, a child or a cat who dies prematurely will be replaced by its genotypic twin. Corporate patenting of all our body-parts and attributes—even our very selves—

apparently lies just down the road.

Other enthusiasts claim sanctimoniously that, since only one parent is necessary, the option to remake oneself (or, more to the point, a replica of oneself) solely from one's own genes, without "contamination" from a partner, is a basic, inalienable reproductive right—as though Thomas Jefferson would have written it into the Constitution if he had had sufficient foresight. But, alas, he did not even perceive the telltale forensics he left in the progeny of Sally Hemmings that would convict him of adultery and racism generations after his death.

Not Enough Genes in the Genome? . . . .

In this climate of product puffery, the mapping of the human genome has been accorded the sort of awe and genuflection that really should have been reserved for the Second Coming or a declaration of world disarmament. Barry Commoner reviews the epidemic of hyperbole:

> In 1990, James Watson described the Human Genome Project as "the ultimate description of life." It will yield, he claimed, the information "that determines if you have life as a fly, a carrot, or a man." Walter Gilbert, one of the project's earliest proponents, famously observed that the 3 billion nucleotides found in human DNA would easily fit on a compact disc, to which one could point and say, "Here is a human being; it's me!" President Bill Clinton described the human genome as "the language in which God created life."[4]

Yet, in light of the multidimensional intricacy and labyrinthine infrastructures of life, two separate genome teams could not find the genes warranting this complexity. In early 2001 after years of detailed chromosome analysis, researchers reduced the revealed number of human genes suddenly and dramatically from a projected 100,000 (based on the number of human proteins) to about 32,000, each containing between 1000 and 1500 letters. A mere 300 or so of these have no recognizable analogue in a mouse (see p. 171). By comparison, a roundworm has about 19,000 genes, a fruit fly 13,000. Many plants have more or less the same number of genes as we do, a fungus a third as many. If genes are the source codes for life and individual species, there are apparently not enough of them to program the full array of proteins and protein combinations for viable worms and moles, let alone intelligent beings who carry out genomic research.

There clearly isn't much difference genetically between humans and mice and not the amount of difference we would expect either, in terms of raw data, between a roundworm or fungus and a man. Modern apes and humans in fact are so similar, sharing more than 99.5% of each other's genome, that biologists are hardput where to begin in distinguishing us at a DNA level from gorillas and chimpanzees. As far as our chromosomes are concerned, we are the same animal. Even baker's yeast and humans share 400 genetic memes. Human genes have, as noted, been transplanted successfully into flies. Plus, the deviations between some groups of fungi surpass those between insects and vertebrates. So how can genes alone be used to distinguish humanity (or rabbithood) or to glean a person from a carrot? "If the human gene count is too low to match the number of proteins and the numerous inherited traits that they engender," observes Commoner, "and if it cannot explain the vast inherited difference between a weed and a person, there must be much more to the 'ultimate description of life' than the genes, on their own, can tell us."[5]

Apostate preacher that he is (see pp. 196–198), Commoner is satirizing the hubris and bungling of modern biology. Fools and charlatans they be who think such a gambit could actually work and we would get to carry ourselves around on CDs.* This white-collar gullibility is at about the level of crocodiles in sewers and the sheikh who

---

*A new development in pure material sciences will apparently soon allow technicians to run a piece of genetic material through a series of microtunnels and tiny filters and come up with its genome in about fifteen minutes. If heredity is all and only molecules, a portable decoder should be able to siphon and read our digits, molecule by molecule, almost as rapidly and capably as a Netscape program browses the internet in a palm pilot.

ostensibly set his RV into cruise control and then stepped into the back.

"In fact," Commoner continues, "an inattentive reader of genomic CDs might easily mistake Walter Gilbert for a mouse, 99 percent of whose genes have human counterparts."[6]

"We had a hard time explaining the control mechanisms when we thought there were 100,000 [human genes]," lamented Francis Collins, head of the American contingent to the Human Genome Project. "Now we have only a third as many."[7]

With just 300 words in the human dictionary that are not also in a mouse, Craig Venter, president of Celera Genomics, admitted the unlikelihood of any simple genetic determinism: "[T]his tells me genes can't possibly explain all of what makes us what we are...." He added, "It's become part of the common language to say we'd like to have the gene for this or the gene for that, but the common language is wrong. I believe all of our behaviors, all of our sizes and functions clearly have a genetic component, but genes only explain part of any process. We are around as a species because we have an adaptability that goes beyond the genome. If everything was hard-wired, we wouldn't have survived."[8]

Where does the stuff that isn't hardwired come from? How can it work so well so consistently without either ball-bearings or wiring? What are genes anyway? Is a "gene" its code written in serial characters or the protein-based expressions of those characters? If the latter, then where is the "gene" located? If the former, then how does it componentially organize letters into traits and traits into tissues and behaviors? What about the capacity of single genes to make two, three, or more different proteins? What about the even greater, more unpredictable potential of macromolecular proteins and the tissues and organs they give rise to?

* * * * *

Richard Lewontin summarizes many of fallacies of genetic determinism, while explaining the baffled letdown that followed the "triumph" of human genome projects:

The major irony of the sequencing of the human genome is that the result turns out not to provide the answer to the chief question that motivated the project. Now that we have the complete sequence of the human genome, we do not, alas, know anything more than we did before about what it is to be human. At the time of the completion of the human genome sequence, scientists already knew the complete DNA sequences of thirty-nine species of bacteria, a yeast, a nematode worm, the fruit fly, *Drosophila,* and the mustard weed, *Arabidopsis.* In each case it is possible to estimate how many genes are present in the genome, using two methods. The first is to compare stretches of DNA sequence with sequences of particular genes already known from a variety of organisms. The other, for DNA that does not match already known genes, is to use certain sequence motifs that are common to all genes. When this so-called 'annotation' of the human genome was done it was estimated that humans have about 32,000 genes. This seems a rather small number when the comparison is made with the fruit fly (13,000), the nematode worm (18,000), and the mustard weed (26,000). Can human beings really only have 75 percent more genes than a tiny worm and a mere 25 percent more than a weed? If, as the eminent molecular biologist Walter Gilbert wrote, a knowledge of the human genome would cause "a change in our philosophical understanding of ourselves," that change has not been quite what was hoped for. It appears that we are not much different from vegetables, if we can judge from our genomes.

The reaction to the discovery that human beings do not have much more genetic information than plants and worms has been to call for a new and even more grandiose project. It is now agreed among molecular biologists that the genome was not really the right target and that we now need to study the "proteome," the complete set of all the proteins manufactured by an organism. Surely the very complex human being must have many more different proteins than a small flowering plant. Although devotees of the genome project kept assuring us that genes made proteins and therefore when we had all the genes we would know all the proteins, they now say that, of course, they knew all along that genes don't make proteins. Genes only specify the sequence of amino acids that are linked together in the manufacture of a molecule called a polypeptide, which must then fold up to make a protein. But there are many different ways in which a long polypeptide can fold, resulting in different proteins. The way in which the folding occurs may be different in different cells of different organisms and depends in part on the presence of small molecules, like sugars, and on other proteins.

Moreover, a gene is divided up into several stretches of DNA, each of which specifies only part of the complete sequence in a polypeptide. Each of these partial sequences can then combine with parts specified by other genes, so that from only a few genes, each made up of a few subsections, a very large number of combinations of different amino acid sequences could be made by mixing and matching. So knowing all the genes of a human being doesn't really tell us what we want to know.

One prominent opponent of the genome sequencing project, William Haseltine, CEO of Human Genome Sciences, has long claimed that the right way to find all the human genes is not to sequence the genome itself, but to go directly to the products that the cell makes when it reads the genome. These products, nucleotide sequences called "messenger RNAs," are then used by the cell to manufacture polypeptides. Haseltine claims to have detected 90,000 of these messengers in human cells, but whether that means there are 90,000 different genes or 90,000 different combinations of bits and pieces from approximately 32,000 genes is unclear....[9]

• • • • •

The expression of a genome is subject to a bottomless web of sub-causes and subeffects; thus, the Human Genome Project was a bit of a wild goose chase. The decoding of our species map may have been a majestic technical accomplishment, but the notion that it should hold the key to life is an absurd misgauging of the actual depth of the processes giving rise to living forms—a confusion of a secondary mechanism for its originary source of design. "Outmoded research funding patterns ... continue to see complex phenotypes as primarily derivable from genomic and proteomic databases."[10]

Scientists set ersatz goals and then declare victory when they are attained. They place the bar scandalously low because they want prizes; they want to get to celebrate. But traits are neither additive nor configured any simple way. They are not reducible to information packed and spooled onto micro-cables. The notions that tissues are the outcome of ordinary messages and that entities can be constructed by depositing ciphers like coins into assembly lines converting raw materials into biodesigns are misrepresentations. They involve counterfeit assignments of numerical values to elementary physical processes.

There is no fixed alphabet or simple algebraic expression of life. The etiology of the biosphere belongs to a pantheon of deep events wherein the word "decipher" is not even relevant. One might as well try to decode clouds and sand dunes.

The human genome map was smoke and mirrors, theater and hype. In truth, scientists dressed a sliver of an utterly insoluble puzzle in a borrowed costume and assigned it one-dimensional pathways with epicycles, while ignoring its preponderance—e.g. how the system actually works.

Paradigms and Ideologies . . . .

I quote often from Barry Commoner's timely and celebrated debunking of current genomic research that appeared in the February 2002 issue of *Harper's Magazine*. My scientific readers have since warned me that Commoner is out of the loop; he has zero current standing among working microbiologists, with no contributions to reputable peer-reviewed journals in three decades. He is not so much a spy in the house of genes as a veteran saboteur, and his tone of implied authority and insider smarts is less than forthcoming. That does not invalidate his article, but it does make it old news repackaged as a scoop.

I, on the other hand, am even further out of the gene game, having never been in it. I face the continual problem of having no informed context for evaluating statements pro or con the relative roles of genes or cells at any number of levels. I can only weigh the public arguments and counter-arguments.

It looks to me like a hung jury: no cells without genes, no genes without cells. At the same time, we are caught inside more than one set of contrary paradigms; we can't let go of the old dichotomies, and we do not have a new one coherent enough to adopt. Commoner is justified in spreading the word that the current model is irreparably flawed, but he is overzealous in behalf of his own solution, guilefully exploiting a semantic and epistemological confusion as to what constitutes a gene. "[T]o a certain extent, [Commoner's ruckus] is all . . . a quibble over nomenclature—is it the 'same' gene that is specifying different proteins or are they really different genes that happen to share overlapping DNA sequences? The fact is there is still one DNA sequence per protein."[11]

Commoner is engaging in a bit of sophistry, inflating the significance of a muddled distinction already identified as such in most freshman-level textbooks.

In 1968, back when he *was* in the game, in an essay in *Nature* entitled "Failure of Watson-Crick Theory as a Chemical Explanation of Inheritance," Commoner declared: ". . . the biochemical specificity which governs the biology of inheritance is embodied not only in nucleic acids but in proteins as well."[12] This has been his pet postulate for a long time and, as we shall see below, that should not disqualify it. While it may not be the gospel truth, through a glass darkly it augurs deeper verities. However, Commoner shouldn't at the same time pretend to be the Einstein of biology. He is more like the dormouse at the Mad Hatter's tea party, having popped up from a long drowse to utter the same peep. We are not anywhere near the dawn of a new gene-protein paradigm at the level Commoner is marketing as news.

The clique of biologists who have objected most strenuously to the Commoner article have piloried it as yellow journalism while ridiculing its appearance in *Harper's* as an indication that not only could its author no longer publish in a scientific venue if he wanted, but he chose instead, as one put it, a popular magazine to link "his views to the activist campaign against plant biotechnology..., still smarting from his intellectual defeat at the hands of James Watson and Francis Crick."[13] This attack on Commoner—entitled unnecessarily but tellingly, "Is Barry Commoner a Mutant?"—is at least as vitriolic and biased as anything *he* is saying.

To summarize: Commoner's critics begot the original problem by pretending genes are fine and that anyone who says nay is unpatriotic—which Commoner then compounded by insisting genes are not okay, but in a self-aggrandizing rant and by "solving" the problem with his own misplaced concretenesses. Since neither party knows what life is, they are engaged in ritual combat of differing allegiances and biopolitical agendas. They are equally at fault in not owning up to the more basic crisis at the heart of science itself—a crisis that is not resolved any better by proteins (or phase states) than genes.

·····

High-profile imbroglios like this expose the self-interest peppered with bravado and sarcasm that characterizes so many debates in modern science. It is hard not to notice that, after one departs the epoch of Newton, Lavoisier, Darwin, and Priestly and enters the modern frame, we have few true scientific innovators and honest oracles in the old style but plenty of pretenders, fudgers, forgers, and casuists—even famous ones. The result is that, while technology has soared, the scientific foundation underlying it has crumbled into expedient gadgetry, having been hijacked and corrupted by special-interest groups. This is forcing science into a dead end from which it can only emerge by returning, one day, to new raw observations—nonideological empiricism. When disputants on opposing sides begin talking to and respecting one another again, the great human "truth" enterprise will move forward—but until then it will remain stuck in petty squabbles and false epiphanies. And eventually technology—as it bumps into the real limits of the present paradigm—will stall too.

· · · · ·

What we are talking about is future paradigm shifts. Meanings do not change from political pressure or p.r. They realign when the old guys die or when new facts become so obvious that they would be disputed only by ideologues or fools—Copernicus and Galileo are signal instances of the whole cosmos turning on words uttered by a man on Earth.

As Thomas Kuhn concluded in his classic book, *The Structure of Scientific Revolutions,* paradigms shift mainly as other paradigms supplant them: "At the start a new candidate for [a] paradigm may have few supporters, and on occasions the supporters' motives may be suspect. Nevertheless, if they are competent, they will improve it, explore its possibilities, and show what it would be like to belong to the community guided by it. And as that goes on, if the paradigm is one destined to win its fight, the number and strength of the persuasive arguments in its favor will increase. More scientists will then be converted, and the exploration of the new paradigm will go on."[14]

Popular beliefs *don't* suddenly shatter and then re-form from one instance of an elevator operator "projecting" murky images from his brain onto film. Science ignores events that embarrass it, and it ostracizes people who keep pointing out those events. Whether thoughtographic telekinesis is possible or whether Ted Serios is a sleight-of-hand con man is a riddle that will likely have to await a future epoch during which science has found a way to measure or at least address mental effects. Until then or until a UFO lands on the scientific equivalent of the White House lawn, this and other anomalous events will have to remain (at best) pseudoscience.

· · · · ·

Paradigms *have* commuted, even rapidly, in recent times. The notion of continental drift—that land masses move upon the Earth's molten core—was in disrepute as Noachian catastrophism for more than a century from its inception. After it was propounded in 1858 by Antonio Snider-Pellegrini in his book *Creation and Its Mysteries Revealed,* a number of scientists extended and elaborated it in the following decades—for instance, Alfred Wegener in 1915 in *The Origin of the Continents and Oceans* and Alexander L. DuToit in 1937 in *Our Wandering Continents,* the latter using firm evidence from paleontology, sedimentology, glaciation, and rock types. It was not until the 1960s, however, that the paradigm flipped in a virtual eyeblink when the discovery of ocean rifts in the context of the development of plate tectonics provided an indisputable subterranean mechanism for events observed on the surface of the planet.[15]

In 1980 in a paper in *Science* entitled "Extraterrestrial Cause for the Cretaceous-Tertiary Extinction," prominent Manhattan-Project physicist Luis Alvarez (with his son geologist Walter Alvarez and two other scientists) drew attention to the unusually high percentages of iridium in Cretaceous/Tertiary (KT) deep-sea boundary clays from 65 million years ago in Italy, Denmark, and New Zealand. With platinum-type metals generally rare in the Earth's crust relative to their cosmic abundance and with no other logical source for a spontaneous spike, the authors suggested a giant meteorite or asteroid impact from which pulverized iridium-rich dust was injected into the stratosphere and travelled in a layer for years, snuffing photosynthesis and leading to the extinction of the dinosaurs and thousands of other species.[16] This interpretation, while initially regarded as a bizarre heresy, was near universally accepted in less than a decade.

We now await equivalent, far more basic and profound paradigm shifts, not merely in calculating the role of genes in the formation and maintenance of species in a biosphere but in assigning a status to consciousness itself in relation to matter.

• • • • •

The following are my lay precepts regarding the failure of genetic determinism. They are redundant and overlapping much in the way the systems I am describing are. Since there is no true entrée into the genetic paradigm, all my distinctions overlap and split hairs upon one another. The enumerated points should be viewed more as motifs than a sequence of logically advancing premises; some are in fact subsets or different ways of looking at others. Also, remember, they are outsider art. I am borrowing other people's scorecards. I don't fully grasp the concepts expressed below because I don't work with things like tandem sequences, multigene families, aneuploids, transposons, etc. I have no idea what they *really* are. I am only trying to get at the background out of which these terms come. Forgive any errors of context and scale that occur in my foreground.

Genes are not blueprints for organisms. . . . .

1. Genes do not have independent meanings.

Genes and traits exist not as wholes but in bits that must be reinterpreted together in the synthesis of messenger RNA and again in the manufacture of proteins.

Insofar as genes are not end traits, their apparent function is to project matrices with potential for wide-ranging expressions. Actual traits occur only contextually in organisms. System development follows not so much from the level of information coded in genes as multi-level processes yielding transgenetic waves of events.

By chemistry (morphogens), codes can transmit messages and even link these together by tiers of information, but they cannot imbed configurations and links in semi-closed, repeating synergistic cycles that regenerate and reproduce whole bodies and meanings.

The regulation of heredity and biological design comes ultimately from millennial events and submolecular dimensions of such profundity and uncertainty (both literally and figuratively) as to overwhelm any simple sense of genetic control of protein structures.

• • • • •

The effects of genes are discontinuous and sometimes epistatic. Genes do not operate independently or in linear causal chains such that a codon has a singular or exclusive function and unchanging outcome; they express themselves only as parts of genetic-epigenetic loops reconfigured by constant feedback from their milieu.

. . . .

Epistasis: a nonreciprocal interaction in which one form of a gene suppresses or otherwise alters the expression of a nonalternative form of another gene implicated in the same tissue complex in an organism but independently inherited.

. . . .

Equivalent results are attained by different sets of genes increasing and decreasing in transcription abundance and reversing roles under perturbing conditions, such that they operate as mirror symmetric pairs, "each coregulating and yet buffered from other sets of genes."[17]

The information bank from which genes collocate proteins has no small print indicating what proteins will form in each context or what types of tissues and traits or creatures will arise from those proteins. From species to species the same genes can elicit quite different traits. "Observations have been made of proteins performing similar functions though with different genetic origins ... evidence at the molecular level of parallel, convergent evolution ... in which redundantly complex genetic and biochemical systems can manifest multiple causal routes to similar functional ends."[18]

The potential expressions of a specific gene comprise a condition called pleiotropy—a situation-dependent multipotentiality.

One gene can give rise to a single polypeptide or to many different polypeptides. Conversely, in rare cases, many genes can collaborate on one polypeptide; for instance, antibody proteins are configured by several genes participating together as their DNA instructions are differentially interpreted during codon translation. The gene that determines a specific antibody polypeptide is for the most part unlocatable and ill-defined. It has neither a fixed geographic

. . . .

Pleiotropy: *the condition of a single gene's regulation of many different, unrelated tissue effects.*

. . . .

site in molecular space nor a traceable trajectory. Delocalized through the genome, its design is expressed only by its dynamic interaction with other equally delocalized genes. Creatures cultivate immunity as they come into contact with new germs and toxins.

Since this particular method of translation is an exception to the general rule—untrue for virtually anything but antibodies, and mostly untrue for them as well—it probably bears more phylogenetic than ontogenetic significance, affecting deep principles of system origination at a sub-blastoderm rather than organ level of implementation.

The usual way that many proteins collaborate on a single polypeptide is by some proteins acting as splice-choice factors for the RNA products of other—and sometimes their own—genes (see below).

. . . . .

Genes express singular and multiple characteristics, which also may be expressed by one or more genes or a family of genes.

Collections of genes (multigene families) with similar or identical sequences arranged in tandem can induce the manufacture of single proteins in parallel fashion—for instance, the nucleosomes of histones which are made of five kinds of histone proteins and help to package DNA in eukaryotic chromosomes. Histone genes are present in multiple copies, but each copy of a given type makes the same histone.

Multigene families also write subdialects of the same messages; one example is the differing embryonic and developmental forms of hemoglobins in red-blood cells.

Ribosomal RNA genes (which don't encode proteins) are present in multipotent copies.

Tandem groupings of genes may have originated from unique biokinetic events setting off chain replications or by repeated unequal exchanges between homologous chromosomes. Such families retain their uniformity by eliding any divergent sequences (mutations and other variations) that arise later in them.

• • • • •

2. Traits are always disperse, multiform, and nondiscrete.

Some properties—phosphorescence, hue, dilation, etc.—can be identified, isolated, and transferred with DNA across organism boundaries and even between phyla, but they still have no unique expression or meaning. They just happen to carry temporarily literalizable design codes in historically discrete contexts. The properties *per se* are not concretely fixed in the genes; they are stabilized effects of molecular interactions.

Eye color, musical ability, or size of claw may be demonstrably influenced by identifiable codons. This is more or less corroborated when biotechnicians transplanting a gene from a luminescent jellyfish are able to create a glowing rhesus monkey. Likewise, the adaptation of Antarctic ectothermal animals can be tied to two or three mutations affecting enzyme function at subzero temperatures—its catalytic efficiency, regulatory efficiency, and structural stability.

It is hard, however, to imagine how maternal love, combativeness, homosexuality, or migratory routes could be stored substantively or synoptically in genes. Perhaps certain hormonal components of these can be attributed to DNA, but the pure phenomenology of comprehension and patterns of behavior cannot.

There is an enormous distance from genotype to phenotype and no simple rules for getting there.

• • • • •

3. The message of any one gene must go through numerous and labyrinthine feedback loops and multiple hierarchies of appropriations and reappropriations of its by-products before it is completely expressed.

Traditional experiments establish entry and beginning points with control outflow, but genes and embryos are not experiments.

They have no paths into them; they do not use exclusive controllers and chains of command. Their designs are nonhierarchical, even democractic. Control, which is always local, is also always global.

Traits that are not stored in genes arise from vast systems with nucleic components; they depend on the innate and emergent complexity and diversity of cybernetic and environmental processes which they themselves partially generate.

Insofar as genes are not "genetic" in isolation nor work as unilineal embryogenic triggers, they represent an enormous range of concurrent states fluctuating among equally multidimensional environmental factors. By transmitting information in shifting cascades of RNA and protein molecules, they generate secondary patterns that perturb other patterns in nonlinear fashion. Genes' expressions interact with proteins encoded by other genes, including those on other chromosomes, to make their own proteins; proteins also organize other proteins. The expressions and functions of a given gene are dependent upon and entangled with the expressions and functions of all other genes in the burgeoning design of an organism. Single genes are thus implicated in micro- and macro-development.

DNA does not exclusively govern "the molecular processes that give rise to a particular inherited trait. The DNA gene clearly exerts an important influence on inheritance, but it is not unique in that respect and acts only in collaboration with a multitude of protein-based processes that prevent and repair incorrect sequences, transform the nascent protein into its folded, active form, and provide crucial added genetic information well beyond that originating in the gene itself. The net outcome is that no single DNA gene is the sole source of a given protein's genetic information and therefore of the inherited trait."[19]

An embryo is a model simultaneously (and almost contradictorily) of, on the one hand, approximation by resonant chaotic feeback and, on the other, meticulous optimization of state. At a certain point the main functional unit is the emergent whole, even before it stabilizes, *even before it exists*. (See pp. 159–160 for a foreshadowing of this.)

When deep systemic redundancy and epigenetic regulation combine, it gives the transient illusion of creature inevitability.

· · · · ·

Messages of both general "house-keeping" genes and tissue-specific ones are intermittently dispatched along pathways controlled by higher-order master genes. Special regulatory genes contribute collectively to a transcription complex consisting of themselves and various environmental and metabolic influences, including fluctuations of heat, metabolites, and hormones. These factors are all singly and collectively morphogenetic. They themselves are organized by confluences of yet other environmental and metabolic regimes (see below). They create a micro-habitat that determines how the core messages of a gene select a discrete expression from the multipotential states it holds in linear/nonlinear balance.

· · · · ·

4. Proteins elicited by the same RNA molecules do not always fold or become active in the same way.

Lay people have generally assumed that amino-acid sequences determine protein physiologies consistently and absolutely. However, proteins are not just mass-produced molds and widgets; they are active structures. The ribbons of linear molecules composing them must get folded up into a very precisely organized irregular solid of one sort or another in order to function biochemically and educe an inherited trait. Some emergent amino-acid/protein pairs tend to become misfolded unless they come into contact with a so-called chaperone protein that guides them into their traditional shapes.

As a protein molecule is plaited into its three-dimensional shape, its outer surface is configured distinctively such that discrete enzyme activities are initiated by the distinctive topography of its folds. Luminescence, hazel eyes, a stubby tail, and the like are synthesized molecularly only if the proper protein surface chemicodynamically requisitions the right enzymatic activity.

The relationship between genes and amino acids, and amino acids and proteins, rests once again as much on hidden protein dynamics as on a faithful, regimented application of a mathematical message.

• • • • •

5. Thermodynamic and topological activity continues to write new design information and to transform in-place molecular structures during gene transcription and protein manufacture even as it did as during the original inscription and assemblage of cybernetic chains.

External perturbations affect the outcomes of DNA transcriptions throughout their periods of expression. Life forms represent not just the information archived in their DNA but mandates expressed when genetic molecules and proteins attempt to restore dynamic equilibrium in their cells and bring the chemical and mechanical forces around them to satisfactory resolution.

The microstructures that determine DNA sequences are not at the same level as the chemical structure of genes and proteins. In general, proteins follow pathways that favor certain mechanical solutions to their tensions, a method of distribution common also in factory assembly lines and sheets of water flowing across land.

Genes must translate their information into phenotypes in strict accordance with thermodynamic laws governing growth and multiformity. Their filaments and tubules shear, bounce, attract, resist, and regulate entropy inside and outside their own bodies. Otherwise, in a molecularly orchestrated universe, they would have no intrinsic logic or mode of embodiment (see pp. 124 and 364).

Molecules acting at a distance translate biodynamic effects into chromosomal effects (and vice versa), as genes transmit signals from sites on DNA helices to other sites separated from them by thousands of nucleotide pairs.

• • • • •

6. Genes change their structure absolutely and irreversibly, so the expressions of genomes change.

Mutations of genetic molecules can occur mechanically from environmental infringements, clumsy copying, failed codon repair, base-pair deletions, random genetic recombinations, structural constraints, probabilistic gene flow, and entropic deterioration. After all, the messages were written, embodied, and preserved haphazardly, mechanically, and fortuitously in the first place. They became semantic after the fact. They can erode or jiggle their own hieroglyphs; they can even turn back into eoliths and scallops. They arose thermodynamically from ringlets and are always on the point of dissipating entropically into ringlets again. They can undo whatever in them was done—though sometimes partial "undoing" leads to greater complexity and refinement than more memory does.

· · · · ·

7. The ribosome may contribute to protein specification during transcription.

The genetic code can be modified as well as corrected ribosomally so that, given the chromosome order, the proteins expressed may differ from those expected. Though traditionally considered a mere information-neutral spool on which proteins are assembled, ribosomes can have dichotomous effects on expressions from single genomes. "[T]he biochemical specificity of the ribosome contributes to the amino-acid sequence of the protein synthesized on its surface and so to the protein's biochemical specificity. . . . '[D]ifferent ribosomal mutants may permit qualitatively different types of translation of the same codon."[20] (See also p. 211.)

· · · · ·

8. Enzymes are as "genetic" as DNA.

Already-completed and in-process proteins and enzymes are involved in editing, deleting, reorganizing, and recoding messages as they are received, then translated and transcribed. Thus, genes and proteins constantly interpret and rearrange one another's handiwork.

Metabolic reactions catalyzed by enzymes engage in circuitous, interdependent networks operating at many levels. As one enzyme manufactures a substance, others process it. These enzymes as much as DNA enforce the exactitude of biological replication. Commoner explains:

> As the single fertilized egg cell grows into an adult, the genome is replicated many billions of times, its precise sequence of three billion nucleotides retained with extraordinary fidelity. The rate of error—that is, the insertion into the newly made DNA sequence of a nucleotide out of its proper order—is about one in 10 billion nucleotides. But, on its own, DNA is incapable of such faithful replication; in a test-tube experiment, a DNA strand, provided with a mixture of its four constituent nucleotides, will line them up with about one in a hundred of them out of its proper place. On the other hand, when the appropriate protein enzymes are added to the test tube, the fidelity with which nucleotides are incorporated in the newly made DNA strand is greatly improved, reducing the error rate to one in 10 million. These remaining errors are finally reduced to one in 10 billion by a set of "repair" enzymes (also proteins) that detect and remove mismatched nucleotides from the newly synthesized DNA.
>
> Thus, in the living cell the gene's nucleotide code can be replicated faithfully only because an array of specialized proteins intervenes to prevent most of the errors—which DNA by itself is prone to make—and to repair the few remaining ones. Moreover, it has been known since the 1960s that the enzymes that synthesize DNA influence its nucleotide sequence. In this sense, genetic information arises not from DNA alone but through its essential collaboration with protein enzymes—a contradiction of [Crick's] central dogma's precept that inheritance is uniquely governed by the self-replication of the DNA double helix.[21]

Enzymes are the fundamental constituents of metabolism, the building blocks of life. The regulatory properties of their cycles cannot be traced to any of their components. It represents multiple levels of feedback self-organized by its own flux and thermokinetic equations (see pp. 127–128). Although the network took shape long ago haphazardly, its haphazardness was subject to intrinsic, non-arbitrary rules of organization and optimality, rules that have no ontological existence outside of the pattern behavior from which they arose. Metabolic-enzymatic feeback is subject to multi-site modulation along every nutrient-output route to energy.

The system is too ancient, complex, and tautological, and grounded in far too rigorous, moment-by-moment constraints of local precision and global synchronicity to be controlled everywhere at every level by rate-limiting biochemistries.

. . . . .

9. The code is edited and interpreted at multiple successive stages of translation and transcription.

Enzymes splice together initially unconnected nucleic-acid molecules and add, subtract, or combine regions of an original transcript, changing the meaning of its message and the proteins cultivated. Individual sequences are converted to others in the natural course of things. Ineffective and dangerous ones are discarded.

Enhancers, promoters, silencers, decondensers, repressors, regulators, and epistatic effects inhibit and activate specific genes, leading to the selective expression of different genes from the same genome. In short, a range of proteins retranscribes, relocates, translates, and degrades cell information, contributing to metamorphosis (tadpoles and frogs) and evolution (tadpoles and crocodiles).

Such RNA-level permutation and control are crucial for functioning organisms. But where, once again, are its decisions made? If genes organize other genes in such complex and multidependent ways, what organizes overall system behavior? What turns a control on and off? If it is simply another gene, what regulates it? What gives chemical cycles an appreciation of existing regimes, and how were the regimes and their savvy molecules established in the first place?

"Where there are no answers there are no questions."[22]

• • • • •

10. Almost 40% of human genes do not generate fixed amino-acid sequences but are alternatively spliced.

In an alternatively spliced gene, each nucleotide sequence is unrestricted to encoding a single protein. Instead it is split into fractions that may be recombined in variant ways to encode a range of different proteins. A gene is usually alternatively spliced at the stage of genetic transcription during which its DNA message is being transferred to messenger RNA. From fifty to sixty specialized proteins combine with five small molecules of RNA at stations of the mRNA molecule for which they have a chemicodynamic affinity, and these form a so-called spliceosome. The whole molecular configuration, functioning now as a catalyst, severs sectors of the daughter mRNA of a single gene and rebinds them together in progressions that have totally different nucleotide sequences and meanings from their forebears.

The proteins predicated by single spliceosomes differ not only from one another in their amino-acid chains but from what their original gene, unspliced, would signal. Though necessarily using the codons of their original triplets and limited to variations of these, they cast them into such diverse memes as to generate totally heterogenous and novel states. Commoner notes "that a single gene originally believed to encode a single protein that occurs in cells of the inner ear of chicks (and of humans) gives rise to 576 variant proteins, differing in their amino acid sequences. The current record for the number of different proteins produced from a single gene by alternative splicing is held by the fruit fly, in which one gene generates up to 38,016 variant protein molecules."[23] Because of spliceosomes, the multiplicity of traits that can arise from similar-sized genomes varies radically. Alternative coding provides a partial explanation as to how the bulk of 32,000 more or less identical human or mouse genes give rise to such different creatures, likewise how hominids and apes diverge psychoculturally—and, to a lesser extent, anatomically—from essentially the same genome. It even accounts for why humans and plants can arise from equivalent scales of configuration.

• • • • •

Alternative splicing was first observed among viruses in 1978 and in human cells in 1981. The genome quest did not begin until well into the 1990s. By then participant scientists already knew that genes were not the sole proximal source of morphogenetic information and that proteins did not exactly replicate DNA instructions.

So, why did they charge ahead with their epic project, make extravagant claims, and then act as though they were surprised to come up short in terms of counts and trait predictability?

It was like the Star Wars missile defense. If you want to do something and the funding and political support are there, better not let facts get in the way.

• • • • •

11. Some genes overlap such that a polypeptide may begin reading its series out of phase by a single base in the middle of a gene.

Codons are usually transcribed in only one direction, without overlap, and with one kind of framing, but there are instances in which base triplets are treated like anagrams or rebuses with slippage of frame such that, for instance, AAC (asparagine) followed by UUU (phenylalanine) becomes CUU (leucine) followed by UAC (tyrosine) or UGU (cysteine). Translational frameshifting provides different readings of the same series of codon sequences such that two or more totally different proteins can be made by one mRNA molecule. Genes become palindrome-like words hidden in other words from which translations generate novel meanings—new tissues and organs.

Although reframed transcriptions have been confirmed only in viruses and retroviruses, the fact that the phenomenon occurs at all means that instances of it may contribute to variability of expression in other kingdoms too, especially during microbial-level formation of the earliest phyla.

• • • • •

12. Chromosomal imbalance (aneuploidy) provides an example of malignant epigenesis wherein cells make phase transitions to metastatic states.

A healthy organism is not just the result of "good" genes and their ungarbled messages; it is an outcome of multiple levels of systems running by operational feedback. Likewise, even so-called genetic diseases are not attributable to bad genes; they are the culminations of multi-tiered energetic miscues. "Human disease phenotypes are controlled not only by genes but by lawful self-organizing networks that display system-wide dynamics. These networks range from metabolic pathways to signaling pathways that regulate hormone action. When perturbed, networks alter their output of matter and energy which, depending on the environmental context, can produce either a pathological or a normal phenotype. Study of the dynamics of these networks by approaches such as metabolic control analysis [MCA] may provide new insights into the pathogenesis and treatment of complex diseases."[24]

MCA departs from models of absolute genetic control while demonstrating other trackable and measurable chemicophysical influences on biomorphology. Its principles have been applied recently with some success to cellular models of neurodegenerative diseases such as Alzheimer's and Parkinson's. Using "quantitative measurements from human tissue (such as substrate concentration, ions, etc.) and standard equations of thermodynamics, it is possible to identify not only the key control points but also the effects of altered genes and environments on the overall bioenergetic processes of cells in normal or diseased tissues. In short, the phenotype—the flux of matter and energy through the system—is predictable from quantitative measurements fitted to kinetic and thermodynamic equations; no additional modeling or computational strategies are required."[25]

The controls and rate limits imposed by genes are critical, but only in terms of one another, terms that are constantly fluctuating in accordance with multiple influences on the system and also, on more than one level of space and time, with one another. Pure information is replaced along the way by nonlinear uncertainty states.

Cascading molecular hierarchies become oscillating phase states. The triggers are not only distributed but situational.

· · · · ·

The accepted etiology of cancer is that "a rogue super-cell ... somehow not only manages to 'disobey' the environmental signals that should check its growth and determine its function, but also succeeds in colonizing other, sometimes very different, tissue environments."[26] Magnitudes of cell mutations lead to carcinogenic protein products and so on. An alternate view, popular for nearly half a century before being abandoned in the 1960s, is that cancers are caused by a more complex epigenetic process rooted in aneuploids—genomes that have been made unstable by irregular complements of chromosomes and thus generate their own environments with their own proteins and their own lineages of cellular offspring. Tissues developing aneuploidally contain neither their usual diploid (double) complement of chromosomes nor any euploid (even) multiple of the normal haploid share from each parent cell. The critical algebra has broken down into a more or less anomalous sort. (See also "diploid" and "haploid," p. 283.)

Aneuploidy suggests a chaotic accumulation of stimuli flowing from chromosomal irregularity until a threshold of these is reached. A new stable state is established in which a malignancy becomes as trenchant and homeostatic as the "normal" condition was beforehand. The mathematics of this event graphs into a sigmoidal wave, which matches both observed pathology patterns and evolutionary dynamics.

The effects become collectively more and more discordant until it is not so much that they are playing a different symphony as that wind is blowing empty bottles and cans across a parking lot—cacophony beyond recapture as a tune. Apostate microbiologist Peter Duesberg, best known for his non-HIV explanation of the AIDS complex, likens transforming a cell by aneuploidy "to transforming the sound of an orchestra by randomly altering the composition of instruments, rather than by 'mutating' specific players."[27] In cells, this leads to deterioration of differentiated function, infringement of other tissue, genetic instability, metabolic deviations, unstructured morphology, and increasing malignancy. With de-differentiation, distortion of both anatomy and metabolism, and burgeoning genetic wobble, the conversion of normal, healthy tissue to a cancerous phenotype requires a massive *nonlinear* shift in expression of gene products rather than a mere quantitative disturbance from single mutations.

Malignant invasions, if aneuploidally sown, are difficult to suppress and resistant to multidrug attacks because the cells in them are not transformed mutationally in a linear fashion; they spawn heterogeneous populations of eccentric creatures, "an autocatalytic process in which each generation of aneuploid cells is succeeded by another with different degrees and types of aneuploidy and, hence, metabolisms."[28]

These are teratologies, abortive evolutions of creatures from a nether zone under the biosphere. "Aneuploid theory explains tumor formation, the absence of immune surveillance, and the failure of chemotherapy,"[29] as the differential products of aneuploids elude the cellular posses because they are operating at a higher exponent of formative disorder, i.e., formative complexity, using the cells' own bodies—and not something else—against themselves.

"Cancer is us because it is derived from our own genome. What makes cancer cells not us is that they have rearranged our genome to differ from their diploid predecessors in both the number of chromosomes and the dosage of thousands of genes. Since there are no new genes and no cancer-specific mutant genes, and no new chromosomes (except hybrid or marker chromosomes) in cancer cells, there is little or nothing for immune surveillance to monitor. This is especially true for the earliest stages of carcinogenesis where the immune surveillance mechanism is supposed to be the most effective but the aneuploid cells [to the appraisal of the mechanisms doing the surveillance] are least abnormal."[30]

Morphogenetic space gained a temporary freedom from entropy by selling its soul to nonlinear patterns arising from chaos. Thermodynamics reclaims its due when a body entropically dissolves, as we all do eventually.

Cancer is the life of a monster disguised as a stranger trying to make itself real, to give itself a meaning on its own terms. Ultimately it creates not so much an alternate regime as a wave of sabotage that spreads by diversifying and fights off attempts to rid affected matrices of its invasion. "The transformed cell is a damaged cell trying, against the clock, to resolve the ever-more complex, irreversible, and unstable gene dosage it accumulated by self-complicating mitotic asymmetries. If it manages to do so, it will establish itself as, literally, a foreign species ... in its host, but the autocatalytic process of karyoptic [chromosomal] evolution will continue and will spawn different cells, which will guarantee, ultimately, the survival of a subpopulation when chemotherapeutic or other challenges appear."[31]

When this theory was recently expounded by an old friend Harvey Bialy (in a forthcoming paper with a colleague), I asked him why aneuploidy could not just be a subset of the more familiar "rogue mutation" he cites. He responded the next morning: "The piece I wrote with Roberto admits of no hybridization between the aneuploid hypothesis and the 'gene mutation' hypothesis, and there is no way that one can be considered a subset of the other.... Mutations in specific genes are not the cause of aneuploidization, and the mathematics of mutation theory is not sigmoidal."

I told him I was also having trouble grasping why the mutation model was any more reductively genetic and less epigenetic than the aneuploidy model. After all, I said, mutations impact living systems in a latent and nonlinear fashion too. His answer was: "You need to read some of the references because you do not understand too much about the gene mutation hypothesis of cancer which says that over- or under-expression of a particular subset of genes (onco genes or tumor suppressor genes or mutator genes or whatever fucking epicycle the morons want to add) is responsible for all the aberrant phenotypes of the malignant cancer cell.

"Try in your mind to distinguish this from the result of the small imbalances caused by many thousands of gene products as in network disruptions."

I take his point to be this: Complex interactive, interdependent life systems not only operate in robust euploid genomes in ways that are difficult to track and mathematicize, but crippled aneuploid genomes develop metastatic potentials and tumorigenicities from quite other, equally multivariant stimuli and attractors. Evolution's malignant shadow is grounded in a different but equally elusive epigenesis with its own molecular mathematics.

I next asked Bialy for further clarification as to: 1) what, if not mutations, causes aneuploidy in the first place and 2) the difference between sigmoidal and other curves. I got this slightly piqued response:

"Most of the powerful carcinogens are not mutagenic. Aneuploidy is based mechanically in [cell] spindle disruptions that come with age naturally. [It is mechanical.] It is a chaotic phenomenon.

"If two phenomena have different curves, they are different. QED.

"I can't believe that you are having any problem with distinguishing A from B. This is the single hottest debate in biology right now. How have all your collaborators in embryology failed to point it out? *Cancer is the inverse of differentiation, but the two follow very different trajectories.*

"Two more points and probably the last I will write today unless you find the hours it will take you to read the papers I sent and have some additional questions. Like an old girlfriend at Bard [College] once told me (only substitute aneuploidy), 'Pushkin is my life.'

"The first is semantic. If you want to consider aneuploidy a mutation, then it is, in Peter [Duesberg]'s words, 'the mutation that makes a species.'

"The second is an equally, I hope, helpful clarification. Logistical (dimensionless) equations are fundamentally of a different character than probabilistic equations."

A day later he more graciously added:

"Here are a couple of other formulations of the primary distinction you seem to be unable to grasp that may help. Regarding 'gene-mutation' and 'aneuploidy,' the difference reduces to that between 'number' and 'type.' It is the equivalent in genome-scale biology to the replacement of the rate-limit enzyme concept in metabolism by that of distributed control."

If I understand him correctly, the control is (no surprise) multiple and multidimensional, not a series of commands along a formal network. Thus cancer corroborates the chaotic, dimensionless basis of metabolic flux and biological differentiation (and de-differentiation), hence both evolution and devolution of tissue, each along respective trajectories subject to their own multiple attractors and nonlinear control pathways.

. . . . .

13. Genetic information, which is not form but potential, moves through systems fluidly and silently prior to expression.

Molecular sublanguages override the nucleic script they generate. The deep etiology of any code, especially one of inherently mutable and unstable signals, ontologically supersedes the temporary terms of that code. That is why dialects like French, Italian, Walloonian, and Portuguese can arise from a common ancestral tongue. Likewise, languages can later share words and merge back together.

More genetic information always resides in an organism than is used in its phenotype. When traits are not expressed, their componential basis remains in the genome, reasserting itself unpredictably or in times of environmental stress. Inactive genes may be converted to active states by a variety of cellular activities.

Not all inherited genes have the same (or even any) effects every generation. Genes can be silenced—in certain contexts they fail to manifest, i.e., to elicit traits, likely because of other cellular imprimaturs. Signal-transduction systems "tell" DNA-transcriptional mechanisms which genes to turn off. Information can become dormant for one or more generations.

A gene's large cytosine residue may have a methyl ($-CH^3$) group attached to it, enough of which can lead to lapses of trait transcription. Orange and black tortoiseshell cats graphically display the off-and-on, chance effects of methylation.

Recessives and unenacted mutations may express themselves after many generations of dormancy. Even traits that were bred out can return, as variations rearise in lines of ostensibly uniform homozygotes.

⋅ ⋅ ⋅ ⋅ ⋅

There are so many variant, interdependent pathways among genes and proteins that developmental potential from mutations, recombinations, transpositions, codon jumbles, as well as overall systemic promiscuity and built-in thermodynamic uncertainty, can accumulate recessively for millennia with essentially no noticeable effect on the organism—no ill effect, no beneficial effect, no phenotypic consequences at all. Then, for mechanical or biochemical reasons, something may detonate, delivering alien offspring.

A simple mutative tweak in the relative timing of sequential gene transcriptions (heterochrony) can shuffle their entire organismic expression, turning a lineage of pre-apes into *Homo erectus* or, over many more generations, some ancestors of small crabs into beetles. The genes don't change nearly as much as their relationships and synchronizations to one another do. What shifts is rhythm and relative chronology—the development of one embryonic organ or tissue node and expression with respect to another. Changes in timing can avalanche into changes in morphology, enough to generate a different species out of the same genetic strings. More often they leave nonviable concatenations in their wake.

Heterochrony: the anatomical displacement in time of the appearance and/or maturation of one feature or organ with respect to another; change in the inception or timing of development such that the occurrence or rate of development of a trait in one organism is either accelerated or retarded by comparison to the occurrence or rate of development of the same aspect in an ancestor's ontogeny.

Revisions of expression of single amino codes in relation to one another are critical in an evolving system governed by its whole. At levels of transcription and protein chemicodynamics, these variables continually adjust gross and absolute scale, relationships of parts, and allometry of organs, which can gradually turn the equivalents of lizards into snakes, of lions into cats, and of magnolias into roses. After the age of dinosaurs, heterochronic mutations likely chose a few "lucky" ancestors of tree shrews to be mice, other even "luckier" ones to be lemurs and hominids.

As Darwin would have it, the creative aspect of phylogenesis is merely trial and error. As chaos would have it, the overall system tends toward intelligent sorting and self-selecting for unknown nonrandom reasons.

• • • • •

Genes can also alter their expressions heterotopically, as a particular transcription series "slides" from one germinal tissue layer to another, i.e., from ectoderm or endoderm to mesoderm during evolution of multicellular creatures (see pp. 287–288). Heterotopic displacements at a gastrulation phase can jumble genetic meanings profoundly, transposing a curl or cavity from one organic landscape to another. Identical genes end up playing quite different roles heterotopically in the fabrication of skin, skeleton, muscles, and viscera in the same life form, as well as new roles in quite different life forms. This process yields not only shifting scales and structures but radical departures of their emergent relationships to other tissue configurations. A mesodermal meaning is quite different from the same motif expressed ectodermally. Despite sharing genes, a brain is never "just another kidney," and *vice versa*. They differ chemically, structurally, and functionally.

Biological design intelligence is always being built up for a wide range of future uses that cannot be predicted. Whether expressed in actual tissues or stored recessively, it is subject to myriad elaborations, specializations, and new metabolic contexts. Organs and functions are modular and flexibly congruent in relation to one another

* *

Heterotopy: changes in the tissue layer from which an organ differentiates during embryogeneis.

* *

at every stage. As they evolve and change heterotopically, they assemble new molds with potential for continued transformation and different metabolic functions. Genetic and behavioral expressions are driven into tissue levels that are in the process of arranging and inventing themselves, thus can be convinced to change their destination. For instance, lungs are offshoots of ancient marine creatures' hearts. Physiological change is invariant at more than one level with genetic change.

Genetic space is at once algorithmic and nonalgorithmic, unruly and pedantic, inventing new parameters while, to all appearances, playing deaf and dumb. Somehow emergent designs are kept cohesive and functional enough amid turbulence to make at least a few ontogenetically survivable transitions from one species to another, from a familiar design mode to revolutionary ones. How they defend biomorphological integrity and metabolic function during this process is one of the mysteries that separate biology from chemistry.

In this context, eugenics makes no sense; populations cannot be restricted to certain preferred traits because the genomes of future generations will invent new traits as well as restore motifs and characteristics edited out.

* * * * *

14. Neither the genetic code nor the developmental process expresses permutations in a linear manner.

Although just about all genetic engineers scoff at the idea that they are naive enough to try to transfer concrete characteristics into genomes, what else do they do? By their own job description they "precisely identify the individual gene that governs a desired trait, extract it, copy it, and insert the copy into another organism. That organism (and its offspring) will then have the desired trait."[32]

However, transgenes may "disappear" into genomes without a splash. Other cells treat them as unnecessary information and methylate or inactivate them in their general defense against surplus or invasive DNA. It is no wonder that, while technicians routinely splice in promising new codons, proposed traits with acclaimed therapeutic or other benefits rarely ensue.

As noted throughout this chapter, genes are not independent of one another; they are subject to one another's continual, intersecting vectors carrying new information and alternate interpretations. Since developmental hierarchies are dimensionless and nonlinear, biotechnologists cannot account for every factor or privatize every transgene after it is set loose.

As long as a disease is identified solely by "a variety of cascading molecular events" that are designated by researchers "as causal mechanisms and as targets for drug design," molecular and genetic medicines will miss "epiphenomena of a primary toxicity of metabolic origin."[33] Try shooting an arrow at a mouse borne by a cyclone. And these are intersecting cyclones in meta-stable equilibrium with one another, functioning not so much like isobars as the collapsing stars of cosmomythology, event horizons that send their laundry to other sectors of the universe or other universes.

Engineered gene transfers lead to unintended and unwanted effects in recipient organisms, not assignable even to a prior understanding of the gene being transferred. While failing to express themselves for generations, transgenes and mutations may dovetail heterochronically with other loci that themselves may then be epistemologically altered — together they combine to manifest spectra of unexpected traits. The ghost says, "There are more than albinos where I come from."

As transgenes continue to shuffle potentiated biological information through genomes, they hold ungaugable perils to the biosphere.

• • • • •

15. Locating inherited causes for traits on chromosomes does not prove expression by a single gene or from a single message.

Bioengineers fiddle with genes and observe results; then, by reasoning through reverse genetics, "single cause" hereditary diseases are traced to codons. However, even when the consistent, reproducible effects of a so-called gene are intellectually isolated in this fashion, its consequences still differ from individual to individual. Other components, many of them unidentified, impose themselves. Even twins or clones, despite apparently identical genomes, have divergent developmental patterns.

According to biologist and philosopher of biology Richard Strohman, the truth is that "almost all human diseases are context-dependent entities to which our genes make a necessary, but only partial, contribution. Molecular biologists have rediscovered the profound complexity of the genotype-phenotype relationship, but are unable to explain it: Something is missing. The missing element was described 35 years ago by Michael Polanyi, who characterized live mechanisms in DNA as 'boundary conditions with a sequence of boundaries above them.'"[34] Levels of constraint (or boundary controls) include, first from genome to transcriptome, the regulation of gene expression including "pathways that detect energy levels* and repress DNA transcription when cellular NADH levels are increased"; next, from transcriptome to proteome, "posttranslational modification of proteins"; then, from proteome to whole dynamic system, "metabolic networks of glycolysis and mitochondrial oxidation reduction ... display[ing] control over all enzymes of a network ... includ[ing] cellular redox potential"; and, finally, from dynamic systems to living creatures (phenotypes), the relationship between localized regulatory (metabolic, hormone-signalling) systems and "distribution over many systems and levels."[36]

• • • • •

---

*These include oxidation reduction (redox) levels. Strohman adds: "Metabolic control analysis may reveal, on a cell- and tissue-specific level, changes in redox potential and in key redox-sensing proteins that in turn are related to changes in gene expression and therefore to disease phenotype."[35]

Most microbiologists are still attempting to derive phenotypes from genomes and proteomes because investors in corporate biotechnology have put their money on agent-based diagnostics and therapies; they do not seek the real playing rules that govern the output of these systems. Interdependent environments—subcellular, cellular, intercellular, organismic, even ecological—play major roles in the transfer of potential from genotype to phenotype. In this progression

> many levels of control are introduced ... each defined by a dynamic system of self-organizing proteins. Polanyi illustrated his concept of levels of control with a metaphor from the game of chess: "The strategy of the player imposes boundaries on the several moves which follow the laws of chess, but our interest lies in the boundaries, that is, in the strategy, not in the several moves as exemplification of laws." Molecular biology, in identifying control levels, has focused on the "moves" of genes and proteins but has largely ignored the strategy used by dynamic protein networks that generate phenotype from genotype. Systems biology is all about finding the strategy used by cells and at higher levels of organization (tissue, organ, and whole organism) to produce orderly adaptive behavior in the face of widely varying genetic and environmental conditions. At the center of this effort is a need to understand the formal relationship between genes and proteins as agents, and the dynamics of complex systems of which they are composed. Much effort has been spent in attempts to predict phenotype, first from genomic, and then from proteomic, databases. But these databases do not contain sufficient information to specify the behavior of a complex system.[37]

And the movements of cells in possums and sparrows are far more complicated than those of rooks and pawns.

· · · · ·

Genes assigned to diseases often entail some degree of misplaced identity:

• Many women with variants of a so-called breast cancer gene do not develop cancer, and even the percentage that does varies by ethnic group and population.

• In the mid-'90s, at least four different candidates were confidently proposed as the single gene leading to schizophrenia.

• The gene on which a transmission error causes cystic fibrosis is 230,000 base pairs long. Its twenty-seven exons, among other affairs, code a protein of 1,480 amino acids. When a deletion of three base pairs eliminates one of those amino acids (in position 508 of the protein), a pathological condition occurs, but only in 68% of the cases, and not identically even in these. As many as 400 other related mutations have been found within the same gene, contributing to cystic fibrosis-like conditions and giving rise to varying combinations of symptoms and absence of symptoms.

Different mutations in one gene can lead to a pathology; other mutations or combinations of mutations in other genes can also lead to the same pathology. Conversely, some whole groups may develop an entirely discrepant disease from the same genetic configuration; e.g. bilateral absence of the vas deferens among Yemenites instead of cystic fibrosis. Each of these patterns can also result in recipients with no disease.

From exposure to a single environmental toxin (like arsenic in local wells or nerve gas from a military attack), individuals may develop quite different illnesses or even remain inexplicably healthy.

We should thank the fates that we are not subject to direct prophecy from a genetic oracle. Only the wildness of nature inside us prevents the lives of our great-grandchildren from being taken over by a technocracy that would make Stalinism look anachronistic. Everyone would be assigned a profile and destiny at birth and set along its remorseless track. We still may approach this in coming generations, but I suspect that cells, organisms, and other rebels will be able to escape full monitoring and launch a counter-coup.

•••••

16. Genomes and the biosphere are maintained in organic rather than mechanical stability.

Biotechnologists presume that organisms are stable because they have tight source codes with multiple utility functions and redundant sites from which to supplant damaged ones and restore lost information. Without mechanical stability, they declare, there could be no heredity, no secure lineages.

Yet genomes do not have real mechanical stability; their source codes are individually unstable at almost every phase of their development and expression. Because genomes are fluid (as we have seen), complex pathways alone survive, while lineages change and evolve.

As naturally occurring perturbations disrupt the stability of particular genomes, over time (through increasing information and feedback) they enhance the stability of the biosphere itself. Genomes maintain organic stability by being not only fluid but mobile within a larger realm of creatures and environments—the whole ecosphere or some habitat within it: a redwood grove, a beaver pond, a stretch of tundra, a glacial bay. Each of these is a vibrating zone of meta-genetic fluctuations and self-regulating feedback mechanisms transferring and reorganizing information beneath a deceptively concrete surface landscape of shedding trees, burrowing bugs, buzzing aviators, and scampering bionts—be they squirrels, centipedes, woodpeckers, penguins, gulls, sharks, or something else entirely.

Genes are not discrete forms, not even potentials to form; they are repositories of multi-origined, temporarily linearized and semanticized information.

The real stability of life forms resides in the cosmological webs in which genes are expressed, not in genes themselves. Something vaster than information stabilizes them. They are stabilized by the entirety of the biosphere—if not by intelligence itself.

• • • • •

17. Natural selection does not work on single genes.

Population geneticists and sociobiologists often follow the idealized mathematicizations of Mendel and his supporters rather than the actual heterogeneous behaviors of genes in systems.

A critter does not "make it" in nature solely by hoarding primo genes. A genome organizes itself dynamically, so it survives only insofar as its characteristics remain both profound and flexible. Remember, genes are mutable and interchangeable. At best, traits subject to natural selection represent open combinations of genes, their protein constituents and interactive developmental processes, and a fluid range of potentials that can change as the dynamic systems incorporating them change.

Creatures and genes fluctuate as they derive themselves from their own and each other's equations. . . . .

There are "not enough genes in the genome" only if you are a pure genetic determinist. There are more than enough multipotential elements to build a complex organism, probably even without alternative splicing. Given emergent properties of complex networks, there is no *a priori* way of saying how many genes would be needed anyway.

Epigenesis in place of genetic determinism makes evolution not just a gene-mortality algorithm or concrete mechanical, statistical event but a multifaceted coevolutionary meta-phenomenon. It is a quadrille of synergy and matter, a flux of multivalent systems — everything happening in the context of everything else, every ultimately realized form in collaboration with every incipient form (organic, planetary, phenomenological). "[T]he inherited specificity of life is derived from nothing less than life itself."[38]

Genes are foci in cosmic webs rather than one-way contrivers of design. Their meanings are solely the contexts of their organisms and the environmental grids in which they come to express themselves, grids which continue to change and thus change their (the genes') expressions. These grids include "a variety of redundant mechanisms: alternative pathways through which normal levels of 'bioenergetic potential' may be restored and maintained even in the presence of genetic or environmental insult ... [and] distributed con-

trol and supramolecular organization of the many enzymes that constitute a given metabolic process.... [T]he control of a given metabolic
pathway is distributed among all the enzymes in that pathway...,
any one enzyme or several enzymes may become rate-limiting
depending on local conditions..., and the interconnectedness of
enzymes is such that it is not possible to change the activity of one
without affecting the entire system."[39]

• • • • •

Life is a vibrating equilibrium stabilizing itself by fluctuations within
continua and generating creative morphogenetic turbulence such
that patternings of tissues and organs arise in waves of discontinuous
variations radiating from potentiated points.

Once we accept that genes are not the patented guardians of
organic form, the sole architects and repair agents of our stuff, we
stand as little more (or less) than elaborate ripples—unified, fluid-tis-
sue matrices resonating with the aperiodic nanocrystals in our helices,
held together not by a formula (though we could not exist without
DNA's circuits to locate and cue us) but by gravitational-centripetal
resonance through chakra-yarn in excited, evanescent field states.

Molecular synergy is the only selectable factor. . . . .

In an email to its perpetrator, Richard Strohman expresses exasper-
ation at yet another public defense of genetic determinism. He is
addressing an old friend from shared "free speech" marches through
Berkeley during the sixties, an evolutionary neo-patriot who is
unaware of either the obsolescence or biohazard of applying avid
mechanism to living systems:

> "[Y]ou have largely, if not completely, ignored, as have the
> neo-Darwinist crowd, the concerns of both classical and
> molecular embryologists who bring a new dimension to evo-
> lutionary thinking. As you know, the Modern Synthesis
> finessed questions of development and the dynamics of

morphogenesis in the vain hope of discovering all the answers in genomes. It never happened, and it never will. The missing dimension is the one of interaction between genes, between genes and proteins, and between both of these and the environment: interactions all obeying dynamical laws of which we know next to nothing and about which natural selection has nothing to say."[40]

In a subsequent article in *Science,* he reinforces this verdict of hubris: "[W]e are still mostly ignorant of the laws governing the context dependency and integration of environmental signals into the output patterns of [living] systems."[41]

Strohman is saying that just because "from gene to message to protein" is a continuous process does not mean that "from genome (total genetic field of an organism) to proteome (total protein field) to metabolome (total metabolic field)"[42] will get you to the same place, i.e., a cornucopia of customized bionts and other concrete results initially promised when "the century of the gene" made biological engineering possible. In fact, Bialy writes, the second series "involves three separate discontinuities, and the obsession with trying to make some version of the first applicable to it is an idiocy."

• • • • •

The structure of biological meaning is as much a puzzle to us now as it was to the Platonists at the dawn of Western philosophy or as the dynamics of the sky were to the pre-Copernicans. Darwin, Marx, and Freud may have driven the great twentieth-century locomotive by their theories of evolutionary thermodynamics applied to life, then to society, then to consciousness, but the twenty-first century must go beyond such a model because form and meaning are not only intrinsic but nonlinear, not only nonlinear but extrinsic.

Bialy and his co-author compare the myopia of a genocentric universe to that of a geocentric cosmos in which everything circled the Earth in spurious cycles and epicycles. They conclude: "After over a thousand years of celestial spheres and epicycles, it needed only the recognition that a circle is a special case of the ellipse to, in a small fraction of that time, do away completely with the geocentric model of the solar system. It may take only the realization that individual genes are just parameters in networks, rather than isolated causes, to do away with obsessive and ever-more abstruse genocentric models of cancer in particular and life in general."[43]

• • • • •

Organisms are generated by an equation with two terms, 1) the molecular activity of genes leading to 2) physical relationships of rearranged molecules in fields—but, as we have seen many times, these are then subject to numerous co-factors, both measurable and beyond measurement. What are called traits are (after all) simply patterns generated by molecules interacting with one another at different mutual phases of dynamic disequilibrium. And molecules themselves are only electron and proton vibrations stuck together.

In a modern world ruled by algorithms and a shifting competitive edge, the Darwinian account of reality is initially compelling. It is now hard to discern, through its hegemonous metaphors enveloping the lineal basis of life, what life forms actually do in order to "become." That nature is a battleground on which the fit and fertile survive and that cells and their genes must be competitive, aggressive, draconian, even rogue are neither accurate nor objective seeings of nature. They are metaphoric displacements, stories that come about because, since there has to be a narrative of some sort, we explain movements and effects we observe according to plots already familiar to us. This allows scientists to experience a dramatic structure behind their equations and to communicate in ordinary parlance as well as numbers.

I repeat: behavior in the wild is not because cells are rogue or, in fact, because they are anything; it is because we want a narrative, must have a narrative, and are good at narratives (especially Westerns and war stories). Life is really about some whole other thing, but we would have to start over from scratch, with new metaphors or without metaphors, to make a description from a molecular-cellular perspective rather than succumb to our own smug moralistic amoralistic projections. After all, commensalism is as prominent as predation in biology as well as nature itself.

It is also possible that cells with their genes are part of consciousness, i.e., unconsciousness too. This could mean explicit ego mind, Freud's unconscious id, Carl Jung's archetypes of the transpersonally collective unconscious, Wilhelm Reich's cosmic orgone field giving rise to vitalized bions, Jacques Derrida's absence-presence of traces, John Upledger's cell talk felt as tissue change, Buddha's "big mind," or all of the above as well as many other things. Thus any model of gene chatter and signification or imagined cell stratagems is a shallow imposition at a level (encompassing an almost limitless plenitude of moving parts) at which any explanation fits, any explanation matches some part going in some direction of the camouflaged design, any explanation is also tautological.

There is no absolute origin of either the "thing" or the psychophonetic phenomenalization of the thing. There is no absolute representation of the gene or final subtext of gene agency. There is only infinite being with openings of exteriority, a chrono-linear logic of strings of signifiers on an apparent one-way irreversible arrow— pasts that never were, presents that can only be lived as modifications or rationalizations of originary events. And there is also mortality.[44]

Like Jung's alchemists, biotech mages in their laboratories are in no position to see the archetypes they are projecting into matter, thus have no opportunity to recognize themselves and their beliefs in them as they appear spontaneously in chemical substances, parading as genetic events.[45]

• • • • •

Old-fashioned neo-Darwinian, genocentric natural selection betrays a pre-Einstein, pre-Heisenberg, pre-Mandelbrot probabilistic bias. On a molecular level where the real action is, there are not only no organisms, no rivalries, no differential matings, not only no discrete and isolated mutations, but no genes and perhaps even no matter. Among charged particles in phase states of dimensionless chaos fields, the meaning of natural selection evaporates. Molecular synergy and dynamic systems are the only selectable factors, but since these are located at simultaneously subcellular and nonlinear levels, there is *no* purely selectable organism or event. There is nothing but transient mirages of whole creatures to submit to the court of differential mortality. And these cannot pass through the eye of the needle.

I agree that you can make mincemeat out of anything by reducing it to matter and then energy, taking away its emergent properties. But, conversely, you can invent a whole house of cards by dismissing the ultimately mysterious, originless basis of all matter-energy systems, e.g. all systems. Just because emergent phenomena suffuse the perceptible universe doesn't mean that we can heedlessly impose scientific totems on any aspect of nature we grab hold of. We *are* only molecules, even as we are more than molecules. Considering the unsteady combination of these meta-conditions together is a prerequisite to understanding life and mind.

Universal evolutionary theory may yet turn out to be a self-inflicted hoax of a few centuries, an oversimplified misreading of the statistics generated by density-dependent population dynamics, survivorship curves, and ecosystem carrying capacities. In its passage out of superstitious creationism while en route to some version of epigenesis, biology got stuck for a moment in hyper-materialism.

As neo-Darwinism crumbles into history without serious damage from creationists, the result will please neither of them, for a vast, indeterminate microcosm will become the new steward of life's grail.

The Earth is the only true genome. . . . .

1. Biological form arises from developmental synergies in environments.

The reality of molecules, genes, cells, and tissues cannot be uncoupled at any level from environmental influences on their development. Morphogenesis is not just a portable computer that runs its same program, whether under a blanket of ocean or travelling near-gravityless between planets. A plant or animal is an artifact of its planet. It is a set of entangled fields interacting with other fields around it. Divorced from worlds, genes would be oddly curdled pebbles. Their mineralogical forerunners (pregenes) are scratched throughout the universe—on Pluto, Chiron, Miranda, and other moons, planets, and asteroids—but they go nowhere beyond hieroglyphicness (see p. 121). This is more than the obvious necessity of an atmosphere, ocean, and landscape for organisms to aggregate and survive. It is how matter itself is planetarized and individuated.

Every living motif depends on both molecules that are temporarily embryogenic and molecules that are temporarily geological. This is axiomatic Earth science: "The chemical elements, including all the essential elements of protoplasm, tend to circulate in the biosphere in characteristic paths from environment to organisms and back to the environment."[46] Molecules of inanimate stone, molecules of DNA, and molecules of enzymes and proteins overlap and change roles and bodies in going from dormancy to metabolism, from raw material to organized text, and back.

Nitrogen in protoplasm, when broken down by bacteria, passes from an organic to inorganic state, escaping into soil and air. Nitrates, sucked out of earth by green plants, then return the nitrogen to an organic phase. Some of these molecules are temporarily informational, some are temporarily embryogenic and metabolic, some remain temporarily geomorphic; all flow nondiscriminately among systems and are impressed reciprocally in a larger astrogeological, ecological, morphogenetic field.

Genetic programming and generic context are bound in such a way that neither can produce a living system without the other. Genomes propagate not single set bionts but a range of phenotypes. There is no DNA expression outside of planetary engagement, in a vacuum without vulcanism and air. This may be obvious from our present greenhouse, but it is not obvious epistemologically.

Tissue is broadly cosmic and microenvironmental as much as tightly informational and semantic. It originates in the Earth's geology as well as in its cybernetics, in its climate and waters as well as in its own genetic logic. Bone is geologically another form of calcium, an inanimate limestone or marble, but containing cells and the proteins (collagen mainly) and saccharides they provide. Unlike mineral in caves, bone is "alive." It accumulates in fields generated by *both* genes and gravity. Note: even after a short time in space, astronauts begin to lose skeletal mass on their heels so, despite her terrestrial genes, a *Homo sapiens* born and raised on Mars would likely grow up spindly and fishlike, reverting in part to the bodies of her ancestors in the Earth's oceans.

Among other environmental factors modulating genetic expression into phenotypes are: micro-temperature, photoperiod (day length), season length, proximity of predators (releasing telltale chemicals), nutrition, ion deprivation, pollution, population density, and cycles of precipitation. A biological milieu originates, develops, and survives as a whole (see pp. 251–252).

There may be *only molecules,* but then, by the same rationale, these become *only cells.* Full-fledged creatures emerge anyway. An active ecosphere makes them alive and gives them a home, a reason for their existence, both before and after the fact. A particular biont may seem lame, surreal, etc., as it swims and quacks like a duck, but that by itself doesn't deter it or its predators or reassure its prey.

• • • • •

2. Every genome is set up to be destabilized and reorganized in states of continual evolutionary flux.

• Genes are altered by the environment (i.e., they mutate); they also get recombined with other sequences.

• Sections of DNA are moved around, reinserted, exchanged, and rearranged inside organisms; genes jump site to site, chromosome to chromosome, chromosome to plasmid, and plasmid to chromosome within individual genomes. These transposons become raw material for spontaneous variations and new species.

• Plant genomes house almost epidemic numbers of DNA fragments (retrotransposons) that skip unsystematically from one part of the genome to another. While disrupting genetic expression, they also participate in the programming of novel proteins.

• Genes can move chronologically and ontologically within embryonic series (heterochrony, heterotopy, heterokinesis).

• Promiscuous entities migrate from organism to organism. For instance, plasmids squirrel into bacterial chromosomes and haul them across conjugation tubes to recipients.

• Codons can be deleted, new ones inserted in their places.

• Genes can be rearranged to form new chromosomes.

• Unequal exchange of material between homologous chromosomes during pairing can lead to one chromosome getting more hereditary stuff than its mate.

• The same lines of DNA code can be amplified and replicated anywhere from hundreds to tens of thousands of times. The resulting enhanced DNA may contain superfluous sequences, but these provide multiple levels of regulatory signals for the gene's alternate configurations into proteins, or new raw material.

• Genomes can duplicate themselves in their entirety (polyploidization), for instance, to allow reproduction by hybrids of species whose chromosomes don't match each other's fully enough ordinarily to allow offspring.

• Random duplication and translocation of DNA sequences iterate entire genes which then develop new functions, as they shuffle coding elements into different proteins. They also expose genes to fresh medleys of enhancers, further altering their expression.

• Genomes can grow to dramatically different sizes for contingent historical reasons often having little to do with the alternative splicing potential of their genes or the behavioral complexity of their organisms. They interpolate and make use of the later strings that earlier strings somehow recruit, a process invariant with neither somatic intricacy nor function and utility in any guise. That is why varieties of mustardlike weeds and mammals have roughly equivalent scales of genomes. Gene fructification is a marker for neither biological subtlety nor mega-complicated architecture. Information can be used, reused, or muffled at different levels for different design purposes. Form can be over-designated, umbel by umbel, with lots of extra scaffolding, or it can be loosely dictated as co-emergent shape fields that interact dynamically out of much sparer information. The latter are perhaps closest to new lines of phyletic transformation capable of phenotypic variation and inventiveness; the former may be more deeply channelled and determinate in their relative shapes, with no particular need or capacity for mutagenic potential or message economy.

Mustard weeds are still more likely to greet visitors to this world following a cosmic cataclysm than politicians are.

• • • • •

3. DNA can be taken up into cell nuclei directly from the environment.

• Unrelated species can mate and exchange DNA.

• Upon the death of cells, DNA is released into the environment, adsorbing onto debris, clay, and sand. DNA is also exuded by living cells. Marine water, fresh water, and soil contain mixed DNA contents, some of which degrade slowly by half-lives, a small portion of which survive indefinitely on surfaces of solids. A meaningful fraction of this DNA may infiltrate genes in foreign species. Although horizontal transfer doesn't happen very often, the vectors used in genetic engineering may increase its occurrences.

Detached genes do not end up at once in the chromosomes of beavers lapping at rivers or bees splattering dust, but they are taken out of water and sediments by bacteria, especially when these cannot find other nutrients. Through such intermediate agents, they eventually find their way into new plants and animals.

• Quiescent microbes also form layers and matrices that gather in colonies of multiple or single species (biofilms) on stones, river gravel, and the surfaces of aquatic plants, and are carried by convection currents through streams and lakes, where they mix with nutrients and metabolites.

• • • • •

4. Genes travel among species and kingdoms, symbionts looking for hosts.

Genes transcend any single creature and even, ultimately, the biosphere itself. Carried naturally by viruses, plasmids, transposons, and other mobile gene-like entities, genetic elements are able to cross species barriers.

Passage of genes between species (or juggling of genes within species to form new species) is fortuitous and opportunistic. Though genes are the greater pool from which bionts draw their components, biological structures emerge only in formal heuristic sequences.

Life forms may not routinely and carelessly share genetic material across nonmating boundaries, but evolutionarily they have done so to such a degree that there are few (if any) species-exclusive genes. Potentiality from any species can hypothetically invade the genome of any other species (including species with which it does not ordinarily mate). Liverworts and amoebas bear the same molecular helices as daisies, moths, pigs, flukes, anthrax. Human genes can function inside insect cells, frog genes in mouse cells. This undermines any privatized notion of a body, species, or self.

If every genome is global rather than species-based, our genetic-embryogenic network is limitless in repetoire and scope as well as fluid in design. Information flows back and forth at more than one level. Each organism ultimately has the entire DNA base of the biosphere to draw on. All life forms are interchangeable and permutable, i.e., coevolving.

· · · · ·

Genes altered or recontextualized through human editing and other mutative processes and reinserted become part of the genomes of the creatures receiving them and (of course) the planetary gene pool. Cell rape is much more invasive than the social kind because it leaves the rapist inside *all* the victim's offspring forever.

Latter-day scientists now play the big game, grafting between species and even kingdoms, introducing fish genes in tomato cells, silkworm alleles in goats, human material in pigs and bacteria, jellyfish in assorted seaweed and monkey cells, etc., thereby spreading genetic promiscuity. As cards are shuffled, dice rattle, and slots roll, players fill casinos and ram in slugs, acting out post-modern theories of nature and its rewards: lemons, cherries, and other hyper-literalized signs representing degrees of temporary equilibrium.

Biotech transgenes, however, are a smidgen of the "genetic engineering" on Earth. The global genome has been self-engineering since the advent or arrival of DNA. Ecodesign is hardly a recent innovation of scientists designating chromosomes.

Transposable elements of human and other genomes may have originally been retroviruses that inveigled their way into cells and then put their own reproductive mechanisms at the service of their hosts. Free microbes, by replicating information and behaving like genes, became genetic, i.e., conscripted, mnemonic, and subject to design hierarchies emerging within semi-closed lineages of their hosts.

Conversely, information nexuses and fragments of organisms have regained primitive but functional independence from their genomes. Genelike elements in plant and animal cells have periodically devolved into "viruses" that escaped their bionts entirely. Some have undoubtedly given rise to new species of viruses, bacteria, and perhaps even parasitic worms, or have invaded other genomes. Whole organisms' genetic potentials have been recruited for design structures in other organisms.

By acting like wildlife, by reinventing aspects of their cellular past, genes are also freelance agents on the town.

• • • • •

As jumping genes and genome fragments with semi-autonomous existences migrate between species, they alter, edit, and replicate themselves in both traditional and new forms. Some of them recombine, then mutate further, reconfigure, and move on to additional species. This process is hypothetically limitless.

Over eons, genes have passed from plants to animals, between trees and vines, and among phyla of animals. Journeying step by step across the planet, they have spread the latent diversity of informational potential in life itself—not plant life, not animal life, not fungal or bacterial life, not even life presently on Earth, but life as open-ended waves. They draw on the raw, superorganic principle that created the biosphere—a precept way vaster (even if disembodied and egoless) than the manipulative designs or protocols of one type of life.

• • • • •

Modern primate DNA is unusual in containing numerous copies of two transposable alien sequences (see p. 162). Catalyzed by RNA-reencoding processes, these exogenous elements have, in effect, recently overrun our chromosomes and altered us in unknown ways. Similar transposable elements have colluded to "modernize" maize and houseflies.

On a grander scale, birds may have descended from dinosaurs by incorporating DNA from insects or from plants. Flowers may have budded from gymnosperms by borrowing bacterial genes. Monkeys may have departed from ancestral rodentlike mammals by appropriating information from primeval salamanders and modifying it to serve a totally different purpose. Pre-Egyptian scarabs may have gotten miniaturized in ants and flies by interpolating fungal and viral coding elements. There is no biological basis for any of this or evidence that would suggest it, but it is hypothetically possible.*

The sci-fi bestiaries evoked by such fantasies suggest a kind of vast pagan collaboration—the storing of both atavistic and futuristic biological information in genotypes of earthly creatures. In this kind of Blakean universe no shape is wasted or lost, no morphology

---

*Cell biologist Stuart Newman comments on this notion: "I know these are speculations, but they are too idle. In any case, they place too much emphasis on genes. Pekinese and Afghan hounds are members of the same species; frogs and tadpoles are different free-living versions of an organism with a single genome. Doesn't that tell you that major morphological changes are possible without radical genetic change?"

(at core) ultimately nonhuman—nonspider, nonconifer, nonbird either.

We may find one day that the Milky Way is the true genome and that mushrooms, viruses, and far more exotic molecular complexes transmigrate between worlds.* This would mean that either DNA or some precursor molecule is endemic in the galaxy or universe at large.

$\bullet\ \bullet\ \bullet\ \bullet\ \bullet$

Codes and alternate morphologies not expressed or put to immediate use are often lost for good but, as noted above, not always and not all of them. The DNA and shattered, respliced genomes and potentials of extinct creatures (our ancestors as well as brigands and scavengers not in our immediate lineage) are indexed and represented latently throughout us—a mute miniaturized record of the biosphere. Fragments of ancestral redwoods, prehistoric birds and fishes, and various crustaceans, rodents, and other invertebrates likely persist inside our cells from eons long before these bionts or we evolved. Aspects of any of them may reappear in our descendants or be transferred to still other species. Retaining the roots of their original expressions and motifs, these vestigial creatures project their source codes, through new organisms, into totally novel tissue complexes. Provisional dodo, horseshoe-crab, and marsupial DNA may end up making body parts and behaviors that resemble aspects of their progenitors, or they may generate totally unique tissues that represent only the dynamic potential—the nodal fields—that gave rise to distinctive morphologies in their predecessors.

This open-ended kind of embryogenic network has enormous potential untapped on Earth. An organism in another solar system with 200,000 instead of 32,000 "genes" might have the option of "choosing" each generation (during an early embryonic phase) whether it is ecologically more beneficial for it to develop as the local equivalent of a tree, fish, owl, or sapient forager, whether it should

---

*Newman writes: "You are going too far!"

become large or small; male, female, or hermaphroditic.

It is not that we don't embody indeterminate capacity; it is that we are not organized to access it.

In fifty million years, new bionts may supplant the majority of plant and animal kingdoms on today's planet. In his haunting book *After Man,* Dougal Dixon catalogues the wondrous inhabitants of a hypothetical future Earth: hornheaded, hornbilled browsing mammals; desert grobbit rodents and leapers (that follow the extinction of camels and primates in the great desert); zarander pigs with elephant-like trunks and tusks; swimming monkeys and aquatic anteaters; tree-dwelling chuckaboo marsupials; and feline gurraths descended from mongooses.[47]

Something like this is probably inevitable if our pollution, wars, and climatic changes make Sol 3 uninhabitable for humans but not for general mammalian life.

. . . . .

5. Biological replication can occur without nucleic acid.

The agents causing such diseases as scrapie, mad cow, and other cerebral degenerations infect organisms by refolding their host's proteins in such a way as to impersonate their own unique three-dimensional molecular shapes. Called prions, these proteins, though lacking nucleic acids, can match and mirror very different alien protein configurations. As each converted host protein becomes a prion itself, its disease spreads through the afflicted organism.

Prions thus cause other prions to be manufactured from preexisting proteins with the same sequence but different conformation. The prion facsimile and the brain protein it mimics come to share the latter's amino-acid sequence, though only the prion is a fatal agent; the targeted protein remains genetically normal. Old-fashioned genes are still needed, of course, to encode the traditional aspect.

A protein's active structure—its folded architecture—and its amino-acid sequence can thus be independent of each other. Something else is transmitting traits; something other than a nucleic acid is passing information from one protein to another and infection from cell to cell.

Prions may not be as crucial in morphogenesis or the evolution of species as the principle by which they work. Transgenetic information transmission represents an entire, more or less unexplored alternative biology leading to species change and cell transmutation by methods other than DNA. This may or may not involve prionlike entities.

Infectious proteins are the tip of the iceberg. Molecules and proteins probably do prionlike things every day that are camouflaged from us in the general subcellular rush hour. The true replicative mechanism may be imbedded far deeper in nature than the gene and give rise to genes as well.

* * * * *

6. DNA may make copies of sequences from RNA by reverse transcription.

Reverse transcription is a method by which somatic cells can communicate new data to germ cells and instill a form of hereditary memory in them despite the apparent segregation of the two. Some mutations thus represent stabs of system intelligence within oceans of random, entropic shuffling.

For instance, immunological knowledge is transmitted among species across time. The mRNA of antibodies (white blood cells) imprint their encounters with foreign antigens in the mRNA of immunoglobulin genes. Endogenous retroviruses reverse-transcribe RNA into DNA and replace one codon with another. They reproduce by transcribing their own RNA into DNA, which is then housed in a host genome where it makes viral genomes.

Since plants do not enforce barriers between somatic and germ material, DNA changes affect all parts of their meristem simultaneously. Cells in meristems undergo consistent, predictable genetic mutations when roots and branches are exposed to fertilizers with different components. If DNA has this intrinsic capacity, it may undergo metamorphoses in a subtler way in animals, now and primevally. Certainly no one was watching when quantal system leaps occurred.

Some organisms actually seem to prepare themselves for plagues and environmental opportunities. Starving bacteria and yeast cells undergo spontaneous adaptive mutations so precise that they could not have come up with them by chance mutations or random variability. In fact, bacteria otherwise quite accomplished at editing out accidental changes to their DNA and chance genetic garbage undergo a kind of natural self-engineering in response to life-threatening external crises, selectively preserving only the most useful and beneficial adaptations:

Cancer cells and other parasites call up dormant genes to aid them in battles against toxic agents.

Insects adaptively transform themselves when attacked by insecticides.

Cell cultures react to drugs by mutating.

• • • • •

The occurrence of directed mutations is obvious to those who make it their business to investigate such things. How else would we know about them if biologists did not report their existence? Yet their implications are mostly ignored. The party line is that somatic cells cannot influence DNA.* Thus nothing that happens in the eventful lives of bionts can be returned to germ material and inherited nucleicly by offspring. Quoting an unrepentent Crick, an unrepentent Commoner twits the axiomatic authority of this prohibition:

> "Once information has passed into protein it cannot get out again." This means that genetic information originates in the DNA nucleotide sequence and terminates, unchanged, in the protein amino acid sequence. The pronouncement is cru-

---

*August Weismann, a biologist at the turn of the nineteenth century, was the first to assert the nontransmissibility of acquired characteristics to the germ plasm or genotype.

cial to the explanatory power of [Crick's] theory because it endows the gene with undiluted control over the identity of the protein and the inherited trait that the protein creates.[48]

Yet protein "information" does sneak back out. Genes and the environment engage in not just multidimensional interactions and gene exchanges but direct impartations of organized text—so-called reverse (soma to DNA) flow. Templates, once copied, trickle back through double helices in the opposite direction. Prints make new negatives.

Spliceosomes are one example of protein-orchestrated reverse transfer of morphogenetic information, but what if living systems more fundamentally regulate and reconsider themselves by the flow of information back and forth between elements?

• • • • •

It is possible that many "random" mutations occur only after "intelligent" transformations in underlying cellular fields orient the genetic code toward change. Mutations may thus be correlated with environmental transmissions such that germ-cell DNA is not just an antique tablet revised by invasive mutation but an ongoing by-product of an active cell talk between the organism and the environment.

Life is more a two-way mirror than a program written and altered aimlessly by blind and deaf ciphers.

• • • • •

7. Organisms aid one another by sharing DNA.

As genes are borrowed from neighbors or provided by parasites and viruses, the genome reorganizes itself to be able to combat diverse threats. This is particularly evident in bacteria under attack from man-made toxins and antibiotics. Strangers become collaborators rather than adversaries; organisms find themselves in cahoots more than competition, acquiring the software they need from one another, even other species. Bacteria thus act less as individualized animals than as a fluid molecular-genetic-ecological network coevolving through configurations of environmental possibilities.

Antibiotics themselves may function as sex hormones in some bacterial communities, stimulating them to mate and multiply; hence, also mutate and develop self-serums. The use of toxic chemicals and engineered genes may turn out to be finally the self-defeating ploy of a limited intelligence (the human mind) against the far greater and subtler intelligence of the biosphere itself. The limited intelligence thinks it is vanquishing an invader in a zero-sum war-game. The invader may not welcome the supposed attack but uses it to increase its informational base and that of its entire bacterial genome.

Humans thus enhance rather than diminish the developmental potential of their supposed adversaries among the viral, bacterial, and arthropod worlds.

• • • • •

8. Biotechnology has the approximate subtlety of whomping on a hornet's nest with a tennis racquet.

Technicians enlist pathogenic viruses and plasmids because these are skillful at invading cells and genomes, by-passing protective barriers and defense mechanisms that would otherwise inactivate alien DNA. To facilitate DNA transfer, mobile genetic elements, viruses, and bacteria that are inherently wanton and predisposed to convey genetic material are then altered to be even more promiscuous.

When a bacterial gene is transferred to a corn chromosome in order to confer an insecticidal protein, the biotech assumption is that the plant will receive only that protein and take on its specific traits. Yet alternative transcription or splicing of the gene in an unfamiliar morphogenetic context could lead to a range of new proteins, none of which bear any relationship to the source gene or its traits. The chance characteristics generated by the strange proteins then affect all future corn plants in their lineage. They are diffused into the environment to alter ecology, agriculture, and health without any sense of what they are or what effect they might have on the farms and societies of the future.

Transposable elements being both manipulable and promiscuous, bioengineering is not only cumbersome but dicey. When supposedly inactivated strands of genetically engineered material are transferred to bacteria, they may still recover in the context of latent genetic recognition and reorganization by other bacteria. Thus, genes created by engineers could eventually travel far from their original soybean or cow stock, change form, change expression, change tissue layer, and (in the future) dwarf anything imagined by their makers.*

Transgenes and crippled laboratory strains of DNA spilled throughout the environment can survive additionally as naked DNA adsorbed on particles, later taken into bacteria. Loose in the biosphere, DNA is invariably self-transforming; it does not retain its historical configuration. Recombining with bacterial chromosomes or other plasmids, its genetic strands transmute the physiology and chemistry of host cells. Bacteria then multiply and spread new codons to other species. Artificial viral vectors and transgenic vaccines metamorphose into genetic parasites, carrying antibiotic-resistant and carcinogenic traits with often-wide host ranges.

Altered combinations of gene sequences, if recklessly cast into epigenetic mazes, lead down the road, through arbitrary recombinations, spliceosomes, prions, and the like, to new infectious elements. Diseases like AIDS, mad cow, and ebola as well as outbreaks of bubonic plague, cholera, *E. coli,* and tree blights may represent incursions of human and plant genomes from engineered DNA or other mutagenic technologies (immunizations, pesticides, antibiotics).

Viral genes from animals inserted into plant viruses later recombine with and genetically reorganize other viruses, architecting even more deviant pathogens. Vaccines produced in plants for introduction in humans and farm animals and (particularly) for use in medical gene-replacement therapies are as likely to generate their own

---

*Harold Dowse objects that the artificially-manufactured complementary DNA he uses has absolutely one expression only and cannot be coopted like genomic DNA for alternate transcripts (see p. 164). He considers this argument against biotech disingenuous.

pathogens as to protect against existing ones; in fact, given the boundaryless complexity of the biosphere, they are more likely to create than to cure diseases. Cross-species epidemics of hoof-and-mouth, mad cow, and AIDS-like complexes may become far more common in the future because of engineered recombinations of endogenous viruses. Monkeypox and hantavirus, transferred from rodents to humans in central Zaire and southern Argentina during the 1990s, have spread well beyond their original rural villages. At the same time, new strains of distemper and rabies of unknown origin have been infecting lions, panthers, wild dogs, and giant otters across Africa.[49] Near the time of publication of this book, SARS (Severe Acute Respiratory Syndrome) arose in the slums of Asia, possibly from civets, and travelled swiftly by car, train, and jet.

Biotechnology is randomly and cavalierly reformulating the biosphere for all subsequent generations.

• • • • •

9. Genomes responding adaptively to the environment more suggest rejected Lamarckian evolutionary theory than a neo-Darwinian paradigm.

If evolution represents not just series of random selections of traits and organisms but morphologies and behaviors generated by feedback loops involving proteins, enzymes, genes, and other molecules and operating throughout the biosphere; if mRNA can resplice and retranscribe contextually; if proteins can acquire environmental information and then shape-change; if information can be reorganized genetically; and if acquired characteristics can be inherited; then the barrier between soma and germ plasm *can* be breached in principle, violating the Crick dogma that genetic messages flow only one way, from genes to proteins.

The relation between life and intelligence is thus quite different from what neo-Darwinian purists have presumed.

The archiving and indexing of multi-tiered changes in homeostatic fields may in fact be the primary etiology of DNA. Its criterion may be to respond molecularly and replicatively to intelligence (from wherever). Far from launching deeds of system origination, it would anchor biological systems from within as they evolved (see pp. 154–155).

Perhaps in the same way that signal transduction from cells leads to selective transcription, mutations at large specify and rewrite targeted genes. Outmoded and archaic sequences are gradually replaced over time by systems transferring information irrespective of species barriers, quarantine of germ cells from somatoenvironmental events, and even of conventional understanding of time and space. Their transactions are opaque and immeasurable by the kinds of experiments that are being conducted. But they explain what trespassers are doing here in the first place. They redeem the "black hole" origin of life.

Modern experimental biology is in the process of startlingly and without notice abandoning the Darwinian-Weismannian view in favor of a nonlinear Lamarckian paradigm* which seemed (before quantum theory) quite archaic. Species apparently do arise dynamically in relation to environments and thus constantly improvise the biosphere and redefine biological identity.

A process that accumulates informational potential in genomes just as easily disrupts development. . . . .

Transferring vectors artificially across species and kingdoms does not take into account what the biosphere—with its permutations

---

*For historical and philosophical accuracy, it is important to distinguish between evolved systemic responsiveness to environmental cues, which is a hallmark of a sophisticated evolutionary trajectory entailing natural selection in a nonlinear neo-Darwinian mode, and a more primitive responsiveness that is a property of excitable matter and not a function of evolved genetic circuitry. Both probably exist, but the second is closer to what Lamarck had in mind when he proposed that evolution occurs by environmental feedback into germ plasm, and the first is what I have in mind. The reality may lie at a level deep enough to combine the two, or at least render them corollaries of a larger canon.

of DNA code, transposons, and transspecific outlaws, etc.—will do with new messages, what Frankenstein (and even Dracula) will become over long reaches of time. Biotechnology presumes that the most feral, promiscuous, and renegade system of all is tamable by one of its parts. It pretends naively that we know its design principles.

Gene engineering is dangerous not because it doesn't work, but because it does, to a limited degree. Biotechnologists are skilled and pertinacious enough that, with their micro-machinery, they are able to transfer live, active vectors into genomes. Unfortunately those vectors then become part of multi-levelled systems technocrats neither understand nor can control—systems pivotal to the vitality of the planet and the future existence of life.

As microbes proliferate, their genes trade hats, contaminate, and spread, breeding new diseases. Mobile strands of transgenic insects and bacteria, released into the environment, cross species barriers and end up in other creatures, including humans. In the form of coding errors such elements jump from insects and other arthropods into primates, mutating along the way. One such vector of exogenous DNA, called *mariner*, causes a neurological wasting disease by hatching an enzyme that cuts human DNA in the wrong place. A few more of these, along with some new immunity-destroying retroviruses, may contaminate the human genome to such a degree that reproduction and survival become far less automatic or things to take for granted.

There is no innocent transfer of traits because there are no innocent traits. There are no fixed traits, either—no taboos or rules in the flow of living stuff. Undercover prions are truer indications of how actual life forms develop and maintain themselves than targeted transgenes.

Biological engineering is tampering with the inherent homeostasis and meaning of the biosphere itself.

Genes get to play in this game because they are not in charge. . . . .

The ontological relationships necessary for jump-starting life forms require whole, borderless systems. There can't be a biology without an integrative ecology, without tiers of organization, for every biont depends on extensive morphogenetic, psychosomatic, socioecological relationships to function, even to exist. A community literally propagates the energetic forms that make it up.

The stability of organisms and of ecosystems depends on interspecies signalling, reciprocity, fluidity, mutual responsiveness, and feedback among levels—in sum, the capacity to derive complexity from simplicity, multiplicity from unity, unity again from complexity. Life itself is at once an embryogenic and a phenomenological contagion.

Ontogenesis and speciation express not merely abstract algorithms but events unfolding in environments. The egg by itself is not computable. A description of its fertilized components ("the total DNA sequence and the location of all proteins and RNA") cannot "predict how the embryo will develop."[50] Yet genetics and microbiology focus on events *in* the embryo at the expense of the milieu that gave rise to it and continues to redetermine it.

Planetary morphology ricochets into developmental morphology. As external inducers are assimilated into genetic inducers, their shifting trajectories particularize bones, neurons, and B-cells as well as anatomical and psychoneural designs. In fact, the emergence of discrete organs like ocular lenses and limbs may have been specified along the way by rotting leaves or microbial parasites turning into enzymatic information.

The developmental plasticity of a genome is characterized either by a relatively smooth range of morphs or by discontinuities of phenotypes (a state known as polyphenism). "[T]he European map butterfly ... has two seasonal phenotypes so different that Linnaeus classified them as two different species. The spring morph is bright orange with black spots, while the summer form is mostly black with a white band. The change from spring to summer morph is controlled by changes in both day length and temperature during the larval period. When researchers experimentally mimic these conditions, the summer butterflies can give rise to 'spring' butterflies.... [Likewise], at one temperature the snapping turtle embryo becomes male; at another temperature it becomes female. Fed one diet, a female ant larva becomes a sterile worker; fed another diet, the same larva becomes an enormous fertile queen."[51]

Alga-eating spadefoot toads hibernate under Sonoran sand until awakened by thunder to breed in the temporary pools left by the passing storm. Maturing rapidly, their larvae then bury themselves until summoned by new colliding molecules of air (or, tragically, the drone of motorcycles, newcomers to the desert). Shrinking ponds also trigger wider mouths and more powerful jaws which allow tadpoles to eat each other.[52]

Juvenile *Daphnia* absorb embryogenic information from predators. When flies release kairomones that betray their presence in a pond, neighboring crustaceans develop large helmets. Carp embryos are catalyzed by kairomones from pike that have already eaten a carp, to form a "pot-bellied, hunched-back morph that will not fit into [a] pike's jaws."[53] Tadpoles of wood frogs sharing water with larval dragonflies are not only smaller; "their tail musculature deepens, allowing faster turning and swimming speeds to escape predator strikes."[54] The gray treefrog "responds to predator kairomones both by size change and by developing a bright red tail coloration" to intimidate.[55] When the eggs of red-eyed tree frogs in "vegetation overhanging ponds" are attacked by wasps or snakes, "the embryos sense the presence of the snake and vigorously shake in their egg cases. Within *seconds* ... (having achieved gill circulation) [they] hatch prematurely into the water."[56]

Ichneumonid wasps manipulate the internal maps guiding the webs of orb-weaving spiders by secreting a hallucinogen the webmakers consume while thread-spinning. Thus parasitized, the arachnids spew drunken zigzag designs that well serve the egg-laying needs of the wasps.

There is no such thing as an antiseptic sequence of genes giving rise to a solitary life form; bionts only emerge and act in planetary networks and ecodynamic fields. An organism is a sequence of spreading collaborations, not a modular federation of digital attributes and alleles. Each gene, each cellular vortex is the result of and contributes to a natural order from which it draws both its reality and its future agenda.

The phenomenon of organelles and cells woven into creatures—the condition we are bequeathed as bodies—is in fact an enormous, vibrating instrumentality stretched across the planet, a hologram resonating through shifting pivots.

• • • • •

Just as life cannot invent itself out of contextless molecules, the human world cannot exist apart from evolving stages of tribal and clan organization. Semeiologies and tropes take over from cells; they organize structures and anatomy as cells do. Knowledge travelling in paraphysical networks does the work formerly assigned to genes and genomes. It induces shapes into open and fractal spaces, creating habitats for life. Bands emerge from organisms emerging from morphogeneses. Biophysical homeostasis becomes neurosocial homeostasis. An organism becomes an identity, then an idea.

In the parlance of Christian biologist Pierre Teilhard de Chardin, the geosphere synthesizes the biosphere; then the life network sizzles into a noosphere.[50] Whorls of stellar dust become helices of DNA, which become spirals of totems and signs. A new, cosmic intelligence fabricates its own culture. This is how galaxies turn into astronomers. This is how matter (mothar) incubates language. It was all originally startalk—ma!

Morphogenesis and noogenesis have been mutually inductive from the origin of the biosphere, though mind hibernated in the unconscious of nature itself. It meandered through cellular and precellular webs along the only paths available—seaweed and amoebas, worms and crabs, tissue clumps, bioelectric currents, cephalopod

• •

Noosphere: the "thinking layer" of the Earth, which has spread outside and above the biosphere and lithosphere since the Cenozoic era.

• •

ganglia, lizards, wolves. Everywhere molecules accreted, mind followed shape and melded with structure. Personality and behavior colonized emerging organic designs according to their morphologies. Each nascent level of biochemistry and tissue patterning bore within it the terms for the next level. As structures coalesced into signs, fresh organs and organisms represented them. Likewise, each level of culture and technology grounded in a prior complex gave rise to more deeply circuited progeny.

Human languaging is an extension of protozoan interactions, coelenterate synapses, schools of fish propelling themselves in synchrony, newts copulating, birds shifting in flocks, stampeding herds of boars, anteaters sniffing sand, newborn possums squealing at their mother's teats. Zooids cannot help themselves; their pagan motilities become supplements and signifiers; their lives propagate glyphs; their cells and tissues become actions; they collaborate on and foreshadow every aspect of meaning as well as every denial of meaning.

A hummingbird among flowers is also generating language and culture. It is just not our sort of culture. Squids and centipedes are carrying out acts. They may be traces, unconscious exploits, but much of our edifice is unconscious too. Acts are the syntax of life.

We who form words can intuit our sentience by examining the myriad levels of incipient signalling in a water droplet. Mind arises from trillions upon trillions of episodes and interconnections across space-time. But this is also how cells accrue from molecules.

Culture is meaning that has recaptured but also reinvented its own subliminal, subcellular basis, and done so while in full metaphysical bloom.

• • • • •

A symbolic social world unfolding from a biomolecular substratum generates artifacts and emblems as well as enzymes and tissues. While genetic morphologies are made of proteins, cultural ones are made of anything. Mind talk thereby extends cell talk into other boundary-less domains; yet both are scripted in raw signals and their subsequent logistical distributions. Both draw upon feedback from each other (and everything else) to survive and then advance.

Morphology, behavior, language, and thought are functionally the same thing. There is no solely linear or mechanical basis for any of them and certainly not for all of them. Life requires multiple interconnected, pluralistic components in order for creatures, symbols, perceptions, ideas, rituals, and communities to arise. Mind is the sum of vast integrations, not genes, not proteins, not tissues (though genes, proteins, and tissues are among the fields that amalgamate in its creation).

⁕ ⁕ ⁕ ⁕ ⁕

It is all still here. The thundering of surf speaks for the ancient, celestial planet. Dozens of gulls aligned along the sand are unconsciousness, eyes of universe, prior to knowledge. Taking sudden flight, ascending as waves, higher and higher, they circle the shore, synchronizing in almost perfect patterns that approach a different perfection. On the boardwalk from the roller coaster: a futuristic screech of terrified awareness. Birds and waves and Mexican rap drown it out. The hieroglyph of chaos is always prying open infinity, threatening to turn into mind.

We evolved phenomenologically even as we evolved phylogenetically, by molecular and molecule-like integration, by significations and metaphors playing the roles of mRNA transmission and cellular context. In fact, phylogenesis and proprioception are faces of each other. The integration of feelings and tissues, layer by layer, makes up the rich textures of a world and a life. This is a critical feature that did not inscribe itself in the fossil record, except as cave paintings and chipped rocks are the residue of an unfathomable event.

7

The Wahhabi Critique
of Darwinian Materialism

Religious Fundamentalism and the Global Economy

"Mere anarchy is loosed upon the world...."
—William Butler Yeats, "The Second Coming," 1920

The Dangers of Biotechnology ....

Within biotech circles, acknowledgment of limits on genetic determinism is industrial blasphemy. The public is treated to a sanitized, Chamber of Commerce view of a coming Golden Age of biology by scientists pretending to understand how cells and genes actually work who want to sell us their plan to ride DNA like a supercomputer. Those who support the cybernetic transformation of the Earth believe that present genomic lapses are temporary and that a meteorically expanding technology will soon fill in the gaps. Just look at what got solved in the last century, which began without phones, planes, television, or (for all intent and purposes) electricity!

Addressing the U.S. Senate in 1999, Ralph W. F. Hardy, president of the National Agricultural Biotechnology Council and one-time director of Life Sciences at DuPont, a major producer of genetically engineered seeds,[1] simplified matters: "DNA (top management molecules) directs RNA formation (middle management molecules) directs protein formation (worker molecules)."[2]

This conveniently capitalist and Reaganesque parable is an absurd infantilization of the actual situation but, more significantly, it bears an undeliverable campaign promise: that a "new top management" of scientists and technocrats will eventually effect a "corporate takeover" of the cell.[3]

Their "massive uncontrolled experiment"[4] rests on the presumption that life is a vast circuitboard-like web in which genes and traits can be transferred routinely between organisms (modules) in such a manner that "in each of the resultant cells the alien gene will encode only a protein with precisely the amino acid sequence that it encodes in its original organism; and that throughout this biological saga, despite the alien presence, the plant's normal complement of DNA will itself be properly replicated with no abnormal changes in composition."[5]

By attempting to unravel the fabric of life as if it were a linear algorithm sewn by a molecular needle, scientists undermine precepts of holistic, phenomenological organization. Genes and proteins work together to generate forms because they have had millions of years of evolution through which to design molecular activities in dynamic, elegant homeostases of trillions of reactions and enzyme loops. With untold generations to fine-tune their products in balance with epigenetic cycles and shifting ecological-cellular feedback, native species have *intrinsic thermodynamic logic.* Operating systems "encourage" their genes to interact synergistically with proteins. Backup redundancies repair DNA damage and errors, as 1) spliceosomes propagate alternate proteins from potential states, 2) chaperone proteins fold emergent proteins in delicate configurations, and 3) gene-silencing agents inactivate or suppress potentially disruptive messages.

Engineered life forms do not share this entire history and thus are innately less stable than indigenous ones. Transgenic corn, soybeans, tomatoes, cotton, and the like are holistically marred because their genomes have been perturbed, sequences of chromosomes disrupted, amino acid/protein manufacture de-synchronized in unknown but key ways that are "unspecified, imprecise, and inherently unpredictable."[6] In their relocated vicinities genes and proteins lack familiar contextual flanking consisting of tissue-specific compasses and positional sequences for optimal polarity and expression.

Even when a gene has been successfully grafted and its desirable characteristics are expressed as predicted, other elements have been thrown into new uncertainty states. Deep changes in molecular structure and biochemical activity have already been set in motion in billions of transgenic plants with little concern for long-term consequences to the biosphere.*

---

*This is Commoner's basic argument in his *Harper's* article—that the permutating complexes that direct RNA and specify protein synthesis in plants will be disturbed, when translating bacterial genes, in a way that produces far more unintended than planned proteins and, by the law of averages, more dangerous than beneficial ones. The rebuttal: plant biotechnologists ask for evidence that this type of genetic subterfuge would be any more of a problem in biotech crops than it is in ordinary horticulture and agronomy. Nonbiotech plant breeding has produced many mutants, some of them by using radiation and caustic chemicals, without any demonstrable poisoning or other ecological or health damage for more than half a century. Technicians claim that their crops "cannot be different from ... [and] must be substantially equivalent to conventional varieties before they can be marketed.... Every single differently shaped leaf of lettuce, every different color of bell pepper, every new variety of citrus fruit is the result of genetic mutations that produce different proteins which were noticed and then selected by conventional plant breeders."[7] Additionally, natural retrotransposons in plants, just as likely to generate randomly altered proteins, have not poisoned the biosphere over the history of botanical evolution. If a protein-synthesis machinery is badly enough damaged by artificial (as opposed to natural) transgenes to produce toxins, then it won't generate the desired traits either, let alone hide the toxins under such traits; in fact, most often, no offspring at all will be produced. The biosphere is too healthy, vast, and resilient for us to damage by our meager intrusions.

• • • • •

Flowing from unnaturally hardy crops to wild relatives, modifica-tions tend to cross-hybridize into super-weeds with invasive traits. Oilseed rape and potatoes on farms in Europe have already spread biotech codons to their cousins.

Gene silencing can also remove the desired characteristics of a transgene, in fact routinely so with species of tobacco. After all, lines of redundant transcription inactivate mutant DNA in nature. Engi-neered plants, either in the first generation or later ones, may not gestate or may not flower.

Organisms secondarily develop toxic or other unwelcome char-acteristics to accommodate transgenes. Supposedly flavorful trans-genic tomatoes turn out to have limited climate tolerance and, after they ripen, their overly fragile skin makes transport to market near impossible.

Scientists declare GMOs (genetically modified organisms) safe after checking them for known allergens and toxins. Yet what about other blights? Soybeans reengineered with a gene from Brazil nuts induce their own pathological side-effects when consumed by humans; likewise, strains of yeast genetically spliced to ferment more effectively.

Although in 2000 the Monsanto Company acknowledged the presence of extraneous strands in their transgenic soybeans, their spokesman concluded that "no new proteins were expected or observed to be produced."[8] Yet in 2001 scientists in Belgium found "that a segment of the plant's own DNA had been scrambled. The abnormal DNA was large enough to produce a new [and] poten-tially harmful protein."[9]

These defects (above) are relatively minimal and fixable; they are the least of our problems. Consider:

Transgenes swallowed in engineered corn, rice, tomatoes, soy-beans, or any "Frankenfood" made from those plants may resist digestion and end up in gut cells, then the bloodstream, and ulti-mately, despite the supposedly impermeable barrier between soma and germ plasm, the human genome as well. Some of these invader genes will come from bacteria, insects, or lichens. They will spread their own biological meanings.

From transgenes inserted into cells, mutant lines of progeny may be launched. This is the legendary parturition of Spidermen, Lilliputian children, and other, more common teratologies. Their images generate fears of tumors and spoofs of sprouting horns.

Descendants of transgenic plants with bio-insecticidal characteristics "select" species of insects immune to pesticides, which will make future agriculture more challenging and also more technological. Engineered by companies to be resistant to their own herbicides, biotech crops undermine economic botany, while more notably creating new markets for stronger herbicides. Organic farms, with the unique potential to restore taxed soils and reduce carbon-dioxide outflow to the atmosphere, face escalating challenges from biotech-fabricated pests and other mutants.

Agronomic biotechnology represents not only a loss of genetic complexity but an increase in environmental toxicity and nutritive devitalization (from miscues of incompatible genes)—ultimately, privatization and capitalization of the food supply with increasing North domination of South. Corporations of developed countries gradually take over the essentials—land, water, food, air. Loss of biodiversity means pandemic famines and—bottom line—reduction in the quotient of solar energy diverted into the biosphere. The depleted cultural landscape then sprouts mines, abandoned farms, razed villages, poisoned wells, etc., followed by sweatshops, stockades, bastilles, child prostitutes, mercenary armies, and boat refugees. Planets elsewhere probably do a better job of organizing life.

• • • • •

As genes are declared the official source of all biological characteristics, medical and social problems come more and more to be attacked by genetic engineering. Indicted codons accused of spawning deteriorative diseases like cancer, asthma, heart attacks, and diabetes as well as less discrete conditions such as manic depression, aggression, hyperactivity, obesity, homosexuality, predisposition to criminal behavior, etc., will soon be identified (or misidentified) for scouring from the genome. Of course, they may return—and they will not even be the real villains.

What is the difference between traits and behaviors? From a sociobiological standpoint, there are only traits, behaviors being their complex coalescences. However, even that fragile distinction is quickly disappearing under the smoke and mirrors of a biotechnocracy churning out new products and licensing them for commerce.

As chemicogenetic control becomes social control, medical acts turn into police acts. The present eugenic package sold to a hungering humanity uncannily resembles earlier promises of industrial, nuclear, and cybernetic revolutions: unlimited progress, utopian machines, improved quality of life, longevity, global wealth, myriad avenues for the pursuit of happiness. Good luck!

Already subject to ethnic cleansing in the service of fun, sport, revenge, business, partying, whatever—organisms and their traits are deconstructed into fictive desirable or undesirable commodities. The campaign to brand and market such characteristics as health, creativity, genius, and beauty (in designer babies by patenting the "best" human genes) is a modernized demonism—finally a pretext (as medicine men and women of indigenous peoples foresaw) to steal souls.

The beautiful society portrayed in magazines and cinema throughout our eBay-Viagra-MTV-porno global village—the illusion of eternal yuppiedom—is attainable only through trait-engineering. By godlike distribution of trademarked attributes, all citizens get to do and be what they want in the virtual marketplace (or, if it is too late for them, they can customize their children). Soon only products will remain, reinforced undoubtedly by identification with the adventures of media stars—no values or feelings to support them. Our inner experience from which our lives derive their meanings will be more and more depreciated and marginalized. In fact, the *central purpose* of present civilization is to discredit individual experience in order to imprison us in matter and entropy (e.g. the expanding market) forever.

As a reign of quantification, hypocritical moralism, and narcissistic materialism washes out texture and meaning, compassionate and ethical acts get replaced by business deals, legalized rackets, fake churches, payola, and enforced jihad, always subject to either local or global corporate, ecclesiastical, fast "food," and sitcom homogenization. In the process we are inviting a whole new range of disasters for which we (as the current ruling species) don't begin to have an answer. The bullied youth of the Earth's wealthiest populace—those who have been deemed ugly, weak, weird, or nerdy by their jock peers—bring guns to school and terminate classmates and teachers. Others stuff down junk food, smoke crack, shoot X, sit in video trances, and browse on internet smut.

Is it any wonder that Islam, for all its own bugaboos of repressive patriarchy and xenophobic fundamentalism, assails this hegemony from outposts in Iran, Sudan, Yemen, Pakistan, Egypt, Afghanistan, Chechnya, Indonesia? Who wants a world that expunges the tribal, the sacred, the ecstatic, the innocent, the epiphanic? Who wants a hard-edged realm made up only of acts, things, and their diversions? What about the calls of birds, the echoes of distant prayers from minarets?

9-11-2001 . . . .

A gap is created in the book. Words that were here on September 10 (my Islamic critique of the West) are rendered fraudulent and hollow by people incinerated and jumping from buildings being vaporized. Nothing will ever mean the same again. Suicide bombers sailing jets into skyscrapers have left holes in every text in the world.

Pyres of gypsum, asbestos, and steel ignited by aviation fuel send a horrifically compounded plume across the immaculate blue of time. Burning and crumbling towers preempt all ideas, markets, philosophies, and sciences. Made of raw materials, the symbols of the West have been shattered and obliterated, the rubble of disintegrated citadels now entombing thousands.

The sixteenth trump is drawn, the Tower struck; hapless figures fall from stories above eighty. "The crown knocked from the tower by the lightning-flash is the materialistic notion that matter and form are the ruling principles of existence. . . ."[10]

The impact exposes a flimsy fortress—illusory, impermanent, spun of toxins to resemble a real temple. "The lightning . . . breaks down existing forms in order to make room for new ones."[11]

Militant Islam said, "Your bastion is a sham. Our martyrs will reduce it to Allah's molecules, to dust so fine the streets will disappear under funereal ash. We will scatter your habitants like mice from a burning barn."

· · · · ·

People along Third Avenue commented on a plane flying lower than anyone had ever seen. But this was no plane. It was a steed from another dimension, from outside history.

Its passengers were already dead. Its hijackers were dead too, but they had been "dead" for years. Aiming their hearse like a Stinger, they laid the dreaded first strike on an American city, a cataclysm outmaneuvered for decades of deterrence and nuclear brinkmanship. So began the war of the desert kingdoms against the computerized metropolises of the West.

The wraith spotted from Third Avenue, not a plane (maybe a Trojan horse) could hardly be witnessed by ordinary sight. Its existence had been transformed by witchcraft and charmed against the financial institutions of the West, the arteries of which passed fuzzily below in the city the ghost projectile loomed upon.

It was about to burst into history with a hideous, metallic shriek.

The echo of that collision crossed the planet in a tsunami. "Be patient," Osama bin Laden told his cheering comrades in a cave on the other side. "The best is yet to come."

• • • • •

"That doesn't look like a rescue plane. Oh, my God!"

A deceptively familiar object crossing the sky until almost the last second … then it was Satan himself.

"Daddy, *they're doing it on purpose!*"

Without warning, the acme of Middle Eastern performance art unfolded before a startled global audience, a Middle East that cares nothing for Stravinsky or Christo, Dostoevsky or the Beatles, that has no use for the mystery or wonder of the West, that serves the cult of death over life, that issues *fatwas* against mediocre novelists, a Middle East that wants to call the shots.

• • • • •

Horrific, unwarranted, absurd, yet at its deepest level these strikes were a biospheric irruption. When goods and markets and money-changers in business suits replace honor and service and warriors, cargo-cult elders dispatch their youth to reclaim the Holy Land from Mammon.

The battle has now been joined. Overshadowing World War II and the Cold One and swallowing our children's and their children's and children's children's lifetimes, the attacks of 9-11 began something even more deadly than pitting tanks, warheads, A-bombs, and military-industrial states against one another. It released the shadow itself—the disparaged doppelgänger stalking technology, kept under wraps by the feeblest of taboos, the most token of patrols.

We can only hope now that the great Buddhas will not stand by while sentient life is thrown into total darkness. Either way, after the smoke has cleared, a different planet will sit in our place.

• • • • •

9-11 was the initial strike of the stillborn Mohammedan theocracies against the corporate dream-factory that has stolen everyone's prime time. There will be many more such attacks and they won't all be Muslim. They will be carried out by the snipers, cells, radical priest-hoods, and self-commissioned militias—human and other—of all those excluded from the game. So dark and benighted is the present time, so drained of spirit, so innured to its own injustices and brutality, that its liberators must be far nastier folks, more brutal and loath-some than the thing they oppose.

Yet they are our karma, our reciprocity, the soldiers of our pre-cise betrayal and neglect—a collective vengeance for how we have treated the poor and helpless among the species of this planet.

You might think that there is no effective argument against tech-nocracy, against genetic determinism, against capital funds and cor-porate graft, against the self-congratulatory, atheistic pseudo-Christian spectacle of the West.

Dispatching a plane full of petroleum and terrified passengers into some very large inhabited towers is an effective argument. It addresses imperious politicians and scientists directly, in a way they refuse to be addressed, by an utterly alien system of logic and action. It lays down an axiom beyond premise: "You kill us; we kill you." The same statement was made by Saddam when, in retreat, he set the oilfields of Kuwait ablaze: "Dig this!"

And: "I'll be back!"

We do not get to choose our punishment; the universe does that.

• • • • •

We evoked these harbingers and then, bought off by their cheap thrills, neglected to heed them. The script was ready-made in Hollywood. We taught them *Mission: Impossible.* Through Dirty Harry, Rambo, and *Die Hard* we exposed the media power of terror and urban detonation. We even gave them our guns—and then protected their right to arm sociopaths (the Second Amendment). By our arrogant dare we all but lured them to our shores.

The youth of dispossessed tribes completely missed the point, the romantic storyline with its faint refractions of Shakespeare, Steinbeck, and Sinclair Lewis. They saw only the vicious, scurilous subtext; the plastique and big-smoke; the torpor of gratuitous destruction, strip-tease, and bored lust—broadcast to Japanese dishes by satellite, to slums and refugee camps. Sitting in squalor before optic displays, the furies-to-be stared right through the nuance and irony of Western entertainment and read only its surface energy and tawdry carnival—James Bond and Batman, Britney Spears and Eminem, the new Crusaders. When they responded, they did what the makers of the videos failed to realize was their prime injunction—build the weapons and use them in unvarnished renditions of the burlesques pictured. Towering infernos, car chases, pipe bombs, and killer asteroids were suddenly exposed as mockeries of real danger. Imitation, after all, is the sincerest form of retribution.*

---

*A friend, Henry Bayman, writes: "Going through your critique, I'm reminded of Morpheus' words in the cult movie *The Matrix:* 'Welcome to the desert of the real.' I believe 9/11 has done something really bad to the body chemistry of America, all the results of which are not yet forthcoming. May God bring forth good out of all this evil. But I tell you, Hollywood taught them everything. Beginning around 1980, the movies escalated to a new level of violence. Everything was shown explicitly on TV sets in the Middle East and Asia even to new-born babes. How to rig a booby-trap, how to make a bomb, how to dive-bomb planes.... I remember something George Steiner said in *In Bluebeard's Castle,* where he attributed the horrors of the twentieth century to the preoccupation of the imagination for centuries with hell, "ovens and stinking air." Imagined too long and too deeply, they could not but be actualized. Well, it was all there on the TV sets of the whole world, in full color. They did not need to visualize it, they could see it with their own eyes. The 'desert of the real' spread across an

Every international terrorist and martyr craved it, to become the heroic instrument of its realization, to march off the big screen into our cities, super-villains we could only keep imagining until they bellowed out, "Stop imagining us. We are here. Not only that, but we have watched all your films, studied all your nightmares. You can't fool us and you can no longer buy us off. We are the ones finally, the ones who don't need stunt doubles or actors, who don't care if you kill us (because we have many brothers, many sons and daughters, many many). We are bigger than your Stallones and Schwarzeneggers because we are really here and can really act."

Even unappreciated dweeb vandals Dylan Klebold and Eric Harris wanted no less than to cap their bloody day at Columbine by scoring a 747 and zinging it into the World Trade Center.

· · · · ·

Terrorism is invasion by jagged, impossible landscapes, obliteration of prior custody. Everything we take for granted is already strange— mouths into which we insert food, limbs on which we locomote, having bodies at all. Cities with their boulevards and balustrades, tunnels and bridges, planes and schedules are the brittle collective assemblage of creatures working over generations like ants to raise something beyond comprehension. We inhabit these domains as unwittingly as fish in water; we could *be* fish in water. Every morning we slip on garments, shirts and pants that enfold our limbs, yarn sliding over feet and toes; we think, we act; as the planet rotates, we dream. Nothing is obvious; everything could change in an instant. Every moment we live is on the cusp of not-being, our vanishing from this realm as sudden and unbidden as our arrival.

The terrorists showed us how at risk we actually are. "This is

---

unprepared globe through the arteries of the movies was even worse than the original thing, where—in its homeland—people had at least learned to cope with modernity in some way, had developed the means to digest it. I'm reminded of Macbeth's words: 'bloody instructions, which, being taught, return to plague th' inventor.'"

what 'real' looks like," they said. They said, "You cannot impose your version on the whole of creation and make it stick."

On this morning, their story was at least as good as ours. Dispensing with metaphors and style points, they chose the most available explosives with the biggest bang, the most efficient delivery system, the most inviolate targets, and brought them together with pitiless, choreographed brilliance. They delivered to our doorstep the fierce planet they live inside of, not the global-trade sitcom or palace America but tinder of astonishing, cruel histories. They revealed how many facets the universe actually has, how superficial and transient any one world-view. They left us alive against a guileless sword's edge.

Yet compassion arose spontaneously too because, this far from enlightenment, no act is diabolic enough to unleash only evil. In one heartbeat of future shock a veneer of alienation, greed, and road rage lifted off the land. People flowed from across America to Ground Zero, improvising kitchens, music, massages. Crowds along avenues cheered rescue workers. Policemen, firemen, soldiers, and citizens, we were all in it together again.

Wake up! was the hidden meaning, the one they didn't intend. They planned chaos, anarchy, jihad. Yet when Satan invaded, Judaeo-Christian civilization lit candles, held out crosses, sang patriotic mantras, and flew banners of stars and stripes to drive the beast out of the circle.

• • • • •

The war between the Old Earth patriarchy and post-agricultural technocivilization began even before history. It was fought initially with bones and arrowheads; it encompassed countless acts of genocide and colonial empires. Centuries later, a many-faced modernism metastasizing out of England and France infested America, Russia, then China, Japan, and beyond; in 1948 it birthed a second Israel, but by then Jews were no longer Palestinians, Apaches were no longer Apache. Its proximal strike (1979) was Soviet paratroopers racing across rugged Afghan terrain.

Then the orthodox mullahs of the Stone Age struck back.

The battle did not end with the expulsion of Slavic troops from Kabul or the fall of the Soviet whole-damn Union. It flared imminently on other fronts—Taliban cults seeded by Russian cluster-bombs, child molesters hanged from the turrets of their tanks; Somalia and Sierra Leone splintering to kidnapper warlords and child armies; Chechnya, Kuwait, Kenya, Algeria, Sudan, Malaysia, Greece, East Timor, Uzbekistan, Ethiopia, Tora Bora, Kashmir, outer China, the Philippines, Mesopotamia, Bali, arenas of the First Bush War, virtual weaponry against the armies of Dune, the dawn of the Mother of All Battles, an ancient Islamic mega-nation against the infidel mercantile states and their post-modern, transnational coalition, a war (contrary to opinion) yet to be fought.

From the perspective of al Qaeda, global capitalism—"the crusader-Jewish alliance, led by the US and Israel"[12]—invaded the Holy Land with mestizo soldiers, unveiled women, Burger King, the Colonel's Fried Birds, boombox rap, John Wayne stride, and tankers bearing off oil. Their reaction was disgust, jealousy, then vengeance—proud victimhood and envious rage, a desert homeboy "fuck you!"

They said, "You may have money and trinkets, but we are pure and we are brave. Despite your goods and arrogance and big-town strut, our culture kicks your culture's ass. And, anyway, Allah owns all this stuff—your skin too." They said, "Give us tanks and missiles and planes and we'll fight you with them. Give them to our enemy and we'll fight you with stones and bodies. Send our martyrs to paradise and you create our shrines. Look for us if you want, but we are nowhere and we are everywhere. You are stalking us, but we are stalking you. Fasten your seatbelts, America. We are about to come where you least expect us. We are about to serve the meal no one can bare [sic]."[13]

• • • • •

Prehistoric, post-modern Wahhabi, teenage and twenty-something soldiers in beards, sporting high-fashion turbans and black holy robes, drag through Kabul in Toyota pickups, "City Boy," "Fast

Crew," "Lion of Afghanistan," and "King of the Road" painted on their sides, machine guns mounted wherever. With *madrasa* students from Pakistan and hitmen from Arabia, high on testosterone and with absolutely no future in modernity, they mimic ghetto hoods and drug lords as they enforce petty puritanical interpretations: jumping out to beat a woman whose veil has slipped; arresting a man whose beard is too sparse; cutting off digits and limbs according to Babylonian tenet; confiscating VHS and audio tapes, smashing their plastic boxes, and tying the ghostly ribbons to trees as shrouds of victory; knifing faces out of signs; hurling thousand-year-old porcelain figurines across the room; and dynamiting giant rock Buddhas with hip-hop glee.

The statuary and lotus postures of the East are to them the same as the turrets of the World Trade complex or the White House—icons of sacrilege. They roll through Persian-Sumerian Mad Max landscapes, bearing the court of males, an eye for an eye, Mohammed the merciless, flying Islamic colors to serve their own sadistic reign.

They want to be bad-ass and cool in a way that has nothing to do with the faggot West or the dharma of the East, that takes up the sword of Saladin and subjugates the masses to fashions of their own whim and will. "These men left worldly affairs," declared Osama bin Laden, "and came here for jihad."[14] Meaning war as spiritual cauldron, as gauntlet, as acidic cleansing from centuries of shame and fear. An exercise both of purification and transformation.

Like Nazi goosesteppers, these pan-Islamic warriors stand against just about everything we are or could be: machine cities, rationalism, humanism, secularism, judicialism, jazz, the human Melting Pot, "nigger dancing," female sexuality, downward-facing dog, gay love, dudes, "my heart and soul, babe"—all of rootless, cosmopolitan democracy with its social fluidity, its shysters in suits.

There can be no solace or truce with them. Their fraternity is Bedouin austerity, muezzins, manly authority, military discipline, divine revelation, maudlin violence, dissociated killing, belts of explosives, subjugation to an emperor-god, bringing it on. They vie for the honor of using themselves as weapons against a craven enemy who jets in quick machines and drops fulminations from 30,000 feet. Kamikaze pilots of World War II, serving a Shinto lord, proved their discipline and unbreakable faith likewise.

We need more than a Bureaucracy of Homeland Security (cops and federal agents) because we are dealing with esoteric events moving faster than tanks or rockets or cyberbits and drawing on a different state of matter. There is no security once the jinni, the shadow of industrialization, is loose. Anything that can be idly created can be just as idly destroyed. Any technology can be hijacked for Stone Age vendettas. In fact, while we're fiddling with tags, why not call Tom Ridge's thing the Department of Defense, and call that other one what it really is, what it used to be called: the Department of War.

• • • • •

Before 9-11-2001, rap could pass for mean and street-smart; now a pimp car at the Oakland Airport sounds tinny, its posturing driver ridiculous. He's not bad enough to tie a belt of explosives around his waist or fly a planeload of men, women, and children into oblivion. High and mighty in his duds and jive, he can't touch the Hamas kid.

He comes from a different civilization entirely—from another planet, one ruled by devil women and recording contracts, imported blow and grandmaster thugs; DJs scratching platters; life, liberty, the pursuit of Vegas; Lotto and high times for all. Al Qaeda regards these not as prizes but laxities, dandyisms, acts of hubris against the universe itself. By comparison they are Neanderthal, Assyrian, Wahhabi.

They mean to exterminate our whole alien species of cowards, sycophants to money and machines, skirt worshippers who saddle themselves into cars and hire Mexicans with blowers to chase leaves off their lawns: "the settled bourgeois, the city dweller, the petty clerk, the plump stockbroker going about his business"[15] who, trembling with fear for his life, is incapable of holy war, of submission in prayer, who is dominated by bitches, and so is brought to his knees and crushed by the bravery of selfless jihad warriors, living missiles bursting into their office chambers.

We better get used to it. They hate us. They want to eradicate us, whatever the cost.

They want to strangle the pulse of secular materialism; to rid the Earth of its liberal, multinational, multicultural anthills; to carry out deeds so daring and irrefutable they rouse a mean old version of Allah from centuries of hibernation.

To the same end, Jesus freaks and their Zionist settler allies conspire to drive Arabs out of Palestine, blow up the Dome of the Rock, reclaim Biblical prophecy, and force Jehovah to bring on Endtime.

Why? Because the Big Guy said so.

"I am as sure of it as I am sitting here before you. Why do you think Rabin was murdered? It was obvious. God whacked him. God was furious over Oslo. He *personally* gave that turf to the Jews."

Nothing like a dedicated circuit to a Supreme Being! Anyone can claim transmission on his or her own authority. Anyone can intercede on God's behalf.

It sure is convenient to say you are a servant of Jehovah, and then execute someone, or occupy his land, or blow up his buses and buildings. As long as God says it's okay.

Is it any wonder that the other guys, the infidels in each other's parables, started fighting back?

War always finds a way, for the losers of the last battle will stop at nothing to win the next. Biblical apocalypse games are a self-fulfilling prophecy if there ever was one. Getting raptured and out of here while the nonbelievers plummet to damnation is both a Muslim and Christian gauntlet. It is about as topical as proclaiming that Hera or Thor is on your side and smashing your enemy's armies and deities, no more holy than a peacock spreading its tail or a cat baring teeth and hissing.

Don't forget: Jerry Falwell, Pat Robertson, David Koresh, Jim Jones, and their ilk are lukewarm versions of the Taliban, removed by about the same degree as cocoa is from cocaine. They equally would love to put the men in charge of the women, the women in *burqas;* not only deny funds to artists but kill them. They likewise declare themselves sole and authorized storm troopers of God.

Too much Bible and Koran is like too much daytime TV.

• • • • •

"Make an oath to die and renew your intentions," al Qaeda's lieutenants told them. "Death is the highest bliss. God had to veil it because if people knew how wonderful it is, no one would stay."[16]

Their recruiting tactic was to sell that to kids in desert camps and shiftless Arab hoods throughout Eurasia. Die, wreak havoc, avenge Palestine and Mecca, make Allah proud, go to heaven. The enemy will have no answer, no technology, no weaponry that can incinerate the atoms of that fine purpose.

One moment of pain, brief but profound, profound but brief— a quick illuminated glimpse through the creation's crack—then nothing, no pain, no hanging around, no more humiliation; then everything: the sudden scent of roses, satori.

They did more than carry the twenty-first-century battlefield into the cities of the West; they dragged America into a war that recognizes no boundary between the living and the dead. Their willingness to go over the edge into paradise as martyrs forced all combatants to claim souls, destinies, and after-lives too.

Before 9-11 the scions of the West swaggered as though the meaning of life and where the dead go were an open-and-shut matter. Americans behaved like the annointed seers of this planet, as though they possessed the sole up-to-date world-map, the only modern armies too, the only war games, the moral imperative of all mankind. But suicide bombers showed us we don't have a clue. We know nothing about the dead—the real dead—or how the living use the undead in acts of war. Most Americans still think that the dead are weird while we're here, drinking Bud, having fun.

(Yet I doubt each terrorist will find grapes, watermelons, and seventy-two virgins awaiting him at the banquet. More likely he will awaken surrounded by unconsolable spirits, his own and theirs; he will travel for untold eons and lifetimes, barely conscious, a ghost wandering in a nightmare, in a fog of the pain and longing, amid the rage and regret he set loose.)

Jihad versus McWorld . . . .

It is said that Timothy McVeigh and Teddy Unabomber, isolated from all other inmates (the mere criminals), became friends in federal prison, confidantes to each other. The latter was impressed by his comrade's social skills; not surprising, given Kaczynski's total lack of same. McVeigh posed that the only difference between himself and the boy next door was that the Unabomber blamed technology while he held the government responsible. Small difference, for they both devised clever, anonymous bombs and claimed the legitimacy of any means, no matter the collateral damage.

It is a tangled web indeed. What moral superiority separates Earth First! from al Qaeda? If we secretly exult when SUV dealerships and luxury hotels are torched or (for that matter) when Animal Liberation guerillas batter down laboratories to free mice and minks, how can we take offense when Palestinians and Iraquis cheer at the destruction of towers of Babel and pentagrams in New York and Washington (hubs of corporate capital, appropriation of global resources, and hi-tech weaponry used against Arab and Aztec resistances), when Egyptians in streets cry out, "Bulls-eye!" at big-screen video footage of jets disintegrating into buildings in distant Babylon?

We are not determined in the biosphere by our ideas or politics, even by our acts of pious disobedience. We are targets and when terrorists, like viruses or wolves, move against the West, they feed on us, amigos. There are no noncombatants; everyone at the mall is an accomplice. To Wahhabi we are all Americans (and Jews), regardless of whether we support a free Palestinian state and Islamic self-determination and oppose the globalization of American culture, Zionist settlements, the enforced impoverishment and marginalization of the developing world, etc. They don't care if we are with the Peace Corps, if we bring food to their hungry, shelter to their homeless, medicine to their sick: "Fuck that shit!"—in Arabic. "We are not dogs."

They care for our good intentions about as much as a jackal or tarantula does. We are our lives (our breathing, what we consume, what images we create and export). It is our very bodies and clothing and other accoutrements they hate. They don't prejudge which of us they shoot or blow up; just like the Jolly Green Giant in Vietnam, any gook is fine for the body count. They despise our activities without exception—bar the lascivious desires and orgies of sheikhs (and, of course, European football. In Osama's dreams it was the World Cup, Arabia against America. And guess who scored first?)

• • • • •

Jihad and McWorld—while at war to the death with each other—want equally to subjugate every indigenous culture, to subvert all local economies and festivals to their corporate plans; they want to end peasant life, games, celebrations, fun; they share being for the bosses, the investors, the mullahs, war, and *against* anyone who won't choose sides and prepare for holy battle. They represent ancient and modern poles of fascistic oligarchy, the old-boys club, the clandestine Masonic conspiracy: the Bushes and bin Ladens, Trilateral Commissions and ayatollahs, new world orders. They certainly have no design to protect the Earth, its biodiversity and multiculturalism.

We now dwell under the slim protection of not only NATO but Russia, China, and a few other secular armies. They alone represent civilization in a world which otherwise would deteriorate (we now realize) not into self-sufficient farmers, Sufi poets, peaceful nomads, and wise monks but Taliban gangs, child brigades, itinerant thugs, mafia chieftains, and, yes—George W.—weapons of mass destruction. But don't forget ozone deterioration, CO_2, glaciermelt, AIDS, budget deficits, drought, deforestation, salinization, desertification, overpopulation, concentration of wealth, global pollution, resource depletion, species extinction, dislocation of peoples, and other, bigger bombs than Saddam's waiting to explode in the twenty-first century.

We cannot defeat technocracy and the global market with rhetoric or weaponry; if we try, we will be buried by those who are so much nastier and more fervent and many than us. In fact, Fundamentalists armed with germs and dirty bombs will snuff out the biosphere far quicker than capitalism because, imbued with divine and sepuchral goals, they don't even see the natural landscape or the point of saving a secular Earth. If Allah wants to restore endangered species, then that's Allah's call. Monsanto, Squibb, and Adidas have about the same attitude toward anyone who gets in the way of business.

Jihad against McWorld trumps the Genome as well. Scientists in labs with hi-tech equipment can be incinerated, their neo-Darwinian arguments terminated. Germs, chemicals, jet fuel, and mutants can be turned against their own belief systems. Any weapon made by a factory can be used, however divergently, however asymmetrically, against that factory.

The global cop is in big trouble. The cargo cult and Ghost Dance have come of age. "There is nothing special about making nuclear reactors, cyclotrons, or rockets...," Mao said. "You need to have spirit to feel superior to everyone, as if there was no one besides you."[17]

• • • • •

Our futures can no longer be assigned to DNA alone. Deciphering chromosomes will not stop suicide bombers from strapping on explosives and going wherever they want, spilling body parts and cells everywhere. If they go enough places, then there won't be any more genome research or gene splicing. There will just be us on the Earth again, as during the Stone Age, having to live our lives in mystery and battle under the unassayable stars. That is perhaps what the one God wants and, in any case, right or wrong, fair or unfair, it is coming. Progressive science cannot stand long against arguments of planes as projectiles and other weapons from our own Pandora's box.

Suicide bombing forces those who assign all effects, all self and value to atoms, genes, and bank accounts to face their own lives, to decide what kinds of existences they want and what kinds of meanings and determinations they are willing to go to the mat for. It makes them *live,* claim lives they refused to acknowledge even exist, to inhabit their unexamined beliefs, or cede their very spots on the planet to those who sprout among Gobi sand and the junked autos and debris of modern civilizations.

Science may be the enemy of humankind when it imposes a reckless, proselytizing ideology, but by comparison to religious fundamentalism, it is our savior. It champions atomic, molecular, cellular spontaneity; its existentiality and improvisational magic liberated us once and supposedly forever from the long rule of Medieval priests.

Jihad wants to return the planet to a direct, authoritarian liturgy, to replace empirical experiment with raw instinct, xenophobic passion, scripture, a grand inquisitor, political torture, and enforced prayer—an omniscient, archangelic deity imprinting molecules with his design, programming the first cells with the entire bible of genes needed for descendant species. Such a macho creator, operating by Intelligent Design, would own the bodies and minds of his creatures, ordering them into existence and then governing their durations like a feudal lord or ranch boss.

Biotech mullahs and corporate moguls are no better. Using science to assign creation to mere chance and opportunity, they want to take it all over as unclaimed property. They figure our lives and bodies are nothing personal, just the universe's junk, so they can rightfully turn them into their property. One side sends ghost planes and giant metallic birds into the other side's commodity vaults.

How we get rid of both of these cults in the next hundred years or so before they terminate life on the planet is our real dilemma.

Entropy is inexplicably deposed by antientropy. . . . .

The barter of energy among predators and prey laid the basis for bands and tribes, hunters evoking sacred beasts who gave their bodies to the totem. Markets are natural extensions of botany, the exchanges of primitive chloroplasts converted to frond commensalism and plant symbiosis; later, primates tended the first farms. Gardens and herds tapped directly into photosynthesis and natural selection and gave rise to kingdoms and nations. "Advantages of trade are found in the metabolic exchange of legume root nodule and fungi, sugar for fixed nitrogen carried in amino acids; [they originated] among the mixed microbial and algal communities along the littoral of the earth's oceans four billion years ago. The trading of the econosphere is an outgrowth of the trading of the biosphere."[18]

When long-term biological and social networks are disrupted or stressed, they lose energy and become simpler, with fewer cycles, less efficient metabolisms. A global economy that abolishes barriers to free trade is like an engineered organism or fundamentalist priesthood. Its previously commensal, plural levels are distorted; cooperative reciprocity is discouraged; and hierarchical organization takes over, with domination by capital-generating, ideological methods. No wonder Filipino divers dynamite their own coral reefs to blow a few fish into the surf to feed tourists. A global system forfeits differentiation in space and time; life itself is degraded.

Successful biospheres are timeless, labyrinthine, and reciprocal; they capture, diversify, convert, and spread energy, allowing kaleidoscopic organization before it dissipates. "... [A]ctivity cycles are nested one within another, like Russian dolls, spanning a ... range of space-times from the very fast (nanoseconds) to the very slow (weeks or months), and from the very local to the increasingly global. In other words, the system has a deep ... differentiation.... [I]ts activities are all *coupled* together in a *symmetrical, reciprocal* way ... entail[ing] *both* local autonomy *and* global cohesion. Energy-yielding activities are directly linked to energy-requiring ones, in such a way that they can readily exchange places.... [P]arts in deficit can draw on those in surplus, and the roles can be reversed, so energy is diverted to wherever it is most needed at all times. The net result is a dynamic balance...." This is also a possible definition of life.

"... [T]he more sub-cycles there are and the better they are coupled, the longer the energy is effectively stored within the system and the less is dissipated as entropy."[19]

Sustainable development and social vitality mean efficiently deploying matter and energy, most notably the incidental flow from our immediate sun-star, sluicing it into molecular grids, transmogrifying and sharing it through primeval biosystems. When events cohere with one another, so-called genes and traits disappear beneath collaborative effects—entropy is deposed by antientropy; the organic stability of open, indeterminate systems replaces the mechanical stability of enforced equilibrium.

Natural and symbolic phenomena collaborate with one another to generate and sustain a vibrant web, a span that includes both language and agriculture, songs and supper. Differentiation leads to structure, structure to interdependent cycles, interdependent cycles to mutually beneficial exchanges. Dissipation is minimized. Macroeconomies and metropolises escalade. This is real democracy, real communalism.

For whatever reason, the modern world is hellbent on replacing ecological complexity with something far less fecund or healthy and far less likely to last.

It is not going to be easy to stop this world-view without a global cataclysm. It is presently a runaway train.

"We have emerged for a moment from the nothingness of unconsciousness into material existence." . . . .

It cannot go on forever anyway, even under the best of circumstances. It never was meant to go on forever. The people in trams and cars do not realize, as they journey between ephemeral places, that just as cells arose from sea foam, creatures from cells, and languages and cities from creatures, something else will arise from this by a meta-morphosis incomprehensibly more radical than what has passed from the Ice Age till now.

There is not enough slate or petroleum, air or potable water to have that many more generations plant crops, expand their settlements, groom pets, consume tacos, and discharge wastes. There cannot be infinite rock concerts, UPS and FedEx deliveries, picnics along the Hudson, or dams and factories in China, before the debris and contamination become greater than the influx of energy and cellular mechanism can handle.

After that crisis passes, not only will Gaelic and Maidu no longer be spoken, nothing resembling English will be known anywhere on the Earth.

. . . . .

People imagine they can improve it all by committees, investments, electric fences, arsenals, and power blocs, but no one stays around here very long. In just a few years, there will be millions of new people and trillions upon trillions of bionts arriving on this world uninformed about past crises, anxious to make a difference, and with lots of energy to rumble. Nonselectively they will replace the foolish and the wise. They will bring their own hip-hop and dreadlocks. Acts and effects will arise from their own direct experience of things. But, as that is happening, those who will replace *them* will already be born.

There are more people in Iran who have come of age since the Ayatollah's Revolution, more Central Asians born since the Soviet Union. Kids growing up all over the world have been casually shaped by the upheavals and binges of the '80s and '90s, either directly or indirectly: materialistic nihilism, marginalization of tribes and habitats, globalization, electronic pornography, terrorism, cyberwarfare, body piercings, tattoos, cocaine, heroin, and other molecular abuse. They will live those out and pass the consequences on, even as they inherited them.

It is not that things can never improve. Everything will be brilliantly and utterly transformed from within the world, but *on the* roller-coaster, and within the minds of women and men.

As long as Sun showers Earth with photons, creatures will come out of the cell. It is doubtful that Gaia itself will be extinguished. Something very powerful will one day emerge, even from us. Despite its present flimflam, ours is a stage in something unimaginable. That is its lone saving grace, for the notion of reclaiming and fixing this in its present form for another hundred years or so of transactions is beyond hopeless.

. . . . .

"Other civilizations, perhaps more successful ones, may exist an infinite number of times on the preceding and following pages of the Book of the Universe." So declared Andrei Dmitriyevich Sakharov in his Nobel Prize speech, radio waves carrying it into the cosmos. "Yet we should not minimize our sacred endeavors in the world where, like faint glimmers in the dark, we have emerged for a moment from the nothingness of unconsciousness into material existence."[20]

True enough. We are the one certain mystery, the single remaining beacon.

Somewhere else in the universe it is possible that life forms have achieved health, longevity, the wisdom of their own nature, and interplanetary travel by other laws and means. They also have arisen, for their moment, like us from nothingness into wonder. We, like they, must now "make good the demands of reason and create a life worthy of ourselves and of goals we only dimly perceive."[21]

8

Topokinesis

· ·

Physical Forces in Development

> There is no pair of wings branching forth from his back,
> he has no feet, no nimble knees, no genitals; he is
> spherical and equal on all sides.
> —Empedocles, 484–424 B.C.

Germs become blastulas. ·

Though shape-changing continuously, a cell tends toward sphericity, a sphere representing the sparest and most efficient organization for metabolizing molecules within membranes while conserving and transmitting energy. Cells are spheres for many of the same reasons that stars and planets are.

Ontogenesis begins as gravitational and metabolic stimuli disturb the fertilized ovum. A sphere deforms into an irregular ellipsoid, its proportions changing as its surface increases faster than its volume. The cellular membrane and nucleus, kinetically bound to each other through their cytoplasm, transmit a range of stimulating tensions. Submicroscopic molecules migrate from the membrane into the nucleus where they are chemically altered, then they return to the elastic boundary. Erich Blechschmidt* lays down the first axiom of embryological dynamics as follows: "The cell's limiting membranes, with their constant generation and regression, their continuous rearrangement with the direct help of the adjacent cytoplasm, are the primary agents of biokinetics."[1]

Heterogeneities progress as extracellular environments encounter the cell's soft periphery in waves and conduct their impingements inward where they are correlated and organized by one another, yielding micro-mechanical echoes and equivalents of external effects. These give rise to new microstructures, in part as elaborations of one another and in part from the nucleus' deep archive of feedback.

In response to disturbances the nucleus bombards the cytoplasm with hereditary information and molecular precedent and, while the cytoplasm mediates these, the limiting membrane forces the system to adapt inside a boundary that is disseminating influences from outside itself. The limiting membrane cannot act without the nucleus

---

*When I asked cell biologist Stuart Newman for help in correcting this and the next chapter, he wrote: "Blechschmidt's descriptions deviate so far from the way I understand things that I can't readily do much with your version of them. It is evocative and impressionistic; some of it is plausible but unsubstantiated; some of it is decidedly incorrect. To sort it out I would need all the citations to any experimental literature that you (or he) have relied upon for these assertions. The intersection of your insights and Blechschmidt's theories with what has been established scientifically is not always straightforward. I know that you are trying to interpret the earlier writings of Blechschmidt, but why bother if they are so intuitive and unscientific?"

Much of the so-called topokinetic material that follows must be read with Newman's proviso that it is of mixed and uncertain heritage. It represents my attempt to summarize and explicate Blechschmidt.

Why bother? Because he is the singular poet and visionary of embryological development. Books like *The Beginnings of Human Life* and *Biokinetics and Biodynamics of Human Differentiation* are seminal in esoteric biology, osteopathy, and projective geometry. They provide the guiding principles of a mostly unexplored life science and aesthetics.

nor the nucleus transcribe without the membrane, and neither can perform biologically without the cytoplasm through which they are kinetically joined.

• • • • •

Waves continue to flow in and out of the nucleus, eliciting successive designs.

Then, for unknown biodynamic reasons, "the increasingly unequal distribution of substances at the surface and in the interior stimulates the centrally located nucleus to divide,"[2] fissioning the cell into clones of almost identical size. During mitosis the zygote's subcellular spindle-fibers in their liquid medium respond to force much as multicellular filaments do in tissue; they bend and twist. Synthesis of intercellular material pushes the daughter clones apart eccentrically even as they adhere by attraction between their membranes.

• •

Mitosis: a sequential process of cell division and differentiation that conserves chromosome diploid number by allocating complete copies of chromosomes to both daughter cells; distinguished from two-staged meiosis whereby chromosome number is halved to produce haploid germ cells that restore the diploid number when they fuse in the zygote.

• •

This is a self-organizing network of networks, with relative cell placement its compass. Only from other cells around it does a cell know what it is and what its destiny will be. Its position is the main piece of news it sends to the dormant nucleus. "[E]xtragenetic information acts from the outside inwards and in so doing acts on the nucleus. This is why the positional relationship of cells to the neighboring cells is very important for differentiation processes. It is also why the differentiation of an organism cannot be attributed exclusively to the genes."[3]

• • • • •

Swelling fission is the first in a series of alchemical transformations that seems to have no physical basis except the force of life itself: first multiplication, then differentiation, inversion, dilation, and signification. Quasi-symmetrically maintaining its ratio between surface and mass while splitting into identical copies, each offspring of the zygote essentially clones its own sphere into two continually self-replicating sphericities. A bubble turns into a bubble of bubbles.

One becomes two; two become four; four, eight; eight, sixteen. A fissioning cell simmers into a hive of cells called a blastula. Its sectors, blastomeres, are not independent cells, but monads comprising a medium, the forerunner of tissue. Eight blastomeres are, to all appearances, equal octuplets. At sixteen, some are inside, some outside. Although they are to appearances still identical, their basic geometric and thermodynamic meaning has changed. The ones inside are not only partially suffocated; they are inside, in the "dark," subliminal—a novel phenomenology for them as well as for those that, by enveloping them, engirdle the universe.

This is a crucial stage insofar as embryogenic patterns are initiated and determined by contextual relationships and necessities. Eventually metabolism arises from the need to nourish and oxygenate the inside entities and clear their matrix. This is also how they "prove" they are not ordinary, static pebbles but a new thing under the sun. Where each multipotential blastomere sits in relation to the other blastomeres determines what layer of tissue will enlist it, and where it falls within that layer will imbue it with functions and specializations, e.g. manufacturing bile, electrically synapsing, or hardening into carapace.

• • • • •

In the oceanic environment of the oviduct the blastula continues to fission, producing more and tinier cells, thirty-two, a hundred and twenty-eight, two hundred fifty-six. Eventually "the nucleus-cytoplasm ratio changes in favor of the nucleus.... Not only the fertilized ovum but also the fluid of the oviduct in immediate contact with the cell takes part in ... developmental movements."[4] The embryo invades the mucosa of the womb, sending out tendrilly villi that exhibit suction behavior. "[T]he metabolic field of the ovum comes into immediate contact with the metabolic field of maternal cell aggregations, leading to an intensive uptake of nourishment ... and therefore to an enhanced polarization of the walls of the early germ."[5] As the cells absorb nutrients, their growth is maximized at the position of greatest curvature—their equator.

The first blastula, a fiction anyway, implanted itself in a womb of thermal vents and drank and drank.... If so, it is still drinking.

Tissue Layers . . . .

As adhering cells attempt to compensate for a spectrum of external disturbances, transformations pass from cell to cell, waves in one synchronized fabric. These developments are read and further reflected by each nucleus, which dispatches RNA in new contexts. Though genes respond as they are inundated, Blechschmidt still insists that they "are not the engines of development; demonstrably, they do not themselves produce the later characteristics of the differentiated organism, not even indirectly via the enzymes they form."[6] Growth itself is Blechschmidt's driving force, not growth as "an endogenous process, but one induced from outside.... [O]rgans and organ systems are formed as orderly components of the growing entity, despite the passivity of the genes."[7] This is a radical notion— that the genes with their mRNA transmission are relatively feeble embellishers operating on a living geometry that molds itself in the hands of gravity and its shear forces like a blend of rubber, electromagnetism, and clay (see also pp. 363–364).

The blastula's limiting membranes—tirelessly rearranged by the activities of adjacent cytoplasm from which they consume energy— accommodate stress, change boundaries, morph, and initiate fresh biokinetic events leading to more advanced nuclear responses. The organism establishes deeper fields; the equilibrium of these fields is tipped by further differentiations; the fields then attempt to equilibrate themselves, leading to even more profound discriminations

and denser, more specialized structuring. "[O]ntogenesis is the totality of responses to stimuli impinging on the primordia of the ovum....
By the steps of differentiation the germ tries, without at first coming to a final result, to compensate for the external influences [while] preserv[ing] its original character—therefore, itself...."[8]

The nucleus, under siege, perturbed, responds in the only way it can—recoiling and dispatching waves of nucleic acid. It does not realize, even mechanically, that this is its main hope of survival and the fate conferred on it by ancient spinners of yarn. Despite its prior stasis, the embryo is forced to evolve. "The organism is trying to preserve the metabolic fields whose equilibrium is always slightly upset in the course of its differentiations. The outcome of each of these attempts is a further step in differentiation."[9]

The colony gradually metamorphoses such that mere mechanical multicelluarity becomes tissue vascularization. The promordial version of this template, by outward appearance likely pretty much the same as the modern one, had unseen potential to matriculate as toads, snakes, blowfish, or gnus. Clams and lions were inhumed long ago in one germ.

• • • • •

At thirty-two cells the human zygote is a fuzzy clump, called a morula for its resemblance to a mulberry. Adding more cells, it morphs into two distinct layers enveloping a fluid-packed core. As this blastocyst comes to reside in a field compressed by two pools of fluid, a spongelike zone suffuses in and among the drupes. The aroused wad is defined by and oriented to its outside—which means, simultaneously, its point of origin in epithelia and the imposition of external tangents (gravity and dynamics) on its changing body. Absorbing the secretions of a deeper interior, the inner layer continues to tumefy, polarizing into two clear-cut plates. Attached to each other by firm membranous ligatures on one flank, fluidic ones on the obverse, these domains become the germinal panels of the animal body.

Laying claim to disproportionate amounts of the yolk-sac's food stores, ectoblast, the bilayer's outside, is the precursor of ectodermal tissue; it includes epithelial cells which will compose skin and the neurons from which the brain and spinal cord will form. Wherever it has space to expand, it becomes broad and thin; where hindered, it thickens.

Reciprocal to hungry ectoderm, a more amorphous layer of cells that is trapped inside sinks further while remaining relatively sparse. This is entoblast and will become endoderm, raw material for gastro-intestinal organs.

Cells within these regions are drawn along by their respective fates.

An intermediate layer of mesoblast arises as the by-product of unequal growth rates of ectoblast and entoblast—mesh-like aggregations churned up between self-defined epithelial landscapes. Some of these will become musculoskeletal; others will contribute to heart, lungs, in fact most of the internal organs.

The transition between tissue layers is maintained by nutrients flowing from the surface of the ovum toward its interior. This establishes zones of intercellular substance and augments the spongelike layer between ectoblast and entoblast. With outer and inner epithelia squeezing out liquid and pushing apart from each other, the zone dividing them becomes a dehiscent gap, forerunner of the chorionic cavity. Membranous tissues thinning here will eventually burst through themselves into a true gut.

· · · · ·

Mesoblast starts out relatively scant. Less well nourished, it aggregates slowly, compressing from the center of the embryo up and along its sides while pulling on the overall mesh and coming loose by liquid astringement. "[F]anning out from the disc upward and sideways,"[10] it exerts a growth pull on the emergent ectoderm, an action that will ordain its role in the embryo's differentiation. As it increases in volume, it takes on a pulpy character.

Some mesodermal cell groups given the name mesenchyme collect in fields composed of loosely organized platelike bodies with a paucity of intercellular substance. These semi-independent zooids underlie the development of connective tissue and cartilage. Aboriginal in mode and style, mesenchymal cells are under no particular orientation; they experience a tension and stress equal in all directions. Combining opposed impulses of autonomy and collectivity, they are truly a mystery factor in development (see p. 88).

* * * * *

Ectoderm is more than just primordial skin; it provides cells plus cellular context for endodermal and mesodermal differentiation. As early ectodermal cells procreate and spread, their mass separates and its layers undergo regional variations, providing reference points for an invagination of the whole blastula. The other tissue fields are then reflections and deepenings of ectoderm—ectodermally induced as back-and-forth currents of information and shear patterns enrich and fold laminae. For instance, the interaction between ectoblast and mesoblast impels dominions of their cells to roll up into a neural tube.

Subfields generated by primal ectoderm play a commanding role even late into embryogenesis.

Gastrulation . . . .

A protrusion along the lateral umbilical region causes longitudinal expansion of the blastocyst and, by the end of the second week after fertilization, the embryo, though smaller than a caraway seed, resembles a stack of circular, flat griddlecakes. At this point, a thin line of lateral ectodermal cells—a sprout or axial process—separates itself and migrates along the layer's surface. Gathering mass and momentum, it scores a seam down most of its axis. A continuous migration follows the trail, cells bunching and then collapsing inward along what is called a "primitive streak." This opens into a groove, then a gash along which cells, while continuing to shift laterally and forward, plunge down into the embryo's core—a fundamental physiological process known as gastrulation.

As the central breach draws invaginating ectoderm into the cell-mass interior, it continues to encounter more resistance medially than laterally, so the expanding aspect of the sheet rolls inward medially as it thickens, while spreading outward laterally against the lesser tension of retarded mesodermal growth. By restraining longitudinal growth, the axial process regulates blastula surface expansion, providing both a fulcrum for the infolding of the fetus and a winch projecting separate regional growth rates and stress fields.

• •

Axial process: an aspect of ectoderm which is invaginated into the interior of the embryonic disc, leading to its global folding.

• •

While the primitive streak roils with chaotic, coordinated motion like a disturbed anthill, cells are being transformed, and the embryo is undergoing the most ancient and basic of its ontogenetic phases: inversion from what would be a terminal jellyfish into the mold for a complex, multidimensional template with wide-ranging potential for elaboration.

In association with the axial process, a dense clot of cells, induced in part by unformed elements of the vestigial notochord, deposits germinal neural plate in its receding wake—seeds for the spine and central nervous system. The nexus spreads into the neural groove, stimulating the longitudinal disposition of the embryo. Cells at the tip of the axial process (the invaginating section) become loosely aligned to form fresh mesoderm.

What is now a tube lays a vertical axis by ascending in the direction of least resistance—caudally from its cusp. Dilating slowly but without surface expansion, it offers a pretext for the formation of dorsal ridges and a neural groove. Waves of pressure along it push out a central furrow—the protoskeletal compass for all advanced organisms. Stretching and softening the embryonic skin, the same compression sluices a superior opening for brain tissue while squeezing radial into bilateral symmetry.

The streak itself is a fundamental line in nature, translating circumferential energy into longitudinally helical designs that express worm or skate, crayfish or squirrel. A polyp or starfish—the under-lying orb of every more complex seed and egg—matures into advanced shapes emergent through intersections of trapezoidal grids and Julia-like sets.

As the body of a primitive fish imprints itself magically in the clay of the blastula, the mythical ancestor of the animal kingdom replaces the universal sphere. Its hinge becomes the mandrel for a neuromuscular tree.

The gastrula is a basic die in the universe for objects from sea horses to brains, likely on all planets, however kinetically initiated and geometrically consummated. Its Euclidean counterpart is dis-cernible in axes of symmetry among trees, flowers, and their sama-ras and seeds, also on any planet.

• • • • •

Open your mouth like a child catching rain on her tongue. Make a soundless, bullfroglike hiccough, holding its recoil in your throat and letting the resonance travel downward. Trace your fingers from your lips along your gullet and frontal torso while imagining the cavity below it, under and around your loins to your anus. Activate this corridor with the energy field of your fingers.

It is your original crease, the seam along which your blastula invaginated, then found and transformed its inside.

An immemorial valley underlies your body, the root of your shape and function.

The topology of all complex organisms begins in a bilaminar disc with an inside and an outside. . . . .

A cell is not a fixed unity. In Blechschmidt's lingo it is "a momen-tary aspect of spatially ordered (submicroscopic) metabolic move-ments. The same is valid for cell aggregations (tissues), for tissue aggregations (organs) and also for whole organisms at any stage of development."[11] They each embody local shear forces and biologi-cal fields. They grow according to innate principles. They stretch and deform their own paragons under external edicts. Calling a cell "a momentary aspect of biodynamic movement" identifies matter as "a spatially ordered submicroscopic alignment of energy."

Prior to chorion formation, the colony comprises an amorphous, functionless ball; its interior and exterior geography are circumstantial, inadvertent. It is not even as elegantly nubile as a newborn dipleurula. It survives by the antiquarian legacy of its individual cells. After the formation of the chorion, the embryo takes on *savoir-faire* and finesse, an animalhood that will elaborate its principle from worms and congers to sandpipers and geckos: a sensitized contact zone onto the landscape around penetralia, a sometimes-ciliated stomodeum (mouth) opening to subliminal chambers for breaking down matter and nourishing tissues. It will enter and define space itself.

Tissue plexuses now reflect that "organism" *means* something, represents biological function; it is a force, a separate topological universe breaching the planet through its own vital physics. Oral and anal portals soon make the deep chamber a corridor for pulsatory, anemonelike swallowing and assimilation followed by peristaltic discharge of wastes—i.e., the recruitment and dissipation of molecular and heat waves. The universe feeds and awakes.

Somehow the autonomy of the cell has been subliminalized within, while the autonomy of the cell colony has arisen from there to takes its place. Accreting and fusing, the various layers and nodes gradually conscript one another into ratios of this new coefficient.

· · · · ·

The topology of all complex organisms begins this way: a bilaminar disc with a gut, a bubble with a core, growing denser, fractalizing. This is probably how multicellular life first differentiated itself in the primeval ocean—billions of such fissioning bubbles. They individuated through the sinking, twisting, and introversion of their vasacularity. They become floating, swimming star systems.

The fundamental multicellular shape, the template of all subsequent motifs is marine, a pulsating, saltwatery lipid or cylindrical bell, a gastrovascular sump pump. From this hydrozoa/anemone evolved interdependence of parts, transitional stability.

We are still basically an inside leveraged against an outside. We think the universe from inside, out.

The speed, location, and trajectory of each developing tissue cluster affect all other tissues. . . . .

The areas of the embryo attracting the most cells are those limiting tissues that adjoin fluids. With more opportunities for exchange of energy and substance, they maintain a swifter metabolic pace. Intercellular substances flowing along membranes permeate and then infiltrate them, stimulating their metabolism and vascular circulation. Accumulations of liquids trigger gradients and set the stage for further vascularization and vessel formation.

Tissues press and squeeze one another into and out of soft architectures, sculpting each other's morphology. Nodes and shapeless clumps will gradually be induced into such unique and specialized entities as liver, kidney, heart, lungs, duodenum, blood, T cells, and glands. The synchronization and metabolization of post-gastrulation movements and patterns of tension slowly endow the embryo with the character of the organism it is becoming. Basic tissue nexuses diverge in terms of size, shape, and relationship to one another from a cow to a whale to a kangaroo. Cell membranes generating tensile stress also particularize metabolic cycles, telling organelles in what chemistry they should specialize. Extragenetic information, dependent on position-based matrices, continues to permeate each boundary, radiating through its cytoplasmic rings to the core. As every cell is a metabolic field by itself, morphogenetic becomes topokinetic principle at the level of simultaneously cells and tissues.

Embryogenic dynamics must ultimately follow the shifting bias of structures in relationship to nutritional sources. While limiting tissues discharge their waste toward the ectoderm, younger mesodermal masses accumulate theirs in between cells, initially fluid deposits that, because they are less chemically active than the cells themselves, inhibit growth and dampen the formative potential of tissue. Yet debris doesn't just stifle; it further differentiates and metabolizes. Transcending simple biochemistry, organ anlagen represent biodynamic tension between cells and intercellular substances.

. . . . .

Without ongoing digestive, circulatory, and locomotory urgencies imposing functional design and organismal meaning, pumps and passages, arms and legs might appear out of time and place (if at all). Eyelike nodes would congeal without legible cables to a ganglion, perhaps in the back rather than the front of a skull; and, despite even the most precise hereditary code, mouths might be vertical clefts (or anything anywhere) instead of transverse gaps, because codons cannot by themselves thermodynamically pin a tail on a donkey, place a slit in a face, or connect it to an esophageal conduit manufactured elsewhere by other codons. Biological necessity engages topokinetic design to make anatomy purposeful. Some creatures may look bizarre (overfrilled, overcolored, overcrouped), but their tissues go together and work from the guts out, cohering in a lifestyle.

As geography and metabolism accommodate each other, events and influences synergize and radiate throughout the fetus. Developing organs attain mature shapes and tasks, initially as a function of their positions and gradually as a consequence of the biodynamic activity of the entire embryo. Systems induce one another as they overlap.

. . . . .

While biological designs absorb new motifs and morph, they remain congruent and allometric. This can be viewed dramatically in time-lapse reconstructions of vertebrate skulls; for instance, speeded-up transitions among crocodiles, ostriches, rhinos, jaguars, giraffes, dromedaries, tree shrews, pandas, gorillas. Going back and forth along imaginary evolutionary and devolutionary paths, the snout protrudes and retracts; the brain case diminishes and swells; horns rise and fall, gnarl and narrow; a bill tapers and broadens; teeth crystallize and dissolve; eye sockets migrate fore and aft while chang-ing size and shape.

• •

Allometry: *the correlation of change of shape proportionately with increase or decrease in size, reflecting phylogeny, ontogeny, or con-tingent variation.*

• •

The speed, location, and trajectory of each developing tissue cluster affect all other tissues—usually those closer to it more sub-stantially—for as mass acquires territory, it transmits stresses else-where and, as it conducts and refines metabolism in each region, it provides ecological context for other tissues and cells. It also becomes rooted in a broader cohesive metabolism serving such bio-industries as cardiac transport, peristaltic assimilation, and neuromuscular motility.

The Human Face . . . .

Cerebral and cardiac prominences dominate the early human embryo, so brain and heart anlagen are initially quite intimate to each other. As the neural tube, fed by invasions of large blood ves-sels, advances in the direction of least resistance, ascending and broadening with the cephalic zone into the expanding stem of the brain, it compels the head, as a relatively mobile element, to curl over the emerging heart. The primitive brain is tractioned upward, while the primitive heart descends, the trunk wall collapsing under the weight of the head. As the cranial portion of the blastula broad-ens, the jumbo head continues to radiate a construct downward through the trunk. The space between brain and heart lengthens,

---

▶ Erich Blechschmidt Collection: At the middle of the second month, differentiation equals developmental dynamic processes in fields of metabolism.

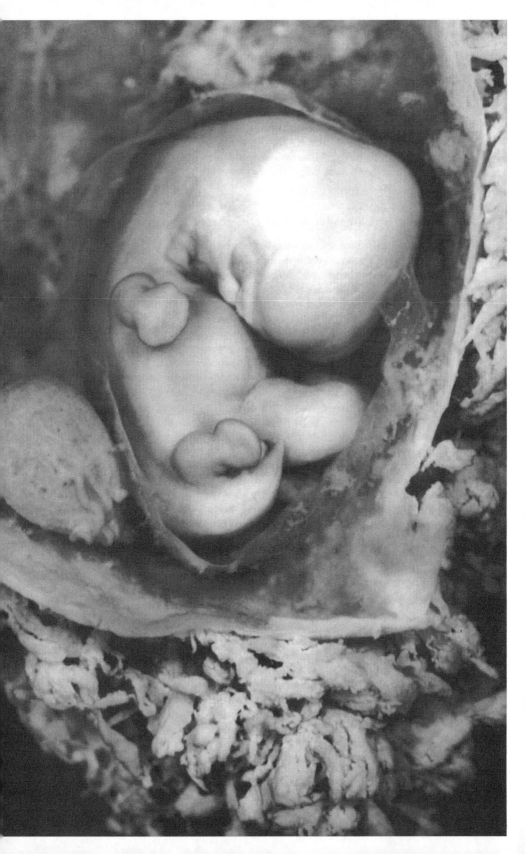

and a countenance protrudes.

The face is not only a matrix between the brain and heart-liver but is twisted and ground from the thickened upper reaches of the neural stalk, the vertebral column of which will later orient neural tracts and anchor limbs. The cranial neuropore widens as the skull bends. Innate growth resistances furrow the head and its ganglia, causing sulci to form between the forehead and nose and between the face and cardiac bulge. As the embryo folds over its heart-liver center, it adopts a characteristic fetal pose.

The mold of the face is literally squashed and broadened between the emergent frontal brain and the heart-liver, rotating from a transverse to longitudinal propensity and crinkling in the pressure of adjacent fields. Gradually narrowing from its transverse alignment, the mask becomes more gracile and expresses meanings sublimated in the brain, heart, and liver. This is the innocent gaze of life on Earth.

• • • • •

The restraining imposition of the skull combined with the formative expansion of the brain forces the head to expand antibasally while arching in a trajectory that is simultaneously convex, superior, and forward. Swelling dilation of the nasal capsule and lower jaw pulls on the skeleton supporting the mouth, further accentuating the dorsoventral plane. Young connective tissue wedged against cartilage at its base is at the same time being stretched craniocaudally between the falx-cerebri membrane above it and the hyoid mass cervically below it. The hyoid tractions the jaw and nasal bridge downward, distending it in a crossplane. The ventrally expanding jaw contributes to the face's depth. Otherwise, it is a blind looker pressed eternally against an invisible glass.

The almost mawkish profundity of the human visage is ordained structurally by the interjection of a nasal apparatus where connective tissue, constricted between swelling optic vesicles and cerebral hemispheres, thickens into an olfactory placode. Ossification of the nasal capsule occurs later when a detraction field (see "bone" in the next chapter) is established lateral to the nasal bridge.

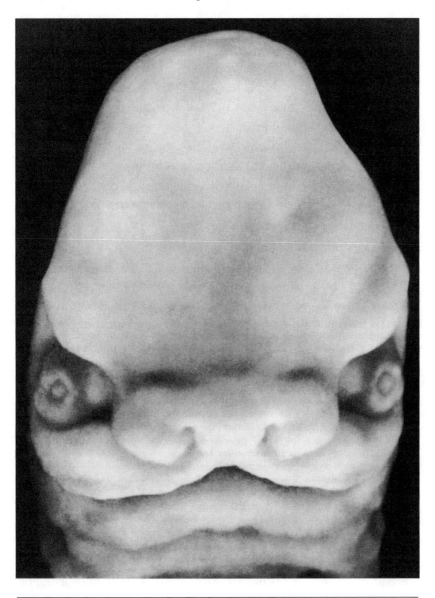

Erich Blechschmidt Collection: Inner tissue between the bubble of the brain and the bubbles of the eyes holds back the skin on both sides of the oral crease and in this way lays the concave disposition of the nose, which is caused by two indentations (the embryonic nose holes). Tissue surrounding these holes helps build the outer nose.

Caught between opposing forces, the simian three-dimensionality of a muzzle is sheared out by the stretching of the nasal capsule in a vertical plane, a tension that also pocks openings in the capsule, aligns them vertically, and pushes them apart. Nostrils then lengthen and taper—embryonic nasal slits, their edges slipping forward, impressing a soft sigil of breath.

In the front of the head, mesodermal protuberances burst through layers of ectodermal sheets, melding into lumps for ectomesodermal organs. Shifts in differentiating epithelium and condensing stress planes generate placodes for light-refracting lenses and vibration-recording catacombs (eyes and ears). These forces originate in part from growth resistances accumulating inside the neural tube as the groove seals over itself, closing the cranial end of the canal.

The head tilting downward and the neural groove widening, ectoderm is compressed at its caudal margins. Otic placodes begin to spiral as their surface thickens. Ears are regional coils of epithelium that is buckling.

Optic vesicles sprout as blind-sac, lateral extensions of the prosencephalon (forebrain), parallel to cerebral hemispheres themselves. This occurs as the brain's tissues broaden while being hindered in their longitudinal expansion by the wall of the cranial cavity. Downward flexion of the fetal head restricts the tissue wall adjacent to the brain. Hindered in its surface growth, restricted also longitudinally, the thickening mass surges outward, bending upon itself as it swells, pushing myopic sacs outward on each side—the optic cup and the first cerebral hemisphere.

Sight is a direct medium of thought, for the eye's sulcus is a swelling and broadening of the brain itself.

The vertebrate face is not an isolate satyr wrapped in a ganglion on a stalk; it is the personification of the emerging nervous system, lineaments squeezed out between swelling cerebral hemispheres and mazes of intertwining viscera. It is also an irrepressible mark of presence that, even among lizards and birds, focuses biological energy, rivets our attention, gives the cell colony an "I," entrains mother and infant to each other, identifies friends, confers charisma on lovers, and hails strangers.

Maintained at a distance from each other by connective tissue compressed between the brain and the nasal root, charcoal eyespots bud spookily inside opaque jelly of an expanding occiput, windows to a core transcending matter. Either a spirit ignited when the first cells went inside the abyss; or spirit is a late visitor to matter and stares through its neurons in amazement.

A similar—though not identical—physiognomy will be imprinted by topokinetic forces elsewhere in the universe.

· · · · ·

The sentinel placement of eye portals, according to Blechschmidt, is topokinetic, not utilitarian or adaptive. While the head expands, the space between the eyes remains relatively stable; yet the floors of the eye orbits approach each other as they are being realigned by dorsoventral traction of the two lateral palatine processes. The bulge between the forehead and the nose induces fresh ligamentous structures which then compel the distance between the eyes to foreshorten in relation to the width of the parietal region of the head. This activity brings the lateral line of sight forward.

Blechschmidt proposes that eyeglasses (with their plastic or metal bridge) represent structural dynamics at the root of the nose as well as the developmental relationship between the nose-ear-eye complex and the convolution of the brain.

· · · · ·

Dilating and shelving convexly above the ocular placode over the outer aspect of the brainstem, two distinct cerebral hemispheres, hewn to a mesencephalon by flattened tissue, assert themselves. As the embryo folds ventrally, mesenchyme at the root of the young mesencephalon (midbrain) becomes transversely aligned and flattened. Meanwhile rhombomeres continue to aggregate at the base of a rhombencephalon (source of the cerebellum, medulla oblongata, pons, and general hindbrain). The shape and functional development of the brain are partially fated by the retarded growth of flattened tissue causing the antibasal aspect of both cerebral hemispheres and the midbrain more or less to cohere.

As eyes expand, mesenchyme along the nostrils stretches into a hammock for an array of musculoskeletal ocular organs. Elongating from nose to ear, they also braid the fibers of the zygomatic arch.

• • • • •

A web of mesenchymal tissue curbs the brain's outward expansion, requiring it to fold and arch convexly in a dorsal direction, broadening the fetal face between the brain and the heart and stabilizing the transverse alignment of the mouth. These biodynamic movements also distend connective tissue circularly in relation to the longitudinal entry of the cranial aspect of the gut.

A separate dilation field around the mouth increases the tissue's tensile resistance, foreshortening its oral cleft. An outer ring of mesenchymal tissue imposing a gradient around the stomodeum induces an orbicularis oris muscle, a fulcrum for the sculpting of upper and lower lips. As these rudimentary structures torque inward, their development is retarded relative to adjacent structures.

At an angle marking the transition of ectoderm to endoderm, the young mouth's corners form. The oral slit broadens, and the tissue surrounding it stretches down toward the heart where it helps induce paired and unpaired aortae, efferent channels, and arteries, as well as their metabolic fields. It also deposits folds that become pharyngeal arches.

As the raw materials of the primitive orifice crimp by degrees, its edges roll inward, closing the slit externally. Behind the sharp-edged dimple a cavity continues to sink along a suction void. Pressure from within causes the embryo to suck, inducing the esophageal groove into a respiratory, digestive cavity. Without such deep inrolling, the tissue would never quaff, and the "mouth" would be little more than a rootless crevice. But then it would also never form.

Soft parts of the oral region come together and seal their opening with a hint of pucker. Accenting the esophageal depths that underlies them, mouth and lips take on biological significance as a tractile

gate through which the mature body will nourish itself—a vestigial wound to which love objects will be drawn.

"By our prenatal development we have prescience of our own
bodies," Blechschmidt notes, "and although we are seldom conscious
of it, we express it in our gestures."[12]

Lipstick, sequins, and facial piercings represent ancient as well as
future crises.

⸱ ⸱ ⸱ ⸱ ⸱

Where growth is hindered in the transitional zone of the lips along
their seam with the floor and roof of the embryonic mouth, a labio-
dental jaw ridge develops, epithelially thickening in spots between
expanding germinal bulges and the oral cavity. As these broaden,
corrosion fields (see pp. 332–333) separate them by growth pressure on
the mucosa, which leaves defects in between them. They intort as
small perforations. Even without solid food, the embryo begins to
teethe. Lumpy primordia later harden into deciduous bones.

The tongue begins as a vaulting ectomesodermal tag that swells
tridimensionally along longitudinal, transverse, and vertical axes.
Dilating crisscrossing cells mold its growth field. Interplay of force-
lines gives the organ snakelike independence, wielding a flexible grid
of musculature into practice speech that babbles silently in the womb.

The vertebrate larynx, emitting unsignified sounds, only incidentally serves as an instrument and chamber of vibration, having
arisen as a sphincter to guard the respiratory passage against the
entry of anything but air. The tongue converts the larynx's unconscious sounds into cries and speech. Without a tongue, no phonemes
could be fashioned on anvils of palate and teeth. Yet the linguistic
organ was hardly forged to gossip; with its crude taste buds it was
primally a sampler of prey and nectars, a spry, sticky limb for snar-
ing creatures.

In hissing snakes and small mammals, the tongue is of little semantic use. In primates, however, it is the spade and staff of the emergent
mind.

Where in their endomesodermal mazes the synapsing neurons

and the brain found meanings to grasp and distend through the oral cavity we cannot even guess, but they soon enlisted mouth, lips, teeth, palate, and pharynx to ejaculate them into the society they created.

Midwife of symbols and science, the tongue forged the alphabet from which hominids derived all others. Australopithecine or Pithecanthropine apes excavated vowels and consonants into the logic strings that now underlie Sanskrit, Apache, Arabic, and Welsh. All languages derive from a single lexicon ontogenetically—a submerged syntactic grid—and differ merely as alternate ranges of sound and breath.

The human world arose out of the animal world. "Unless we had been physically capable of gestures," declares Blechschmidt, "no speech, no verbal thinking, and no social order would have developed.[13] ... What we call instincts are direct continuations of prenatal developmental processes (i.e., ontogenetic but not phylogenetic events)."[14]

Culture is an expression of organs, even as mind is an expression of coalescing cells.

The Landscape of the Brain . . . .

The rapidly enlarging cerebrum dominates the early human embryo. Phalanxes of newborn brain cells feed and migrate by amoeboid motion, absorbing nutrition from the pia mater, a gauzelike vascular membrane shrouding other nascent tissue. Pushing one another apart, these clusters of mesenchyme cause a top layer of themselves to detach from its base, making space for other cell processes to emerge from deeper within. It would appear that membranes attracted to the pia's presence creep toward it.

Where mesenchyme around the emerging brain becomes more thickened ventrally than dorsally, the primitive meninx—an embryonic dural envelope—begins to congeal from it. As the pressure of the neural tube increases against it, the meninx dilates, ultimately forming a tough sheet that spreads multidimensionally along a friction plane to enwrap the entire brain and spinal cord in its integument, while generating hydraulic pressure from fluid trapped within it. Its unequal growth pull also contributes to the curvature of the embryo.

Pathways of blood-vessel formation irrigate loosening tissue inside the early brain and spinal cord. Interstitial components expand and vesicles fuse with one another to lay down aqueducts and a lacustrine network. As the neural tube materializes, additional neural-crest cells are induced out of it by the meninx and its wrapping. These cells, clarions of the nervous system, multiply and then travel atavistically, as mesenchyme is wont, reamassing regionally throughout the organism. Blechschmidt and Gasser describe this migration as tropismic, impelled as if by an inner agency, "not truly mechanical but very much alive ... as a cooperative action with several partners, especially the neural tube, primitive meninx, and blood vessels."[15]

Meanwhile, the thicker ventral leaf of the dura mater, the ectomeninx, continues to indurate and seal fluid between itself and the embryonic pia mater. Blood components percolate out of the tube through the interstices of the pia, ooze arachnoid fluid, and fill the first cisterns of the arachnoid membrane simultaneously along both the ventral aspect of the spinal cord and the basal aspect of the brain.[16] Named for its cobweblike appearance, the arachnoid (literally "spider") manifests like a delicate frost between dura and pia membranes. Accumulation of fluid projects mechanical dilation forces and differentiation and afterwards hinders further development in the same area.

The cerebrospinal waterway, filling with seepage, stores its accumulating reserve within impermeable dura, pia, and arachnoid layers. During embryogenesis it matures into an internal river, greater (in scale) than the Nile, as pure as underground aquifers, and as vital as the juice of a noni fruit. It becomes a channel along which purl the carnal aspects of dreams, inspirations, and healing.

The pulse regulating these hydraulics supports mammalian identity at a deep, rhythmic level. The skeletal-visceral axis responds to cerebrospinal hydraulics by rotating inward and outward every eight to twelve seconds, a cycle that continues over a lifetime. This cranio-sacral metronome assimilates and resolves psychosomatic events and harmonizes autonomic functions.

Internal liquors, generated in hemato-aquifers, brew a native medicine akin to the mercurial seeds sought by Parcelsus and John Dee. The wellspring at our center, they support our original mirth—which is why cures that ride the dura have such remarkable effects. Life, as it is being lived, imprints itself continuously and synopti-cally in the cerebrospinal fluid to be mirrored, cleansed, and trans-muted.

The various biorhythms and tides (the collective inhalation-exha-lation/flexion-extension/respiratory/circulatory/craniosacral cycle) comprise a quantum system transcending simple physiology or physics. According to William Sutherland, founder of cranial osteopa-thy, the cerebrospinal fluid—potentized and transmuted by an invis-ible bioelectromagnetic field originating *sui generis* in the ovum and blastula—is the chief ordering principle of the embryo, the blood-based, blood-transcendent partner of DNA in the vital sphere.

Universal intelligence infused in the brain's "blood" also implants its blueprint as an organizing hologram in stem cells throughout the blastula. These embryonic coins, the multipotential progenitors of cells that later specialize to construct tissues and organs, absorb the therapeutic potential of the bioelectromagnetic field—continual uni-fied flows of "liquid light"—and discharge its Breath of Life through all the tissues of the body.

· · · · ·

A tug on the neural tube from its own membranous coat induces the formation of mesenchymal crests with their own sensory gan-glia and dorsal roots. Harbingers of a full-blown "fiber-optic" net-work, the neural-crest cells, having been summoned ventrally by the thickened and restraining meninx, come eventually to align in a ser-rated band on both sides of the tube. Representing primordia of gan-glia and nerves, these ridges elevate themselves out of the primitive spinal cord at spots where capillary veins also contact arteries (see

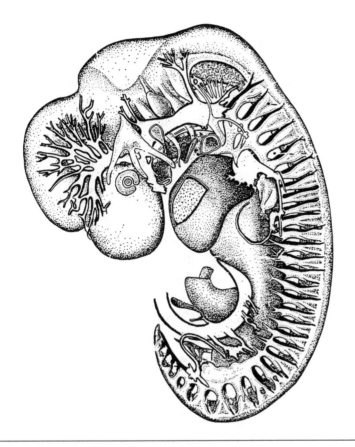

Erich Blechschmidt Collection: relationships among veins, nerves, and spinal ganglia, culminating in the mesencephalon at the caudal end.

pp. 338–340). These missions are carried out with stunning synchrony—blood, nerves, and spine coordinating one another's appearance and integration as if syncopated by an outside conductor.

Nerves are thickened skeins dynamically pulled out like taffy. As the membranes of their fibers are drawn by longitudinal and circular shear forces, the filaments string electrical pathways. The optic nerve, initially stubby and thick, attenuates as the expanding brain forcibly extends the distance between the nerve's point of origin and, at perigee, its insertion. Fields of mesenchymal tissue over the cerebral hemispheres and between the emerging mouth and eardrum similarly guide ophthalmic, maxillary, and mandibular forks of the trigeminal nerves to sensory insertions remote from their original sites. Chemical stimuli synapse along these pathways. The upstream flows of their signals are coordinated miraculously in the brain.

Distance and context, projected telescopically to the remotest reaches of the universe, are generated in ratios of cells comprising mere millimeters.

• • • • •

Further enlargement of the brain is induced by its irregular adaptations to the functions of mesenchymal tissue. A variety of structures stimulates its differentiation locally while restraining it globally. Parts of the dilating central ganglion collide with each other and fissure in such a way that two components of the dura mater, the falx cerebri and tentorium cerebelli, individually balloon. Biomechanically the falx is a girdle between cerebral hemispheres against the growth tug of descending viscera. Enveloping the cerebellum, the tentorium braces the cerebral occipital lobes. While the brain both bends and folds, the spinal cord—its nerves rooted peripherally in the body wall—stretches into a narrow vertebral canal along the vestige of the primitive streak.

As components of the primitive dura impose restraining cinctures on them, lobes of the brain explode convexly and antibasally into one another. Under its corset the cerebral surface bulges and fluffs unevenly, interthreading fontanelles between unfinished bones and differentiating in subtle layers. Where mesenchymally orchestrated eccentricity of growth is greatest, mysterious gray matter collects in cerebral hemispheres, its field of quickening intelligence sown perhaps by the atavism of migrating cells, perhaps by cerebrospinal fluid bearing the Breath of Life.

• • • • •

The young ectomeninx matriculates into a wafer of connective tissue that divides into twin sublayers, responding to the ossification of an aspect of its outer surface with the pull of the orbital septum in the direction of its inner aspects. The upper plane yields to the attraction, condensing, hardening, and losing growth capacity; the lower one merges with a field equalized by the expansion of the arachnoid.

Tissue forming a mantle around the meninx proliferates from ossification points and provides a grid for protective structures surrounding the brain and its spinal nerves. Lines of force radiating outward cause young cells of the bony skull (osteoblasts) to tilt forty-five degrees against a surface of osseous substance, itself the result of massive biomechanical shearing. As their pressure is resolved in contusion and calcification, they kern a helmet around the ganglion.

Blechschmidt interprets our fetishistic attention to the head, expressed through hairstyles, jewelry, and ceremonial crowns, as reverence for the life-sustaining activities inside the cranium.

Creatures Elsewhere . . . .

By as-yet-unknown methods of propulsion, terrestrial life forms may someday cross the 4.4 light years to our nearest stellar neighbor, Alpha Centauri, find there an Earthlike planet orbiting the star at a temperate distance, land their module on its surface, and scout for habitants. No matter how alien this world, how unrelated to our own, astronauts will expect to find flowering and vegetating stalks, various gnats and mites (some of them airborne), arboreal crawlers, quasi jellyfish and lampreys, sharklike predators, as well as facsimiles of whales, walruses, rodents, simian hunters, and amoeboid membranes in dew—because it is unimaginable how else cells or their equivalents might matriculate, metabolize, and organize.

They will expect to meet phalanges and faces, for biodynamics and metabolism provide limited occasions for somatic hierarchies, conduction of signals, sense organs, and proprioception. Placodes ripening into eyespots, ear canals, nostrils, and legs guide embryogenic masses everywhere, through thermals and across perilous ground. Tissues orient by fore and aft, left and right, extending themselves over gaps, bearing spring-loaded tools. Without limbs and feelers, animals could not discern where or what they are.

There may be other ways of packaging molecules but, until we encounter them, we will expect heads and eyes, gullets, metabolizing tracts, and limbs with digitalization. Critters are biomechanically inevitable.

One day we may have a cosmic biology in place of our provincial one; then we will know how universal not only are heads and intestines but DNA and its helices.

"Human dress could be called the oldest anatomical atlas ever published." . . . .

On Earth cytoplasm, nucleic acids, and membranes are the first components of an ecosystem, interpolating through epithelia, tissues, and viscera into organs and nerve nets; then minds, cultures, ceremonies, cities, and civilizations. Biomorphology may begin in a quantum realm of molecular particles but, as it compounds while spiralling and splitting, it engages vectors of resistance and avenues of possibility and responds to them morphogenetically on (relatively speaking) macro-scales. Biocomplexity is ceaselessly projected into escalating domains of exterior geometry that it also invents.

Unpredictable designs occur on planets and diverge in membranous fields they themselves are organizing. They convert tissues into artifacts. Their expanding substance comes to include cells and symbols both. They manufacture a) fissioning cells; b) the meanings generated by communications between cells; c) morphogenetic feedback; and d) transcellular dialects that pull consciousness out of matter and dispatch it through tissues such that embryogenic networks shape objects and artifacts above the level of themselves.

In other words, organic shape reflects the world the embryo is emerging into as much as the semi-pristine nucleus it is arising from. An embryogenic field spins out metadimensional tissue webs which reverberate and thicken by contact with terrestrial obstacles and occasions, turning calibrations of these into even more complicated structures, as nonlinear information ripples back against the cell membrane through the cytoplasm into the oracle at the nucleus.

Oak trees, horsetails, snails, and llamas are multicentric projections of the same clusters, the same basic nucleic acids and cytoplasmic fields encountering unique resistances and opportunities. They are the individualized fruits of deep attractors, nonequilibrium basins, and fractal gradients.

· · · · ·

Organisms made of cells evolve transhistorically; develop organs, nerves, and connective tissue; swim; leave the sea; sniff at shoreline; enlist in colonies; fly; manufacture honey from nectar; stalk; steal eggs; shuffle up trees; hide; crouch; lope; split rocks and bones into tools. The biological field continues to disseminate (outside in, inside out) from limiting membranes to tissues and organs, fur and claws, from viscera in and out to clothing, houses, roads, cities, and *these very words,* none of which existed on this planet before cell complexes began to flow out and radiate back and forth through tissue masses.

Human philosophies are ongoing, deepening extrapolations of molecular mass and structure through outward-spreading morphogenetic waves travelling from cytoplasmic beakers and tissue networks at their core. They are mind-centered, yes—but they percolate collectively unconscious, subliminal acts.

After all, a limiting membrane long ago did not halt the exponentialization of the nucleus of the cell.

· · · · ·

The tissue boundary of an organism is sealed in an epidermis, but the body projects its neurovisceral designs into objects of nourishment, protection, power, and healing; into artificial structures that are noncellular but designed by cellular intelligence, cellular protocol, cell aesthetics: handaxes and yurts, blankets and jugs. These are proxy, prosthetic organs. Garments and sanctuaries of furs, silk, wood, stone, leather, and textile continue to reflect developmental resolutions of extrinsically rippling fields. When organs and their neurons encounter snow and winds, they incorporate plant and animal materials—vines, seed fibers, leopard and beaver skins—into haberdashery, even as they unconsciously generated and embodied mesodermal and ecto-mesodermal layers in self-organized contexts during their evolution. The protective layers that a primate pulls about himself/herself unconsciously reenact the cell and tissue evolution at his/her basis. A symbolic field ultimately takes over as the vehicle by which a morpho-genetic field expands and hegemonizes matter.

Blechschmidt identifies the origins of culture when he says: "Human dress could be called the oldest anatomical atlas ever published."[17] Indigenous peasant costumes are extensions of embryogenic waves as well as analogues of conch shells, pods, pelts, manes, coats, epidermises, feathers. The fillips and frillery of human wardrobes are not unlike the gaudy, multicolored raiments of underwater animals, the feathery accoutrements of birds. From a cosmic perspective, clothes represent increased biodynamic layering—a further molecularization of the environment through the nanomachinery of the cell.

The cuts, fabrics, angles, and colors of high fashion, along with their risqué and titillating combinations of what is revealed and concealed (a slit, for instance, from ankle to thigh in a woman's dress, a low-cut bustline, etc.), are further refinements of the metabolic and reproductive (i.e., erotic) accession of space in grids of protein. They represent nostalgia for the enigmatic shapes designing and underlying bodies. Alluring fetal spirals are projected into equally winsome couture.

* * * * *

Mind is a phase in astro-embryogenic evolution, not a break in the continuity of mass and energy and not a flight from biological destiny into abstraction. Blechschmidt identifies "correlations between the micropopulations of cellular aggregates (therefore of smaller organisms) on the one hand and the macropopulations of human groups (socialization in larger dimensions) on the other." This suggests to him that "the modes of behavior of macropopulations are not completely different from the modes of behavior observed, *inter alia,* in the cells of the human organism even if they are never the same. Unless we are able to react somatically with living processes within our bodies, we could not take our place in society to act there." Then he makes his favorite point: "Without living, bodily processes, no higher human achievements leading to professional differentiations and cultural developments would have been possible."[18] Human bodies are necessary for human culture to occur. That is obvious in the cardinal sense, but it is also the rule in every actual instance of a specific human act and the tissue complex underlying it. Embryology provides the seams, grids, and generative syntactic layers along which creature artifacts and activities emerge.

Morphogenesis turns bands of hunters burning scapula bones into magicians in antler headdress round-dancing under a full moon, into farmers digging channels from rivers, into smiths working hot metals until they become bracelets and bells. Rows of wheat and rice are planted by cell clusters seeking nourishment through organs for tissues trapped in the cellular deep. Castles and cathedrals are raised by cells to garnish and expand themselves and their realm. Factories, villages, and roads are offshoots of thousands of cell clusters using tongues, breath, blood, neurons, and symbols and musculoskeletal tools in collective metabiological enterprises.

The manufacture of bark, hide, and horn into fabrics and dwellings leads to the derivation of levers and wedges, pulleys and gears, and cells adapt these for lifting and measuring. The smelting of iron out of bog ore in coal-fired ovens lays the basis for cooking vessels and coal-pot laundries; the molding and alloying of ferrous and non-ferrous metals into manifold devices; the ontogenesis of coins and money; the compositing of aluminum; the forging and hot rolling of shapes and massive design structures from metals; the fusing of silicon into glass; the erection of edifices of steel with reinforced concrete; the synthesis of rubbers and plastics; the extrusion of nylon and rayon bolts from petroleums; the chemical transformation of fabrics; the pulping and suction of vegetable matter into paper and typography; and ultimately the resurrection of the coal-pot as the four-cycle, internal-combustion engine; then the generation of electricity from steam; and the wafering of electronic transduction circuits for automatic machine control and the assemblage of electronic circuits in central processing units.

All of these events are collectively and fundamentally embryogenic. They translate tissue layers into mineral and textile layers; they go from the iterative exercises at the nucleus of the cell to molds of mass-produced plastic designs. These could not be achieved by individual cells, individual organisms, or even generations of bionts; they are accumulated incrementally epoch by epoch, as each new life form and birth class inherits the activities and ever-unfinished designs of its predecessors and translates them into new ones. They are embryogenic not only in their cellular derivation and intrinsic application of metaphors derived from tissue dynamics and designs but in their own systematic, layered, introverted, exteriorized industrial development molecularly and seamlessly out of one another.

Tissues and symbols, tools and machines are expressions of one embryogenic field, as autonomous agents spread "molecularly, morphologically, and technologically in untold, unforetellable ways persistently into the adjacent possible."[19]

• • • • •

It looks like "invasion of the body-snatchers" here. Pods everywhere. Pod City. Movie theaters filled with pods. Pods locomoting bipedally down streets. Little pods running the gauntlet of big pods. Pods building pod factories.

Are not cities incredible things for molecules to have assembled— massive nodes of glyphs and thoroughfares in layers and labyrinths? Red Square, Times Square, the temples of Lhasa, the plazas of New Delhi and Mexico City correspond to tissues and viscera stretched across vast, psychosomatically differentiating fields. Aerial photography makes their biogenesis perceptible. We see shells, mesenteries, metabolic centers, limiting walls, arteries and, at night, neural clusters and synapses.

Topokinesis

Walk down the paths of any modern metropolis or ancient city—even a small village. What you see are the warrens of creatures, lit at night, cubicle by cubicle, in degrees of shadow and lamp or fire glow, each filled with the accoutrements and prostheses of that creature and its kin, its hive-mates.

Why are they all here, living where they are? How have they ended up in these circumstances, luxurious and dire, all within a stone's throw of one another?

They have to be somewhere.

They exist as continuously expanding and bifurcating cellular events, stealing molecules for their masque. They have been, each of them, assembled, melded, forged dynamically in layers, and then born out of space into space.

Consciousness is its own explanation. So is body. So is society. They make themselves.

Creatures have to frame a way to live and pass the time, while using the resources at hand, in this state in which they have found themselves. If they were not here, there would be no "here."

It is all topokinesis—the ceaseless migration of stellar hydrogen outside/in traceries of originary designs.

9

Tissue Motifs and Body Plans

. . . .

Coordinating Form

> For wool-carders the straight and the winding way are
> one and the same.
>
> —Heraclitus, 540–480 B.C.

Tissues transfer energy and shape into other tissues. . . . .

The organs of the living body are not solid objects but the outcome of reciprocal forces interacting semi-liquidly. Like sea squirts and sponges, they require an interior, watery milieu to exist at all, a sealed space that is created embryogenically solely by their emergent relationships to one another.

A sun jelly stranded on the beach by the tide quickly eviscerates into sedimented salts. Its umbrella has no intrinsic integrity, no macrostructure. Likewise, tissues have no organic unity outside of their interior habitat.

Suspensory and buoyant factors in the body reduce the weight of all tissues by half the amount of equivalent structures outside. Subatmospheric and structural forces push each organ up. If there were no other pressures diffusing into the spongeous mass of the lungs, they would collapse under their own weight; yet the hard frame of the thorax impels them out, establishing billowing breaths by its resting state, imparting a subtle rotation that moves the lobes apart and brings them back together.

Morphogenesis establishes "being" as one transformation after another; life is ceaseless compensation and adjustment. Nothing in a body is stationary; every encounter with proximate mass elicits responses right down to subcellular levels. During their fetal development, organs not only have the capacity to change shape and position to accommodate the shear forces impinging on them, but that shape-changing is the biodynamic basis of their very existence. This capacity is not lost but reduced as the organism approaches parturition. The morphogenetic trajectories of organs like the liver, lungs, stomach, and spleen become their active metabolic motilities.

Our tissues were spun not so long ago, and we still experience their torque, the long axes of their phylogeny and ontogeny. We can feel the cerebrospinal cradle rocking. If we are quiet enough, we can track the neural groove's rustle from the third eye up over our skull and down our spine to our sacrum/tailbone. Its vestige runs just beneath the surface, pliant enough for shamans and yogis to rekindle its seeds.

• • • • •

When organs impinge on one another in a mature body, each tries to occupy maximal space. Mass and gravity press against the boundaries of cavities. Hollow aspects sagging toward collapse are pushed outward or fill with fluid. Relationships among tissues transfer energy and tension into other tissues, for morphogenic patterns beginning in embryogenesis impose themselves throughout life. Each organ seeks to glide over its neighbors with a minim of friction but invariably is restricted by, restricts, and deforms them too.

As forces in the external world impose their own bumps, torques, whiplashes, and other insults on mobile organs, their shapes flow through cavities to lodge in soft tissues. When viscera absorb distortions, they develop and then project energetic cysts and other blockages, scars, and irregularities that restrict their ranges of motion and the ranges of adjoining tissues. Strains that cannot be resolved kinetically gradually line up in bunches as immobile fibers, while still-elastic tissues try to translate force away from any knot and dissipate it.

Everything attached to an organ must follow that organ's axis of movement. The range of growth and rotation of a lung or intestine must accommodate all its adhesions and relations to adjacent organs insofar as it transposes its own stress patterns into them and is the recipient of *their* stresses. These various tensions, fulcra, and waves continue to interact in the dynamic recomposition of life forms throughout their existences.

Mechanical tractions synergize with psychosomatic currents in the genesis of emotional and energy channels, determining relative health, pathology, and ductility. Viscera under protraction, in order to release pressure, redirect unwanted forces along any available vectors, rotating and side-bending, shortening distances between each others' boundaries, pushing up convexities, taking stress off unduly compromised components. This is also how they attain shape, position, and orbit in the first place.

• • • • •

Resembling sashes and other ballasts, tissue pathways pull tight and kink over distances. Deflected progressively from fiber to fiber in a morphogeneticlike pattern, an intestinal toxin or cyst can end up in a shoulder, a kidney fixation in the knees. There it becomes an ordinary part of anatomy, habitually registered in daily kinesthesia, integrated as gracefully as possible into the life plan. A musculoskeletal congestion may be only a secondary line of self-protection. The parochial result may be restriction and limitation, but at another level it is function, albeit hampered or reduced function. The same energy, if unanchored and unabsorbed, might threaten the organism more systemically.

Disease usually shows up in the body not at its actual genesis or site of pathology but where the field is most in danger of losing its capacity to compensate. For instance, fixations that reside in the musculoskeletal system often are phantom pains, transposing distortions from more vulnerable tissues inward along contusions and detraction paths, triaging damage to sites well removed from the activating injury or pain.

To heal pathologies one must try not to break through survival-based defense mechanisms, at least not without identifying and attending to the thing they are safeguarding. What seem like dysfunctions in an organism are often highly designed, multiply redundant systems incorporating layers upon layers of feedback and backup functioning. Any attempt to "cure" or normalize tissue without accounting for the vulnerabilities it is protecting can lead to deeper resistance and more intractable knots.

The properties of proteins transcend simple physics. . . . .

The medical system of osteopathy originated during the 1870s as an attempt to apply engineering principles to the shear and fluidic fields of living organisms. In osteopathic terms the skeleton is a structural frame with a network of soils, levers, dikes, waterways, and springs. Lungs and heart are bladder-pumps; meninges of the brain and spinal cord are closed fistulas; other viscera are lobules in constant motility, rippling and semi-rotating, balancing competing forces.

A remedial medicine rather than a mechanical intrusion to preserve life and/or function, osteopathy conducts a sympathetic embryogenic operation, using a patient's own tissue resistance to guide the soft, layer-gathering knife of the healer's palpation. By contrast, severing viscera surgically with a scalpel compromises dynamic homeostasis and introduces scar tissue—though sometimes it is the only antidote for advanced pathologies.

Osteopathy performs a purely energetic replica of surgery, using the tissues themselves to rearrange their own mass and equilibrium. Palpating at a subtle level, the trained hand learns to melt into another's tissues and sense and track their own self-organizing principles. A living structure has a direction, a rhythm, a cycle, a vibration, a texture, and tendencies of sliding and adhesion, all of which can be supported or steered, even through the skin, musculoskeleton, and other organs. The lungs can be gathered up in fatty tissue, ribs, and pleura, and then rolled gently in their orbits. The liver can be rocked back into place. The heart can be irrigated by stacking and gathering the tissue forces around it (dermal layers, ribs, mediastinum, pericardium). The hand melds through these envelopes until thump-thump-thump is skipping along its surface.

The goal, always, is to restore embryogenic position and topo-kinetic potential.

• • • • •

The osteopath's strategy is to follow biodynamic vectors and release them, one by one, both directly (torquing back against the direction of compression) and indirectly (tightening compression until it releases itself). Despite the many-levelled torques and knots of somatoemotional cysts, only an extremely light touch will track deeply enough and disperse the forces holding them together; only guided attention will convince tissue itself to respond embryo-genically.

Practitioners gathering membranous layers while stacking and following force lines leading into energetic cysts need add only five grams or so along the way to standing tissue buoyancy. First the fingers find, match, and gather the resistance; then they contribute their own increment while continually pulling in the slack. Any more force will insult the structure and compel it away from the innate currents that support its core pattern of organization, rendering the therapist's application an additional intrusion which must be armored against. Nonmelting touch that lacks sensitivity to visceral texture and standing energy will actually increase the types of congestion and adhesion the physician is trying to relieve.

• • • • •

Beginning historically in the nineteenth century with simple inspection of the dynamics of tissue, blood, neuromusculature, and skeleton, osteopaths soon came upon a life current, i.e., a rhythm and cadence that distinguish animate from inanimate matter. Manual doctors were palpating something they could barely hold, something that did not even concretely exist.

Within a few decades, practitioners of a cranial branch of osteopathic science, following their mentor William Sutherland, ventured beyond simple matter-and-energy mechanics. They conducted an experiential inquiry into congeries of subtle energies and self-organizing patterns that compose life fields. To identify the principle of life solely with dense anatomy and its immediate metabolism of blood and air overlooks many layers of neurovisceral energy—the cohesive charge holding tissue masses together.

In the 1970s John Upledger shifted the focus of therapeutic touch from correcting mechanisms and adjusting structures to unprejudicially following energy and visceral disposition. His intuition was that an "inner physician" in the body/mind of the client would naturally lead the practitioner to a potential cure by pulling the hands along, instructing them in how to undo its encysted pattern, eliciting from them exactly the combination of support and activation it needed. It would use touch thermokinetically.

The popular name of this modality—craniosacral therapy—certainly does not mean that its denizens work only on the cranium and sacrum; instead it honors the discovery that the small bones of the skull are motile—a property refuted or ignored even today by most allopathic doctors. Blending along pathways that tremulate with cerebrospinal rhythm draws the practitioner's touch to the heart of viscera where she can palpate inertial blocks. Riding the cerebrospinal pulse— placing her hands gently on a bird taking flight—she kindles or enhances the transmission of the Breath of Life to all the organs.

Craniosacral therapists carry out out a repertoire of dynamic techniques: palpating fascial webs, finger-milking cerebrospinal fluid, enhancing the flush of lymph rills with gentle strokes of fingerpads, adding grams of momentum to tissue resistance, and following organ motilities to the peripheries of their embryogenic ranges where they anchor and support them before releasing to stasis. They engage the pyramids of life in their own terms, along their innate grain.

Once healers experienced and followed the resonance of myriad currents of "Liquid Light," they realized they were inside the vital principle, handling creation itself. Some imagined a biophysical current or identified the pulsation with Wilhelm Reich's orgone, Taoist *ch'i*, mesmer, Vedic *prana,* etc. Others felt they were grazing against pure spirit, figuratively shaking hands with the soul.

Here medicine meets religion, foreshadowing that in a true utopia or on some other world, medicine *is* religion. Without abandoning the skeleton and mechanical physics of tissues, esoteric surgeons had encountered a vibrating grid of life permeating everywhere—Fourth and Fifth Laws transcending ordinary thermodynamics.

Manual medicine gradually evolved into hands-on shamanism. What started as a manipulation of structure became reinduction of the forces underlying the growth pattern of its embryo.

Now osteopathy (whether by name or, more often, under other ensigns like manual medicine, Polarity, Craniosacral Therapy, Zero Balancing, Fascial Regulation, and Myofascial Release) operates as a phenomenological, empirical science of biodynamics—perhaps the forerunner of an embryogenic science of the future.

Heart, Liver, and Other Glands and Cervical Structures

The human interior is molded by a series of multi-levelled topokinetic responses to suspensory forces impinging on cell masses and by interactions of shear planes crossing and plaiting.* Coelomic fluid exerting less pressure on the outside of the cardiac wall than blood with its higher osmolarity does on the inside, a primitive heart motif is stimulated to expand and broaden more or less spherically in relationship to the longitudinal axis of the cardiac tube. The tube then lengthens and widens from the variance between pressures inside and outside it; it is the only embryonic capillary in

---

*The embryology that follows, like that in the previous chapter, is mostly Blechschmidt's topokinesis, investigative science as biodynamic theory.

the thoracic aspect of the coelomic cavity having basis for such expansion and morphology.

Adhesion under stress to the thoracic walls at each end twists the twin buds of the formative heart in an omega pattern, as the connector yields first convexly to the left and then ventrally to the right until it dwarfs its inflow and outflow valves and realigns left to right in relation to the dominant axis of the coelom. A bilateral shearing motif also underlies the separate formation of left and right ventricles, with their connecting rod warping from its adhesion to the primordium of the diaphragm, which separates as an intermediate layer compressed between the heart and liver.

Pumping of blood through this system becomes the engine of bodily circulation. A hematic river soon provides nutrition to every tissue-bound zone, spreading a radius of empathy and compassion— literally essential flow and phenomenology from core. A pericardial sheath imposes a bilayered, fibrous boundary around this irrepressible muscle to keep it from hegemonizing the whole fetus.

Blood is mammalian will. Heart is an ancient, primary organ disposition, brain's correlate and other. Its atavistic fury defends animal existence unto death. "Take a bone, take a foot, take my sight, take my memory, but I, I, I beat. I am."

Blood vessels emerge within tissue wherever a positional relationship to a nutritional source develops in the context of a thirst for assimilatible, energy-bearing molecules. The first veins arise as fluid vacuoles coalesce in intercellular matrices, while primordial arteries germinate as endothelial sprigs that penetrate interstices between cells. Once a blood vessel inhabits a tissue, it functions as a restraint. In general, vessels tend to "cause a growth resistance that is directed against the organs they supply."[1]

Bridled by dorsal aortae lying closest to the neural tube, the brain bends in a concave hood over the young heart. From the burden of this flecture, ectoderm and endoderm on the ventral aspect of the head and neck fold into pharyngeal arches (see p. 333). Between adjacent arches at sites of mesenchymal activity, cells that are trapped where pouch and groove are in partial contact corrode.

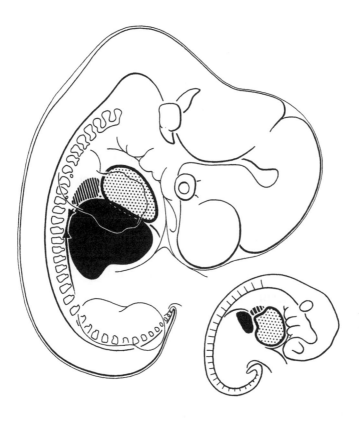

Erich Blechschmidt: development of the lungs showing the restraining function of the diaphragm (liver black, heart stippled, pleura striped).

In general, where voids occur within limiting epithelia, suction fields form, loosening tissue and allowing fluid to seep in. As cells from proximate zones invade these newly opened territories, glands are fashioned—the partially recapitulated descendants of marine invertebrates. These range from tiny sweat organs blistering throughout the skin to whole oyster-like invertebrates: lungs, liver, pancreas, and kidney.

The primordia of glands tend to exploit the direction of least resistance, their cells diverging in wide conical or club patterns, while percolating into loose surrounding tissue. Without cell crowding, which is relieved by gland formation, there is no basis in the same vicinities for the formation of additional structures.

Mitosis in a gland's epithelium occurring more intensely than in its excretory duct causes a lumen to winch interiorly. As fluid is released by the epithelium, the gland tip recedes toward the lumen, and congestions of intercellular substance crumble the organ into a dichotomized, comblike structure with its own particularized chemistry. Similar branched formations fashioned by limiting tissues around the lungs and kidneys become the bronchial and ureteric trees. The same essential topokinetics is imprinted in the soils of river systems.

. . . . .

As the fetus expands, most of its ectoderm dilates caudally, gliding on underlying tissue in the direction of least resistance toward the apex of the trunk. Across this spreading layer, protrusions of basal epiderm begin to rise and tilt, leading to rolling bumps—hair primordia. These fine sprouts are drawn into unamalgamated tissue, ultimately inclining forty-five degrees against the plane generated by their own rolling movements.

Hair is made of animate cells too; as Thomas Hardy wrote of Eustacia Vye: "Her nerves extended into those tresses, and her temper could always be softened by stroking them down. When her hair was brushed she would instantly sink into stillness and look like the Sphinx."[2]

That Sphinx dwells inside all adult postures—the mute bodyself.

Undifferentiated tissue between the epidermis and each emerging hair follicle compresses while loosening in the obtuse angle on its opposing side, giving rise to the arrector pili—smooth muscles attached to hair follicles and dermis that contract to lift hairs. Meanwhile, tiny fields disengaged by fluid in the angle between the arrector pili and the epidermis develop as sebaceous (sweat) glands perpendicular to the epidermis.

• • • • •

The heart, at first no more than a capillary fold protruding from the dorsal wall of the cerebral section of the neural tube, attains its mature shape almost solely as a function of its position, expanding as the neural tube dilates, and bending asymmetrically. A capillary channel emerging in the cardiac mass protrudes ventrally along a trajectory of least resistance.

Because liquid under pressure begins flowing through all the tissues of the body, the heart materializes where it does, at a critical juncture of emerging channels. Vascularization of the rapidly growing brain also forces the muscular pump to augment itself, its gush waxing as the basis of an expanding circulatory system. The surge of these primitive creatures through its matrix—blood—partly induces the heart's shape and texture. The organ then attracts more and more guests until they are a torrent to its beat.

The trunks of large blood vessels feeding the neural tube accrete tissue more slowly than the tube itself, so these structures, balancing their own visceral tension and enclosed by the body wall, add to the force that causes the swelling head of the tube to bend over the cardiac prominence. A huge exocrine "gland" budding from the caudalmost part of the foregut and subsequently induced by intense hemopoiesis coopts the adjoining region, extending its unique adsorptive capacity. Topokinetically the liver forms where three tissue vectors come together: the visceral arch artery, the heart pushing against the thoracic wall, and the gut. It then receives nutrients from the gut to process in its labyrinths, synthesizing and degrading their chemistry before secreting and transferring their altered components to the blood for transport to cells throughout the body.

The titanic liver is a hologram; unlike most other organs, it regenerates itself. All hepatocytes specialize identically, secreting the emulsifying agent bile to break down large globules of fats and complex proteins and metabolize foreign matter. If two-thirds of a rat's liver is pruned, two weeks later a full organ reconstitutes itself from the remainder.

Accumulating under local exigencies of cell growth and differ-

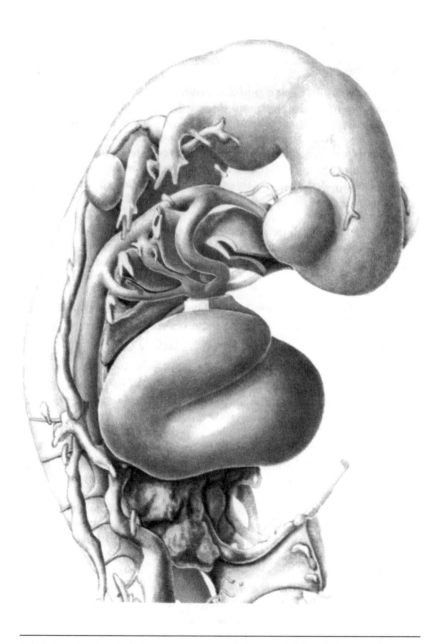

Erich Blechschmit Collection: role of the early folding of the embryo in the formation of aortae and arteries.

entiation, the liver and its storage chamber, the gallbladder, initially form a *cul de sac* on the caudal foregut with the bud of the pancreas. Then, as food pours into the cell colony, strands of liver cells, interwoven with blood vessels and sinuses, begin rapid expansion and production of bile for its digestion.

The liver, pancreas, and gallbladder define themselves in respect to one another as different topokinetic pressures shear out ducts, stalks, and lumens in them. Local chemistry and the chemistry of the globules passing through them help induce their specializations, as they spill their corrosive hormones into the stream of the intestinal tube. Not only does metabolism always have a mechanical and energetic as well as a biochemical basis; on a molecular level, these are the same.

Proximal to the embryonic heart, the liver is substantially defined by its position. Insofar as it directly receives blood that has just absorbed nutrients from the gastrointestinal tract, its hepatocytes filter as well as transmute the thickening broth before it can clog circulatory nodes.

It must grow for morphodynamic as well as metabolic reasons. Its massiveness and specialization are dynamic responses to both the fetal heart and the brain. The cerebrum's mushrooming taxes the circulatory capacity of the heart to nourish it, so the heart pumps out more blood and enlarges itself too. Cerebralization impels cardialization which induces hepatization.

. . . . .

Mesenchyme between spinal nerves attenuates as the heart-liver expands. From the rapid, disproportionate dilation of the cardiachepatic bulge, migrating mesenchymal associations are pulled unequally and longitudinally to generate ribs of different sizes as well as absences of ribs across emerging cervical and lumbar gaps. As the ribs diverge ventrally and caudally, lungs and thoracic wall are tugged into position.

The swift but irregular growth of a brain and spinal cord stretch and distend the body surface dorsally; meanwhile, emergent cardiac and hepatic organs extend it ventrally. These relationships underlie other regional thickenings and thinnings and cultivate a mesodermal field with a lateral cell mass from which the skeletal grid, peripheral nervous system, and locomotor apparatus will emerge.

After the thin central tendon of the diaphragmatic plate congeals as a slat of constricted tissue between the two expanding organs (heart and liver), continued maturation of the liver flattens the diaphragm caudally and ventrally, depressing it from behind the heart. The same compression also sculpts the nasal and bronchiopulmonary aspects of the respiratory tract. As the heart tractions down from the cranium, it collapses while foreshortening the wall of the neck relative to the brain, thorax, and abdomen.

The collective descent of viscera countervails a beanstalk-like ascent throughout the central nervous system. The neural tube and notochord vestige uncurl synchronously in a rising path. A homunculus/dwarf shape is fashioned by semi-cyclonic streams resisting and molding each other in counter-longitudinal flows.

Out of this baseline semi-symmetry, cauldrons of cells, surging against their own accumulating density, gradually forge pathways for broad neuromuscularization, leading to longitudinalization through the nervous system and limbs. The organism thickens while allometrically reformulating its fishlike grid. So a hominid homunculus fights its way out of a "paper sack" into the cosmos.

The Lymph System . . . .

Throughout the intercellular milieu, submerged cells become inundated with surplus nutritive materials. In response to this landlocked metabolism, featherfine unorganized pathways sprout everywhere like rills in rained-upon mud. Without such conduits composing a primitive lymphatic system, cells could not survive their own edemas and autointoxications for more than forty-eight hours. Fluid pressure from emerging connective tissue gradually causes protolymphatic tributaries to link up, forming (parallel to veins) a second organized network back to the heart.

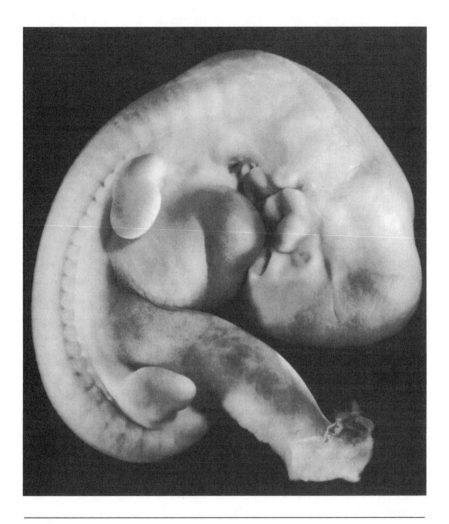

Erich Blechschmidt Collection: Budding blood vessels coming from the heart restrain the disposition of the extremities and divide them into upper arm, lower arm, and hand; upper thigh, lower thigh, and foot. The blood vessels also restrain the skin at the free edge of the tissue, thus providing the occasion for finger building.

As drainage vessels emerge, their fragile one-cell-thick capillaries, anchored by microfibrils, weave a tight net. Ducts covering most of the interstitial environment of the body's organs then coalesce into a closed overflow network that cleanses the intercellular environment of the toxic by-products of cell activity: germs as well as foreign bodies, protein debris, and general pathogens. Immersed in mild pressure fields of lymph, lumps of matter are gradually irrigated out of connective tissue. Immunity and lymphatic drainage develop in concert.

Oozing through its rills in a slow rhythm at low compression and velocity (at least by comparison with blood), embryonic lymph is channelled into purification centers where it accumulates, forming nodes in which its fluid becomes plentiful. In four hundred to seven hundred such bean-shaped sites throughout the body, waste particles begin to be broken down and prepared for flushing in other tracts.

About half the body's lymph nodes bud in the abdominal region, many of the remainder in articulation folds where their delicate structures can be sheltered by the curled fetus.

A furrow between the face and the cardiac bulge causes tissue there to act as a collar on the ventral side of the neck, impeding lymph flow from the head. Local congestion aligns mesenchymal cells tangential to the network, biodynamically proliferating sites in the neck. These drain into the chief vessel and gateway for the entire lymph system, the thoracic duct, which exits into axillary nodes under the arms. Organs flushing themselves through this watershed include the right and left lumbar trunks, intestinal trunk, intercostal trunk, and convex and posterior aspects of the liver.

For Blechschmidt, shirt collars represent the narrowing of the neck and broadening of the lymph-drainage system after the descent of the heart into the thoracic cavity. A necklace marks sternocleidomastoid muscles rotating and proferring the old reptile head. The male larynx is additionally reinforced by the knot of a tie.

Shrugging, Blechschmidt says, is the precise psychosomatic expression that "'there is nothing to be done, it is impossible.'"[3] It signifies the intersection of cranium, heart, face, and thoracic lymph duct in a gesture where meaning is both affirmed and negated. Not any particular meaning but meaning itself: "This message is independent of whether or not the meaning can be formulated logically in words, judgments or concepts, verbally or in writing, or whether it can be substantiated factually."[4]

The Renal System . . . .

As paired short trunk-wall veins detach and travel ventrally away from the neural tube, the mesonephric duct—which adheres to them—must follow until the cranial aspect of the duct is wedged against its caudal aspect near emerging bladder tissue. This consolidation provides an irresistible occasion for drainage.

The caudal end of the mesonephric duct is aligned by the curve of the neural tube. Tissue nourishing itself in the direction of least resistance collects in a metanephrogenic diverticulum where the duct bends the most; it then follows that line, parallel to the growth of the entire embryo. One of its surfaces is liquidic, the other solid.

Here geography becomes destiny, for—where the kidney will eventually mature—positional, morphological, and structural requirements impel particle movements along excretory pathways, opening a fan of calyces. It is the only place in the emergent organism where development and differentiation allow such a structure and, because biodynamic and metabolic opportunities have fused there in need of waste relief, a urinary network forms on cue.

The renal organ functions excretorily from its inception. As fluid wastes rarely stagnate, fetal tissues begin diffusing them through a primitive ureter into the top of the bladder. Preurine trickles through the wall of the bladder into umbilical vessels, a process which also enhances kidney differentiation.

The kidney, like other organs, is shaped and fated by its position and emerging use. "Functions can be exerted postnally," Blechschmidt reminds us repeatedly, "only when they have begun as growth functions in the prenatal period."[5] Permeability of the embryonic renal epithelium together with its strategic placement and preurinary activity requires it to release catabolites, thereby particularizing its metabolism. This local transfiguration, in turn, stimulates additional renal-cell proliferation, tissue enhancement, and organ growth and dichotomization. Reciprocally, the development of a drainage bed encourages increased production of embryonic preurine that would otherwise imperil the fetus.

Even taking into account these interrelated activities, it is hard to understand the near-perfect development of a finished organ—any organ—from raw tissue primordia. Knowing the answer in advance always fools us into thinking the embryo also "knows" at the level of cell conduction, hence dispatches the right docents. The premonition that an archetypal organ shape meets an emergent metabolic crisis (in a fractal landscape) is both accurate and misleading. An organ equally amasses and shears itself from scratch out of the materials, forces, and urgencies at hand. It develops work and dissipates heat out of simple thermodynamic principles and cell kinesis. Convection and gravity help establish the excretory specialization of renal cells, while the cells themselves differentiate according to the nuclear and topokinetic dispensations imposed on them.

• • • • •

Initially the outermost section of the embryonic nephron, trapped between the ureteric tree and renal capsule, differentiates into a glomerulus, a tuft of capillaries that filter solutions passing through them. A portion of the primitive nephron, hindered in surface growth, thins, while the remaining segment folds on itself into a vascular lobule. The narrowing end of the nephron lengthens and is squeezed against one of the branches of the nascent ureteric tree. A corrosion field is generated—the epithelia are compressed so tightly that they cannot feed their vascularized inner tissue, the cells therein perishing. Wherever limiting tissues (two epithelial layers in a thin double-layered membrane) are pressed too close for vascularized inner tissue to nourish itself, adjoining cells starve or suffocate at varying speeds. Their atrophy leaves behind a gap. As the tissues around these spots spring open, either fluids come into contact with each other, carving interior reservoirs, or adjacent tissues flow into the gradient shaped by the defect, while other shapes bend to accommodate.

Equivalent corrosion fields develop between layers of the cloacal membrane, between pharyngeal arches (as noted), and between seminiferous tubules and the testis. These also express irrigative functions.

Insofar as metabolic movements influence the shape and function of the organ, the urinary plane of the kidney becomes spatially oriented outward toward an orifice. As its gaps coalesce, the lumen of the nephron colonizes the mesonephric lumen of the ureteric tree. Waste-bearing fluid is released from the wall of the diverticulum into the ureter-kidney. When renal tubules fuse with calyces, metabolic necessity validates biomechanics. Implicit principles become explicit activities. Primitive urine trickles through the emerging tract.

Respiratory and Excretory Organs . . . .

The head and the cardiac bulge push out mandibular, hyoid, and laryngeal folds. These inner tissues furrow against either side of the esophageal aspect of the emerging intestinal tube. The fishlike pharyngeal pouches, according to Blechschmidt, are not recapitulated— or *not just recapitulated*—from oceanic ancestors but represent the only possible topokinetic solution to the relentless longitudinal expansion of the neural groove. They also participate in the bridling and bending of the brain and broadening of the embryonic face and mouth.

Endoderm of the intestine, dilating on its surface, invades the lateral pouches. Lung primordia are pulled into the thoracic cavity by the expanding chest, which is adapting as well to the thrust of the hegemonous liver. The expansion of the liver pushes against the dome of the primordial diaphragm (see p. 322), forcing it out of its placement behind the heart into the thorax along with the heart and then changing its angle between the heart and liver, making room for pleural sacs and lobes of lungs (see illustration, p. 323).

Later movements of the thorax, as it expands and exposes the pouches to shearing forces, will also slide and mold the lungs until they become pulmonary. The same motility reinforces the descent of the heart. Spatial changes radiating from the aggrandizement of the heart collapse the wall of the neck region in relation to cranial and thoracoabdominal brattices.

· · · · ·

Breathlike flutters kindle fetal breathing and incite life in a saclike float-bladder that materialized long ago alongside the ancestral heart among fish, though not for respiration. Lungs anteceded by many epochs the tentative migration of amphibians onto land, so represent an exaptation. The pulmonary float-bladder eventually became the pump of a second circulatory system, providing buoyancy while propelling a torso underwater and later transmitting atmospheric oxygen to cells in land-invading trespassers. Amphibians, with both lungs and gills, emblemize an intermediate state between oceanic and reptilian styles. As nascent lizards and dragons gradually committed to lungs and land, they sired both birds and mammals.

Blechschmidt adds: "The so-called first breath of the newborn infant is but a stage in the long development of breathing actions that begin with the origin of the lungs."[6] Inspiration is a by-product of the dynamic interplay of brain, heart, circulatory system, and pharyngeal rudiments (*vice versa* too). We reenact its archaeology every time we take another breath.

Directing mind and attention into pulmonary-cardiac cycles while deepening their range is a power-generating and curative exercise in yoga, *t'ai chi ch'uan, chi gung,* and other inner developmental systems. The respiratory aspect of breath melds with cerebrospinal and circulatory pulses to form the Breath of Life which harmonizes the cells and tissues of the body in a vital current (see pp. 303–304).

The excretory primordium originates at the beginning of the second month as a longitudinal bulge or growth adduction of the lateral abdominal wall—a midgut tugged by distal enlargement and proximal shrinking of the yolk sac until an early intestine herniates into the umbilical cord.

The primitive intestinal tube differentiates out of endodermal tissue, caudal aspects of other descending viscera underlying and stabilizing both its shape and developmental pattern. Growth, dilation,

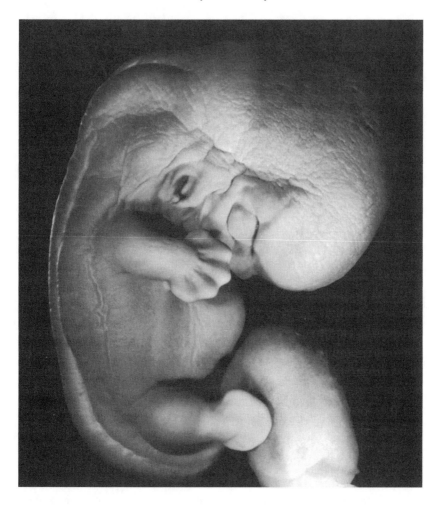

Erich Blechschmidt Collection: The lateral body wall that is visible at the sixth week remains thick between the rapidly growing skin of the stomach and the mighty liver as well as the also-rapidly-growing (and therefore thin) skin at the neural pipe. This also shows some of the dynamics underlying the vertebral column, vertebral arcs, and ribs.

and elongation of tissue inside the tube and around it enlarge its central lumen while burnishing it spherically.

Pressure from its own pulsating vessels helps vascularize the intestinal wall and engenders its distinctive metabolic gradient,

which manifests percipiently in sensations of hunger, parasympa-
thetically in digestion. As the gut lumen widens, adjacent mesenchymal cells, dilated by the surface expansion of endodermal
epithelium, derive circular trajectories by counteracting vascularization endodermally. Continued growth of endoderm adjacent to
the intestinal lumen induces a circular musculature and opens the
conduit. Later torque causes the gut itself to dilate in a trellis striated
by mesenchymal tissue.

The caudal elongation of the embryo itself imposes a lengthwise
extension on the abdominal region; i.e., under a different topokinetic regime the cephalic region of the body broadens. Meanwhile,
rhythmic dilations from multiple shear patterns continue to expand
the intestinal-duodenal tube and twist and fractalize it like a jet's
wake in high winds. This process also helps initiate peristaltic motion.

As a liver-pancreas complex matures and spreads under the force
of hepatic dilation, it dispatches prefood secretions from umbilical
vessels into the primordial tube. Endoderm here thickens and extends
a layer into the intestinal opening—a process similar to the merging of renal tubules with urinary calyces. Because bile and pancre-
atic juice spill through the cranial aspect of the gut, that portion
widens more than the caudal aspect, molding a narrow embryonic
colon along its trajectory. Blood vessels percolating through the
folded lining (mesentery) of the intestine invade the wall and stifle
further growth there.

As intestinal surface pressure still exceeds its total volume expansion, a pressure gradient develops within the embryonic organ, sat-
urating the thickening tissue with tiny protrusions of mucosa laced
with blood and lymph capillaries. These villi immediately transport
the absorbed nutrients of digestion. Compressive forces here result
in vigorous cellular responses rather than "crush" because this is a
hungry cell colony with organismic integrity. The overall shear force
is not narrow-banded enough to smother rather than stimulate
metabolism.

• • • • •

As components of tubular mucus-producing glands in the mucosal intestinal lining (crypts) multiply, their enhanced surfaces cause the epithelium of each villus—packed among the multiple circular folds (plicae) of the intestinal lumen—to detach from its underlying medium. Myriad more villi germinate fractally, spreading and exponentializing the geometry and chemistry of digestion in thickets where they sprout. The zones around and between the villi and their microvilli become so dilated that the separate digestive tubules are forced apart from one another and develop their own cytoplasmic extensions. These little tentacles hungrily suck in raw material, expanding and refining the metabolism of the tissue they populate. Soon they comprise an ecosystem of bellies rooted polyplike in the intestinal wall.

Contacting the face at the perforation of the mouth/esophagus, the front of the intestinal tube spreads across a transitional zone and extends an avenue through which creature appetites are drawn and discharged. Self-identity declares itself in twin passageways, for lungs and guts, in whales and frogs as much as in lions and gorillas, roar esophageal self-recognition. A rough endodermally-accordioned squeak, a wheeze is converted by breath, tongue, and palate into a bark, howl, or pagan call. The same sound is later projected by the primate mind into languaging.

At the other end of the tunnel, the esophagus and intestines sign only a scatology of farts.

Where endoderm of the intestinal tube and its ectoderm impinge too closely, inner tissue that would feed the region is dislodged and corrosion fields develop. As their epithelia disintegrates, these perforations will become mouth and anus.

• • • • •

Our reality is formed as much by our inside, our endoderm, as our ectodermal and social nervous system. Selfhood is derived from tissue layers, their weave contributing a vast semi-liquidity we feel as ourselves.

The tight hierarchies of intestinal components—duodenum,

jejunum, and ileum, as well as auxiliary stomach, liver, gallbladder, and pancreas—recapitulate a marine, amphibious journey, lengthening and coiling in invertebrate stages and organisms. These contiguous tissue complexes provide a fulcrum, literally "guts" whereby animal designs get turned into organs and put inside a body wall. Their templates phylogenetically reformatted, they establish a mechanical and proprioceptive basis for innervating alien matter and propelling toxins out of cell space. A muscularized, cavernous density squeezes extraneous debris in pressurized zones exquisitely along its metabolized linings.

Conversely, nausea is profound endodermal lethargy and toxicity.

We think in our guts as well as our brain. Without an intestine to ground abstractions, the airy synapses of our neurons would evaporate into gobbledy-gook. Certainly our unconverted wastes would drive us mad.

The Origin of the Musculoskeletal System . . . .

Muscles, tendons, ligaments, and bones take on identity in concert with arteries, veins, nerves, and the organs and life plans they serve. The overall musculoskeletal signature of a bear or walrus is different from that of a mole or owl, and that difference is established by small, accumulating embryonic strokes that proceed in waves from the neural tube and somites.

As the human embryo bends and its neural groove closes, the floor of the tube forms; then it spreads in both longitudinal and transverse planes. Because initial manifestations of spinal preganglia are rooted ventrally in the dense and viscid meninx and then anchored in its successor, subsequent movements of the tube must increase the distance between its dorsal aspect and the preganglia in such a way as to induce the development of neural processes. These will assemble an organized nervous system using spinal-cord structures.

The vertebral canal imposes a narrow lengthening corridor on the spinal cord, so the latter is embodied as a thin neural network by comparison to broadened, folded cerebral tissue. Spinal nerves

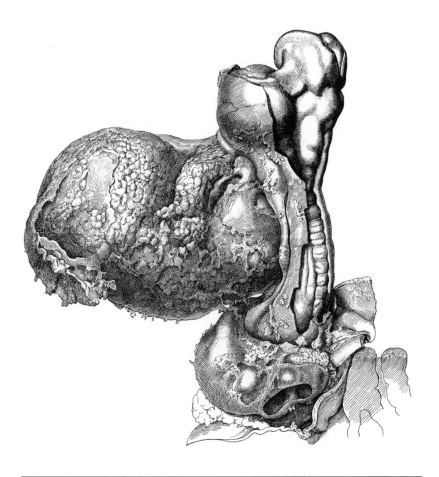

Erich Blechschmidt: early development of ten pairs of somites.

are pulled along by the stretching of the embryo and, as the fetus becomes more erect, they reorient almost 180 degrees from a perpendicular to a parallel alignment.

The neural tube also exerts a general growth resistance and, on only the twenty-third day after fertilization, out of unsegmented mesodermal tissue, thick-walled vesicles—the somites—rise up along the central furrow, broadening dorsally more than ventrally and aligning perpendicular to ectoderm and mesoderm in concordance with the developmental movements of the embryo as a whole.

Capsules continue to amass upward from the sliding that this causes.

According to Blechschmidt's topokinetics, differential speed of growth in adjoining areas of tissue gives the musculoskeletal system its morphological basis, shape, girth, and positioning. The somites' enlargement outpacing surrounding tissues, vesicular primordia must strain against ectoderm, shearing up the capsules that ultimately function as their restraining vessels. An increase in the volume of intercellular fluid prompts mesenchymal cells to adopt a position of less resistance perpendicular to both ectoderm and endoderm. With the ectodermally proximate surface taking on more mass than its endodermal counterpart, mesenchymal cells and nuclei must travel ectodermally, their processes gradually tapering off endodermally. There is simply more space for their bodies near the ectoderm. As intercellular fluid collects around the tapered aspects, their domain becomes depleted. The thicker ectodermal portion, by contrast, swells with metabolic connections.

The entire embryo bending, the neural groove closing, the edges of the dorsal aspect of the somites stretch cranially and caudally along the trajectory of least resistance. As ectodermal surface expands more than its endodermal counterpart, these give rise to dermatome, a neural layer under the skin. Conversely, cells lying on myotome (nascent musculature) form a vortex such that its dorsal strands, following the longitudinal axis of the expanding neural tube, differentiate relative to dermatome.

From the shortening of blood vessels and their realignment by enhanced blood pressure, each vesicular somite develops as a folded prominence lateral to the neural tube, separated from its neighbor by furrows, their presences inducing nerves which adorn the spinal pearls.

• • • • •

As the restraining function of the meninx helps pull neural-crest cells ventrally, they contribute to the further migration of blood vessels, the crest cells themselves ultimately falling into serrated bands along the neural tube, their teeth marking the formative sites of spinal gan-

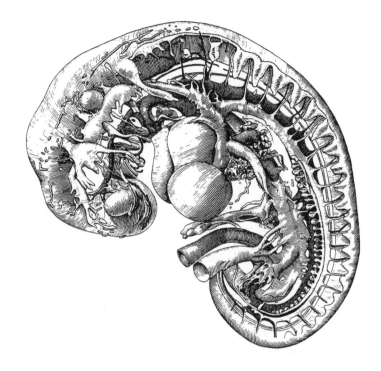

Erich Blechschmidt: relationship between dorsal intersegmental vessels (guiding structures) and neural-crest segments (spinal-ganglion anlagen)

glia (see above). Where contact is made between capillary veins and arteries, neural filaments begin to grow out of the spinal cord. Dilation with transverse compression imposes a vertical gradient on the upheaving somites, and flux of intercellular liquid within mesenchyme contributes to their matrix becoming denser.

A whole new median system wraps itself around viscera, fusing with every level of ectoderm to create the shell, skeleton, and body mass of all but simple invertebrate forms. This musculoskeletal grid propels life jellies throughout seas and, by sprouting limbs, pulls them onto dry land; by feathering arms, lifts them into air; and, by jointing and digitalizing limbs, enables them to manufacture vehicles that depart the solar system.

Muscles and Bones . . . .

The neural tube serves as a passive-growth activator of neuromuscular tissue. Myotomes propagate when surface cells proliferate faster than those underneath and stretch out in layers that wedge segments between them. Mesenchyme, as noted, generally experiences growth tensions of comparable intensity in all planes and directions. Even prior to stresses imposing themselves, many of its cells flatten and accrete as cartilage. In fashioning young cartilage, these fields exert growth pressure on one another and become flattened radially at right angles to the long axis of precartilaginous skeletal tissue, following the course of least resistance.

With insufficient oxygen and with limited avenues of diffusion, premuscle cells from the myotomes—called myoblasts—have difficulty disposing of their own wastes. Their osmotic pressure increasing, the composite tissues take up so much liquid that they swell and, compressed from opposing directions, form large-vesicular young cartilage by contusion. This material is the basis of sinew and forerunner of bone. Its swollen growth gives its cells a piston-like capacity which helps induce muscle dilations in neighboring tissue. Surrounding mesoderm *loses* water as the large vesicular cells around them function as sponges, and they harden and form capsules. Thus, the forerunner of bone is differentiated from the forerunner of cartilage; as one swells with liquid from surrounding tissue, the other dehydrates.

· ·

Contusion: *flattening of cells in a radial plane around a longitudinal axis of pressure squeezing liquid outward from the interior of their zone.*

· ·

In the interior of cartilage zones, cells resorb nutrients among themselves and lay a grid for skeletal structures. Growth resistance continues to compress and flatten them. Wherever proto-muscular tissue exerts tensile resistance on neighboring cell masses, restraining structures develop, transverse pressure imposing a gradually narrowing counter-pull. Additionally flattened and then broadened in the plane of least resistance—transverse to the longitudinal axes of their emergent limbs at right angles—some cells press together and squeeze out their remaining liquid. Leeching hardens them, and capsules indurate around their cartilage. Swelling meanwhile causes adjacent tissues to dilate as musculature.

As new fields establish themselves within the mesenchymal extensions of embryonic muscle bulges, their protrusions stream from their respective sites to points where their fibers cross and diverge. Attaching at inserts, they become primordia for tendons and ligaments as well as for the interstitial aspects of blood vessels.

* * * * *

Locally dilating fields—not where the body would pragmatically seek to exert force—determine the sites of muscles. Early myotomes establish themselves only where spatial conditions allow unidirectional longitudinal growth as well as expansion outward and attachment, i.e., where they can load dynamic forces for muscular action. Myoblasts, positioned by their continued response to tension and dilation, narrow as they shift transversely. Their longitudinal plane broadening, they start to bulge.

Topokinetic forces may shape muscles and ligaments, but creatures accept and learn to manipulate the unique torso and recoil mechanism they are given, often with remarkable ingenuity and talent; witness a hawk or ocelot operating its machine.

Mature muscles cultivate long, slender domains, wriggling into newly available space as the embryo undergoes growth spurts while mesodermally thickening. As their fibers lengthen, their cell nuclei line up in rows causing regional expansion of the exterior surface of each fiber and helping to induce and guide corresponding nerves. Sensory strands are drawn toward substances beyond their tips, sucking into their molecules which they use to accrete membranes and extend. By contrast, motor fibers use intracellular materials to load into the

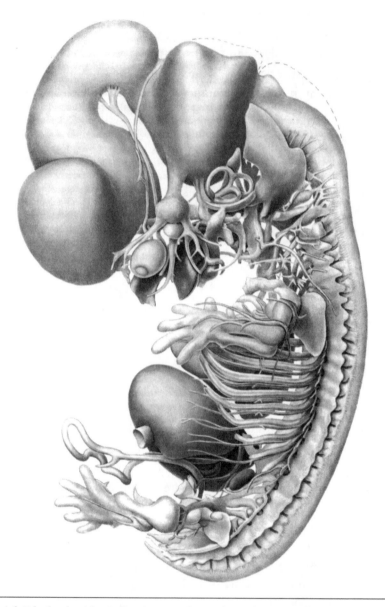

Erich Blechschmidt Collection: In the early embryonic phase reached at the middle of the second month, the growth of the embryo is particularly intense. At this time it is three million times heavier than the sprout form of the first week and has one three-thousandth the weight of a newborn baby.

springs of growing fibers as they are passively pulled toward them.

Muscular organs develop fulcra for later shortening by contraction. The circumstance of their passive formation—as elastic tractiles with multiple attachments rather than as inertial threads—leads to their later active functions (e.g., slinging rocks and cocking wheels). Their uses literally mirror the dynamics of their development.

• • • • •

Where gliding movements wring out fluid, tissue hardens in calcified zones. Bone is produced as underlying interstices are covered by strands of fresh connective fabric. This happens only after the forging of pathways for nerves and blood vessels. "From the developmental aspect ... the skeleton surrounds the nerves and blood vessels."[7]

In general, where strong friction is generated, as at the opening of a bone cavity, new detraction fields with osteoblasts (bone-making cells) materialize.

Bone also crystallizes where mesenchymal cell aggregations slide along the hardened barriers of prior osseous tissue; any remaining liquidity is wrung from the ground substance. A mixture of fat, blood, and connective tissue (marrow) in each bone expands its cavity toward a fibrous membrane (the periosteum).

• • • • •

Bone is mineralized water in skeletonized geometries, but it is also a dynamic response to life's requirement of integrity and resistance, of structure as well as fluidity. Crystallization appears to be the natural biodynamic state of mesodermal cell complexes. All protoplasm, even the cerebrum, is potentially bony; context alone determines which cells will crystallize, which will manufacture serotonin, which will swim as blood or lymph. Where protoplasm becomes trapped, it hardens into gem. This partially petrified tissue formats the body prototype.

It takes bony anatomy to get creatures travelling through space. Most fishes undulate with light musculoskeletal filaments; their collateral land-dwelling cousins hoist denser ossicles as purchase and fulcra for bearing their mass on locomotory levers and even taking flight.

Though skeletal chassis, bone is active cell life in a condensed state, its tissue still consuming itself in its own ripples and flowing toward gravity's well. Where no fluidity at all remains, bone turns into dead rock: the calcite shell of a once-living mollusk, the skull of a gopher. Likewise, where mind and culture overly rigidify and condense, we see the equivalents of brittle, hardened bivalves in thoughts and institutions. Symbols also need to maintain their watery, electromagnetic basis in order to thrive; otherwise they are relics and skeletons, marking where energy once passed.

Limbs . . . .

Four human tentacles begin their development as tiny placodes lateral to the angular zone between the neural tube and the coelomic cavity. Limb buds similar to eye and ear primordia sprout as ectoderm along the dorsal edge of placodes is elevated indirectly by pressure from the spinal cord's eccentric and spiral growth contemporaneous to the retardation of contiguous tissue. Dermatomes coradiate toward each postcardinal vein, while dorsal metameric veins converge toward limb placodes. The upswelling of the placodes tilts their underlying substance medially, forming such deepening depressions that the body wall must collapse into them. These concavities foreshadow the axilla of the upper limbs and the inguinal recesses of the lower limbs. Corresponding elevations become buds of limbs themselves.

Developmentally, limbs are dermatome and myotome— appendages of periderm, muscle, and bone. They differentiate early in the second month of development as "the surface growth of skin is impeded on the flexor sides of the extremities but favored on the extensor sides ... which at first represent ... the palms of the hands and the soles of the feet."[8] They become adjunct sense organs— implements for the fine bioelectric transmission of touch, even as

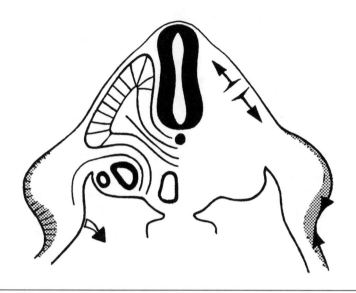

Erich Blechschmidt: transverse section through upper limb-fold region

eyes are instruments for light, ears for sound, and tongue and nose for chemistry.

Limbs uniquely work on the objects they sense, enfolding as well as recording them vibrationally. They connect animals intimately and tenaciously to the world with crowbars, hooks, levers, and gears. Replacing the locomotory shimmy used by water-dwelling hydro-zoans, worms, and fishes, their jointed machinery makes it possible for just about all nonaquatic animals (except maybe limbless snakes) to get themselves from point A to point B, whether on terra firma, burrowing underground, hopping among branches, or soaring across the sky. Tarsiers, eagles, gophers, and macaques each inherit unique ambulatory tool kits, so adapt to different landscapes and habits. Their locomotion is the source of their identity.

• • • • •

Gradually limb buds crease into longitudinal folds with flexion and extension surfaces, the tissue stretching longer and thinner on their outer face. Biodynamic forces induce these plicae as limbs. As ectoderm expands, the ventral portion of each fold sinks. The entire plait then trails its collapsing portion ventrally, initiating growth adduction with a tilting trajectory.

Likely following some asymmetrical head-tail axis from the templates of limbless fish, though with radically different results, posterior limbs thicken and their bases widen in order to equalize top-heavy mass of bipedal transit.

The ascending central nervous system and descending heart precisely position the limbs, two pairs of loci extending asterisk-like from the eddy of the primitive streak. Each one—right and left, upper and lower—has a distinct position and potential related to the other, based on its dynamics and location in the body wall. Upper limb buds gravitate toward the liver, lower ones toward the root of the umbilical cord. (Birds of course have limbs but no umbilicus.)

• • • • •

Limb templates are molded as if a tube under low pressure were being compressed from the outside. When expanding ectoderm draws mesenchymal tissue from the trunk wall, each primordium becomes stumplike and then flattens. This opens both the shoulder and hip regions to penetration by nerve pathways and blood vessels during the second month of embryogenesis. Making contact with one another, nerves knit plexuses. The ascending central nervous system and descending heart help to guide them to positions within each limb, the former providing the dorsal axis, the latter the ventral component.

Neural rudiments converge toward swelling limb folds until they make contact, lacing plexuses throughout the shoulders and hips. The extension side of each emerging limb continues to lengthen relative to its flexion surface. As the ectoderm and stroma on the former become longer and thinner, their complements on the latter thicken. These emerging folds then transpose pressure simultaneously in three dimensions—craniocaudally, proximodistally, and mediolaterally—but at different rates of growth and retardation, sculpting distinct laminae and edges. The distal portions of the upper limb buds taper into elbow zones, whereas the proximal portions broaden into embryonic shoulders. The lower limbs become corresponding shanks with knees and ankles. In birds the upper limb buds differentiate into wings.

Da Vinci's hominid shape in its basic proportions and symmetry is a shear-plane insignia of the Earth—a robe donned by Pithecanthropus, Jesus on the Cross, Adam Qadmon the magician, and the yogi in lotus—all reaching outward into worlds unknown. Though geomorphic in origin, the torso is transcendent in meaning.

• • • • •

Basal ectoderm spreads as its cells send up palisades that assimilate nutrients and release catabolites; by contrast, peridermal dermatome accretes slowly. This disparity of layers sets a stress pattern whereby inner tissue materializing more rapidly peripherally than it does in its interior induces cells to compress into one another instead of becoming aligned perpendicularly. In accordance with the angle of the neural tube and the serosa lining (which is widest caudally), the future shoulder becomes thicker than the future elbow. Likewise the caudal aspect of the lower limb—its gluteal region—becomes thicker than the emerging knee.

As the entire limb primordium develops into sheets of periderm, basal cells, and intermediate strata, global thickening induces its structure to extend outward and then thicken even further.

• • • • •

Mesoderm provides an ontological and functional basis for neuralized phalanges, extroverted implements that denote endomesodermal concepts. Relationships and functions we activate unconsciously and autonomically (heart, lungs, intestines, kidneys, spleen, etc.) resolve themselves mostly by streaming along neurons into gesticulations of limbs. The highly mobile, sensual musculoskeletal system provides the only tissue that is shaped in hollowed fields and made pliant as signal rods. These appendages derive their own intrinsic anatomical meanings and, like the larynx and tongue, forge institutions and rituals that translate the urgencies of cells outward. As Blechschmidt notes, without substrata supporting them, signs could not occur.

Watch any two bipeds in discourse or challenge, their endodermal layers alternately extending and flinching, accentuating and damping, clasping and yielding, activating phalanges in ectomesodermal pantomime.

As topokinetic shapes become acts, acts engender rituals and rituals bear messages.

• • • • •

Blechschmidt considers the limbs an axiomatic instance of the inseparability of structure and shape. The distal limb border sharpens as a direct function of its marginal ridge of ectoderm. From longitudinal growth of epithelium, the acute border of each limb primordium's ridge of extremities becomes undulated—the crests of the waves as fingers and toes. Their troughs thin out (in some vertebrate species a webbing is left behind between digits). Remaining mesenchymal tissue between the primordia of the phalanges and the cutis is flattened from surface growth even as it is stretched longitudinally on both volar and dorsal surfaces. One axis continues to be pulled unilaterally to the end of each phalanx, with capsules forming along skeletal vectors where the neuromusculature is distended bidimensionally.

Joints arise where unequal resistance displaces mass to its concave side, stretching volar-dorsal tissues and causing them to shear. Joint gaps align in the dorsovolar plane because of the stronger growth resistance of the flexor primordium. As a radioulnar trajectory expands to a greater extent than its dorsovolar counterpart, fingertips are sculpted—a topokinetic process similar to the formation of optic sulci and pharyngeal arches under the head's folding.

Each palm cultivates dorsal flexion, while its fingers rotate in a volar direction. The curling of primitive digits induces tendinous mesenchyme to travel away from bone, distending adjacent volar tissue and forming fan-shaped harnesses over what will become tendons. This process also initiates sliding movements, shearing, and clefts; where fluid accumulates in the clefts, tendon sheaths develop.

Stretched inner tissues function as restraining nets. Dynamic tension imposed embryogenically allows thousands of discrete biomechanical applications. This is what hands and feet, fingers and toes express—spatially ordered, tugging planes congealed to termini at digits. Their uses arise concomitant with their morphology, improvising alternative efficacies as the body itself relocates from culture to culture—i.e., hurling sticks, regulating notes on a flute, or playing a video game.

A tracker with bow or lance imposes his *ex post facto* definitions on phalanges, tendons, and ligaments as he generates symbols. Because viscera accommodate only certain activities efficiently, it is not unwarranted to imagine that the occupations of the band and tribe were fetally prefigured and recapitulated psychosymbolically. Even small mammals, birds, and reptiles incipiently gesture, are "cultural" within the provisions of their anatomies.

As an Acheulian tool-making complex diversified among the clans and guilds of late Ice-Age hunters and spread to Mesolithic fisherfolk and Neolithic farmers, handaxes and arrowheads became arrows, hooks, and hoes. Hominids worked the forge and cast helmets, pipes, and engines. Wheels, pistons, circuit boards, and optic fibers continued to evolve from neuromesodermal currents. During this evolutionary process the hand, like the genes, hardly changed at all. But it changed the entire world around it.

• • • • •

There are no moves, styles, or gigs for humans that do not originate in and get organized by tissue complexes, most of which are millions of years old, some of which were established nearly a billion years ago. The shapes and movements of cell clusters, channelled through sinews and synapses into modes of conduct, are the basis of all behavior. No motion has any other cause or foundation; nothing can elude or outfox them.

The bricklayer, the boxer, the Navy SEAL, the opera singer, the shaman, and the lap-dancer are cultivated embryogenically. All of them balance, twirl, jab, leverage, skim, and speak with limbs. Orgiasts in Castro Street catacombs, basketball players slam-dunking in midair, dancers in disturbed equilibrium are not inventing new territory or proceeding on their own recognizance; they are responding to and expressing constructs from inside the tissues of their bodies. Dadaist painters, al-Qaeda trainees, biotechnicians splicing chromosomes, strippers, skateboarders, and rock-climbers are each in their own ways translating viscera, muscles, bones, and neural circuits into actions.

It doesn't matter how extreme the mind tells the body to be. Tupac Shakur rapping, a suicide bomber pressing a switch to blow his body to shreds, a crack addict butchering his girlfriend's remains with kitchen knives and then running the pieces through a wood-shredder, a hang-glider, a supermodel, an options trader, a clown, a transvestite—none of them can conceive and carry out pristine deeds; they cannot escape the repertoire of gestures that nature has allotted them through the ontogenesis of circuits and figures in their bodies.

• • • • •

As a result of locally bridled surface growth, embryonic thickening of cutis molds palms and soles on the flexion sides of emerging fore and hind limbs. This living push-pull system intensifies metabolism there and confers a feeling/grasping motif as well as a stable shape for the organ. The infant will test its implications very soon after birth.

The corium of developing forelimbs, crammed with cells, continues to differentiate itself, while adjacent wedge epithelium spawns crest-like folds in large numbers of cells that undergo intensified mitoses. The epithelial valleys between these crests experience no comparable activity. Folds radiating outward as well as inward provide a papillary system of ridges in the fingers as well as ground material for further mesodermal and endodermal differentiation.

Inner tissue continues to be compelled to grow faster peripherally than in the interior of limbs, so thicker, deeper structures develop below the periderm, and their pressure forces liquid from vessels into surrounding mesenchymal tissue, detaching cells around them and molding lobules (or cushioning sites) which become distended in ratio to the intensity of the pulsating fluid.

A similar biomechanical cushion forms in the subcutaneous layer of the heel. In its embryonic state the heel actually pushes against the angle between the lower abdominal wall and umbilical cord, which develops its flexion capacity before the ankle joint expresses its mobile coupling function while crawling.

· · · · ·

Blechschmidt traces the origin of mathematical operations to the shaping of fingers and their tendons. When we grasp solid objects with both hands and bring them together in a space cancelling the operation, we invent zero. Moving a hand to the right along a rod and clasping it at equally spaced sites signifies positive numbers. Opening the hand, moving it to the left, and removing it from the rod creates a minus sign. "By gestures," he declares, "we shape our own bodies and decipher what would be completely incomprehensible."[9] We invest relevance in blank events and, by them, engender our own transactions with values.

Numbers are pure tissue creations. They exist nowhere at large in the universe. We alone add, subtract, multiply, divide, take square roots, derive pi, and measure substance and energy inside and outside our bodies. The elusive meeting point between thermodynamics and gravity is uniquely human, statistical. The basis of numbers in gestures and gestures in topokinetic folds should serve as a *caveat* to physicists and cosmologists who think that we of all monkeys can solve the universe in a grand intellectual leap of unification.

Identity . . . .

A global nervous system does not develop as such in order to conduct electrical impulses; instead, a thickening of the whole mesodermal layer invites bands of ectodermal cells to initiate multidirectional membrane processes, as many and as extensive as they can. Even as these synaptic primordia creep along, other cells in their domain and between them develop as glia to mediate the transfer of nutrients from blood vessels to them and to dispose of their waste products.

Afferent neural connections are initiated as branched extensions—electron-conductive dendrites—burrow into tissue to be activated. They proliferate more on the concave inner portions of cell membranes between the dendrites where pressure is lower. It is here that substances are conducted across the synapse toward the nucleus. Spreading their surfaces isometrically, nerve cells feed, dilate, and transfer chemico-electric charge. Autonomous cellular aggregations follow these filaments to the boundaries of tissue.

Meanwhile motor pathways are formed by the transposition of cells into muscle fibers pulled passively into cohesion.

Bioelectric currents infuse the entire cell mass, bestowing sensation and inventing time-and-place awareness, astonishing life with its own existence. How the dream mind is spontaneously lit is a mystery. Clearly, a creature knows (beyond knowing) what it is, forgetting everything else.

· · · · ·

Molecules are pulled into swarming cyclones of cells; those cells accrete in tissues; tissues meld as organs; organs and vital fluids kindle metabolism and the phenomenological life of an organism. Mind takes over all networks underneath it and pilots their sum, establishing and deploying identity.

Consciousness is molecular and not molecular, cellular and not cellular, a product of tissue etiology and an epiphenomenon of tissues.

▶ Erich Blechschmidt Collection: At the end of the third month, one talks about not an embryo but a fetus; its individual differences are now perceptible to the naked eye.

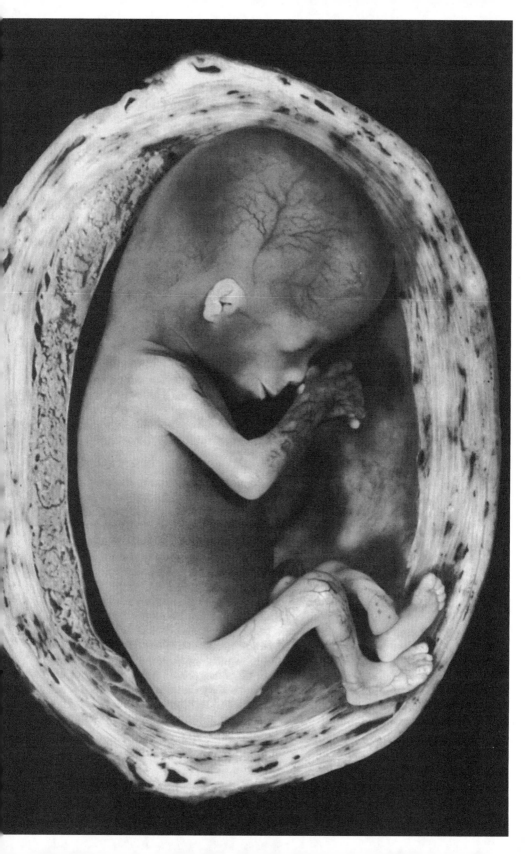

Yet identity is based irretrievably and inseparably in organs such that the majestic vapor of mind itself seeps away when their network is ruptured.

When a surgeon makes an incision and pulls back skin and subcutaneous fat to pry beneath, the human landscape dramatically changes. Soft anemone-like organs—luminously colored, moist, stippled, cradled in saliva-thin membranes—oscillate in deep confederacy and communication with one another, oblivious to thought which is also oblivious to them.

The social being, the self is a mirage. A visceral factory is running silently and undeterred in another space and time—an osmotic machine composed of snails, sponges, jellyfish, and nacreous, speckled bags transpiring in sullen autonomy, sustaining philosophy and will. Appendix, intestines, ovaries, and spleen maintain iridescent, bilious gazes, while the laporoscopic eye invades their formerly covert realm.

The beloved—anesthetized, dreaming beyond memory—is a cabal of bestial objects and their pagan acts.

As epiphenomena, we are shocked to visit our basis in guts, in mere substance. The living cannot view the dead without a ceremony, intended or unso. The living do not even really view the living.

Meaning is rooted in a biochemical, embryogenic labyrinth that tapers out of atomic clusters condensing and enwrapping on a scaffold of cosmic forces. Bodies are processes, histories elapsing in present things. Organs and identities are atavistic, fluid, interchangeable. Even a face is a molecular membrane that would change dramatically if fitted over other bones and vessels.

"A differentiation of something already existing in its essence" . . . .

Blechschmidt's topokinetic premise is that, no matter what chromosomes say, embryogenesis must proceed by physicomechanical divergence, progressing from phase to phase, according to unforgiving requirements of space, appetite, and prior organization. A creature assembles and unfolds as it responds to fields emerging out of its cytoplasm. Hominization, insists Blechschmidt, is inherent in the singular proclivities of a human embryo—whatever those are, wherever they come from (if not from cell nuclei), however they express themselves (if not by mRNA). He is trying very hard not to be mystical, yet he is eliciting a first cause.

What organizes each fissioning germ cell and its blastula toward a phylum and species, if no exogenous genetic plan, no phylogenetic law runs the show? What makes a cow a cow, tansy tansy, a blowfish a blowfish if its seed has to start from scratch and extrapolate the occasion for its tissue complexes and overall shape each time anew?

Shear force and liquid dynamics must be specified at some prime and profound level in order to discriminate a man from a mouse, an octopus from a snail, even as a smidgen of ontogeny recapitulates a smidgen of phylogeny. That smidgen is the quantum leap that catalyzes and guides development, each egg, each organism. It is the confluence of myriad tracks of dimensionless specificity. And the tracks change as suddenly as an electron jumps orbits.

The threshold at which humanity or cephalopody is imposed on tissue is rooted somewhere in gene-space. Despite topokinetics, information must impose and discriminate designs. Blechschmidt hedges on this point because he intuits a superior force to genes, though I see no reason why genes could not also be a manifestation of that force. They may seem cybernetic, but they could be telekinetic and topokinetic *and* cybernetic. If animate shape is vital, it must be regulated in networks and circuits to come to exist in physical space. Where two primordial streams—one paraphysical, one physical— meet is the Rosetta stone of biological development.

· · · · ·

Heredity is as much structure as data, because, in kernel, structure is information, and information is nothing but another form of structure.

Even software is made of something; thermochemical forces underlie the genesis of genes at both classic and quantum dynamic levels. At a scale we can at least see through an electron microscope, the deployment and redeployment of a cell's hard molecular components—microfilaments and microtubules—in response to external pressures translate highly discrete vectors into the cellular nucleus where they turn genes on and off, giving information, releasing other forms of information, and designating the structure of information.

As Silicon Valley proved, you can get a lot of ideas into a very small space, but you need *some* raw material; you can't build machines out of thin air.

This was a planet long before it was a library or temple.

· · · · ·

Since genes (as we know) are not unique or specific to proteins, their memory cache cannot anyway be the full inventory or singular point of reference for creature manufacture. The noncorrespondence of genes to life forms explains the digression of quite different creatures from similar genomes. A whole array of new genes is unnecessary, for nature's basic method is to reuse the same core information in different alternative states, tissue layers, stress planes, and shear fields, often with different rates of developmental timing, for new structures, meanings, and modes of existence—but how?

A "fertilized egg [is] a mere speck of protoplasm and DNA encased in a spherical shell. [How can it] generate such complexity?"[10] How can design which is unfolding kinetically also be anchored by nonlinear streams? How can codes which are devised algorithmically, stored chemically, released disjunctively, and disclosed anagramically—but in meticulous sequential progressions nonetheless—be distributed allometrically and construed with consistent structural and functional results? How does gruel receiving barrages of quasigenetic requisitions rearrange and even improve itself? How do milky beans in pouches become spiny echidnas?

Blechschmidt says simply: "The cell's limiting membranes, with their constant generation and regression, their continuous rearrangement with the direct help of the adjacent cytoplasm, are *the primary agents of biokinetics.* . . . In comparison, the nuclei have only a relatively passive function."[11] (italics mine)

Tell that to the genome teams! What is a "relatively passive" function? How active is a diffident gene allowed to be? How can coordinated development occur without stringent chemiconucleic control? The Sun and Moon don't curd puddles into organisms by convection and tides alone. A cell can't differentiate or collaborate with another cell in making an organism without instructions. It is just as helpless as genes are without context.

If we picture genes as founts spewing out information from source codes, then we must ask not what is providing the blueprints (which we more or less know), but what is receiving them and, more significantly, what is coalescing, mooring, and perpetuating them so that more or less the same offspring arise along a lineage of zygotes every time? What hierarchicalizes and indexes the distribution? Where is quality control localized? The answer remains: everywhere and nowhere.

. . . . .

It is a paradox that organisms, which are so much less concretely specified than the products of assembly lines (at least by our rules), are mass-produced with incredibly more congruency and uniformity. From one embryo to another, the fidelity is astounding. Embryos continually recorrelate and self-correct. Flaunting friction and degradation, they turn their semi-liquid stuff into muscles, nerves, brain, digestion, and other traits, and then integrate them seamlessly. They flirt with and achieve the temporary demeanor of perpetual motion, e.g. in the beating of the heart and the flow of unconscious thoughts. No one expects an embryogenic machine to grind down, diverge, or obsolesce. Yet there is no machine.

Once again the big question is—creativity aside, genius and originality of design aside, fluidity and transmutability of motif aside—what factors are *stabilizing* embryogenesis such that, defying the quixotic variability of the DNA code and the random motion of billiard-ball molecules in trillionfold chronologies of different-scale entropic interactions, each newborn is fabricated with absolute perfect synchrony of circuits, almost flawlessly?

• • • • •

While the physical basis of organization has been overlooked by a genetically biased establishment, to imagine that the hypothetically limitless ova of a species follow a near identical itinerary of development each time from purely extrinsic, exigent forces, shear and pressure fields, and nonteleological actions of cells, teeming, filling cavities, metabolizing boundaries, and twisting out organ motifs—without DNA decreeing what proteins to synthesize in what order and dynamic relationship—strains credulity even more. How could paths of molecules be kept so precise, organism after organism, clone by clone, not only in conducting unfathomable design features but assembling metabolizing structures of great intricacy and delicacy from scratch? How could germs fission, congregate, make original protein complexes, spread coherently, form laminae and vacuoles, introvert, irrigate, sculpt massive interdependent organs, and regerminate identically all their basic thermodynamic potential in offspring via single gametes just from molecules, proteins, and tissues bumping into one another and exerting contingent chemistry, pressure, and strain? Is this even possible once, let alone routinely?

How do blastulas become functioning, sentient bionts from the offspring molecules of a single fissioning cell solely by spontaneous series of cause-and-effect, gravity-induced, fundamentally fallible motions and collisions? How can cells replicate this process (for all intent and purposes exactly) in termless series by mere sequences of interactions in tightening membranes? How is molecular diversity converted into metabolism and knowledge and transferred between structural states? How does droll entropy become comic organism?

Any Rube Goldberg perpetual-motion machine with its absurd links of cause and effect (involving mice and scales, candles and falling coins) looks utterly plausible by comparison.

Blechschmidt can recognize this paradox as bluntly as a high-school biology student. What he is pointing to beyond the genes is a prototypal shaping construct which has a life of its own and its own method of animation into which genes must cascade. He is identifying that construct with a living rather than a historical component of embryogenesis. He is attributing development to a natural and vital ecology and the tireless rambunctiousness of the fertilized egg and its offspring.

"[D]ifferentiations," he reiterates, "begin on the outside of the cell, setting to work here first before gradually penetrating towards the nucleus ... this state of affairs applies to the whole living body as well as to each of its cells...."[12]

Raw but intrinsic directionality is already present *even in a single cell.* Protozoan predatory "intelligence" becomes mesenchymal and then hematological volition, immanent tactics to congeal in a life plan. This same inertial force becomes—without *not being* chemicomolecular—vitally and psychodynamically oriented.

"An ovum has an extremely high capacity for original development (potency), and this capacity decreases continuously with its ontogeny and therefore with growth and aging. The mature organism, although developed, is not a higher entity than the egg from which it grew."[13]

A baby is no more, no less than a smooth, immaculate effigy of a cell.

. . . . .

While topokinesis is clearly a critical and overlooked aspect of development—a gravitationally guided molder of protein clays—there must be some way in which not only an exemplary template, a biological insignia, is transposed into organic matter but a life form changes its design self-consistently, complexifies (over eons), and extends a lineal pattern invariably while remaining organized enough to square with nature and carry out survival activities at every stage. The mechanism must be beyond passive and ulterior.

No one would assert that creatures are possible without genetic molecules collating and unionizing billions of separate molecular actions. Wind, mud, gravity, compression, cytoplasm, and shifting boundary densities obviously don't just come together under archetypal edict and invent plant, animal, and human systems on a windy planet without some mnemonic, nanocybernetic device etching their molecular products and phase states. Topokinesis may have been neglected by gene-hypnotized technicians, but it can't do it all, even if it rides on liquid light.

DNA is also real. Its exquisite thread throughout nature denotes the calling card and prompt of life more than any other object large or small, and that cannot be ignored even if genes do not contain templates for finished forms (something they are not supposed to do anyway and *could not do* in a nonequilibrium framework). In fact, genes are nothing if they are not opals of liquid light. A cell can no more exist without them, finally, than it can without metabolism.

There must be text. There must also be *raison* behind the manufacture of a structure as sophisticated as a chromosome. Micro-mazes and networks of ciphered grids are not scattered liberally throughout nuclei and chemical grids to no end. Nature does not etch, subtilize, and string baubles for decoration or as demonstration of wit. DNA is not only functional; it is core. It is far more deeply core *than it would be as an architectural blueprint for life forms.* Instead it is the primordial molecular collating principle, the great memento, the nonatlas of atlases. Its complexity is a precise and inevitable figuration and, at the same time, a remnant of *our* complexity, not as its originating cause but as its signature.

There is certainly a lot of genetic material everywhere now, from mites in the air to kelps along the shoreline, from comb jellies to hens laying eggs in straw, from bees to betel palms, from bacteria and viruses to scorpions and skunks—a fathomless range of interlinked potential states answering to molecular and environmental orchestration.

Genes are biogeochemical events functioning as sequence-specific domain controls. Long, thin crystalline ridges in moon-stirred pools of cuneiform carbon and oxygen, they jell as hierarchical positioning beads for molds already beginning to form *in* nature. While they play off natural selection, they are not imprisoned by it. They have their own loops and escape hatches. They receive design that is already happening and siphon it according to meta-systemic laws. They store information in order to function as a hyper-primitive, yet incomprehensibly advanced, thermo-quantum index. They are the sole and discrete molecular bridge between physicochemical morphodynamics (bubbles, boundaries, and crystals) and bubbles, boundaries, and crystals fused into tissues and set alive.

. . . . .

DNA does not vie with topokinesis; it represents its bare initial microminiaturization—the letters of a deep cosmic alphabet through which the activities of the geosphere are funnelled and organized into intelligible constructs for tissues to birth.

Cosmic runes were carved long ago in molecules organized in long strands, precisely as cells emerged from the deep, expressly inside the nuclei of those cells. Becoming-animate forms surged through physicochemical states along the ragged borderland of matter and energy. Amidst this incipient activity, genes arose naturally and were imprinted with fossils of primordial forms, which they culled and tabulated into creatures in environments.

This is how we get the bodies we have, the histories we encapsulate, those shaggy pandas in which we chart our way across precarious geographies in a mortal condition. They hold us in their life-and-death thrall. So far as we know, they *are* us—except that such pronouns always have two forms—subject and object, self as self, self as other.

"*Biodynamics serves to unburden the genes....*" . . . . . .

Blechschmidt does assign his own critical role to genes: they mediate between hereditary factors they store (the vestigial and diachronic) and the embryogenic occasion of the embryo (the topokinetic and synchronic).

> There is no doubt that the genes are of great importance for heredity and that they therefore play an indirect role in differentiation. However, even in the young germ, neither the chromosomes nor the genes are dynamically active. They

are not the motors of development and therefore do not themselves evoke the characteristics of the differentiated organism. Genes do not work, even indirectly, via the enzymes they help to form. To believe this last view, only shifts the whole problem of development and differentiation from the chemistry of the genes to the chemistry of the enzymes. Even the most important enzymes are only aids in the process of differentiation, a process that has many other prerequisites apart from the genetic information itself. A cell is much more than the genes. It functions effectively as a whole (cell nucleus, cell cytoplasm with organelles, as well as cell boundary membrane). And it is only in the framework of this whole that the genes have any meaning.

There is no direct connection between the genes in a nucleus and the form of the whole cell. Just as it is unlikely that a sandbank, by itself, can cause ripples at low tide on the basis of its chemical nature alone, but rather needs the strength of the wind and the surge of the waves, so it is unlikely that formative materials, rather than formative forces, are the direct motors of phenogenesis [the genesis of the phenotype]. It is the task of *kinetic morphology* to provide an understanding of these motors. In any phase of development, changes in form and structure must result from the complex movements of particles of a molecular and submolecular nature. At all times, such movements, which are the manifestations of physical forces, are the direct causes for those changes in position, form, and structure that lead to differentiations.... Since we are concerned with living organisms, the most fundamental forces associated with these biophysical components can be described as *biodynamic.* Thus, the form of the organism differentiates directly under biodynamic forces, not chemical-genetic information....[14]

The mission of the cellular nucleus is to regulate ongoing biochemical reactions and provide molecular stability *but in a system that is already embryogenic and biodynamic.* That is, gravity + matter + vital force (or liquid light) + breath = life. Likewise, genes + cytoplasm + microenvironment + life = kingdoms and species.

Blechschmidt concedes that genes take on ever increasing importance in the late phases of development. "[G]ene effects often appear only ... when all basic structures already exist."[15] As organisms differentiate and approach their "final" forms, their molecules and tissues become more deeply indexed by DNA, trading developmental potential for positional integrity. "The totality of the genes, as the *genome,* does not constitute a homunculus that resides, say, partly in the nucleus of the ovum and partly in the sperm.... The genes do not cause differentiations. Rather, they are the chemical constants of metabolism and as such, are especially stable components in a cell. It is certain that genetic material serves to preserve the individuality of the organism whereas, *vice versa,* the *extragenetic material,* particularly the cytoplasm, brings forth the changes in appearance that are observed during development."[16]

It is as though a phenotype is an expression of cosmic (i.e., extragenetic) forces impelled through each cytoplasmic hologram into its germ cell wherein they stimulate nuclear and then multicellular differentiations. These then activate the genome, which is an aspect of their metabolism and not its cause.

Genes first accord specificity to broad phylum, class, and order templates and then imprint individual characteristics through enzymatic and cytoplasmic feedback out of parental genomes. After the basic plant or animal prototype has been established by gravity and shear force, DNA sorts creatures from one another by eliciting their ancestral traits. That is why a marten never gives birth to a badger or a badger a mink.

Blechschmidt admits the importance of a tabular function, though only as a secondary response to living topokinesis:

The genes—recognized carriers of heredity—therefore take part in the preservation of metabolism as centers of reactions. If the metabolism is changed by the superficial growth in the layer of the cell's limiting membrane—this external stimulus is necessary for normal development—this type of predifferentiation determines the later direction of development by a defensive reaction of the organism to the change, resembling that to disturbances. Genes constantly react to disturbances by compensation. According to the different position and shape of the cells in young and older germs the genes are compelled to react ... differently, playing a very different role in the development of the skin, the skeleton, the muscles, etc. This is shown by the germ and each of its cells assuming new topographical relationships in small steps, in an almost steady flow, but never ... any purposeful reactions aimed at particular objectives.... [D]evelopment is not a phylogenetic recapitulation, and ... it cannot be explained simply by genes ... in terms of a sequence of inductions but rather ... differentiations originate dynamically as a partial process in the framework of the whole ontogenesis.... Biodynamics serves to unburden the genes to a large extent.[17]

The activity of tissues, in other words, relieves the even tighter nuclear-space kinetics of genes (helices, supercoils, microfilaments, microtubules). But nothing is recapitulated.

• • • • •

On the plight of how this entire topokinetic shape construct arose and was initially made inheritable in biological kingdoms—if not by DNA runification—Blechschmidt is silent. He refers to "a differentiation ... of something already existing in its essence"[18]—i.e., the human form, life movement itself.

What is life movement? Blechschmidt implies that it is the result of an intelligent entelechy that has imbued chemical, crystalline aggregations with its essence and given them a pulsatory awareness, however rudimentary, and that agency has then conducted its own molecular interactions and regulated flow. It is a not only a vital but symbolic expression of matter itself.

Biologists do not necessarily recognize events originating in waves of feedback from outside organisms and travelling by unknown methods of signalling and transmission. But then, despite the name of their discipline, they do not even really acknowledge life itself. In trying to avoid letting it become an epiphenomenon, they do not allow it to be a phenomenon.

Protean tissues moving against mechanical resistances on young planets imprint an intrinsic life force or animating principle in tide-pools and spume and then *in utero* through the cerebrospinal fluid and cytoplasm. These then grave living tissues with organic functions. That would represent an epistemological entirety not requiring *prior* DNA archiving. The universe is operating outside-in and inside-out on both galactic (gravitational) and molecular (quantum nonlinear) scales, from the cytoplasm of nebulae into the coronas of cells (and, maybe, back). Currents flowing out of and into stellar fields—and into and out of submolecular wells—are threshing hyperdimensionally to yield the vortices of worlds, bodies, and creature consciousness.

• • • • •

Look at a sow bug, a panda, a sea urchin, a thistle. The sublime precision of any worm or bird, flower or fish is not just the critical grinding of chance nucleic factors and algorithms through equally fine selective parameters, extracting functional, even cute designs; it is also the distribution of informational networks through one another's languages, coalescing according to one another's patterning and feedback. Whether a spider or rabbit is a random cascade arising from gibberish or is "ingenuity embodied" depends *at only one level* on whether the universe has any intelligence in it at core (if it doesn't, then everything is finally random and hollow). At a more profound level, exquisitely honed feathers, palps, gills, and eyes evolve on worlds according to whether innate intelligence—and its antipode, innate accident—assemble creatures by trial and error or by dimensionless, self-elaborating networks. If the latter, the universe might as well be intelligent.

I have said it many times in many ways: The depth of this system is revealed solely by its products—any average pig or pigeon, but also the collective thoughts, meanings, and artifacts of lives and nations. Reduce it to cells, genes, or molecules and atoms; reduce it to a particle, trace, or industrial metaphor; assign it to any cause-and-effect series (even nonlinear); any hierarchy of laws and design principles—and you won't capture it; you won't come close.

You can chase more and more subtle ball bearings to the roots of their orbits, but at the point they would betray the motive force driving them they vanish or leave arrows pointing in all directions and nowhere simultaneously. Biotechnicians can move traits and tissues, but they can't activate or get at the *unity* of the system. They can't even escape the prison of false metaphors, the "machine" and "commodity" imposed by this civilization on nature.

Yet remove any scientific hard factor (like genes), and the whole house of cards collapses faster than you can say, "jack sprat." You can't build the system from its parts, though you can kill it that way. You can't find any starting point from within tiers of emergent mechanisms from which to imitate or rebuild any major aspect of it.

It is not a system. Its parts are not parts. Its cause-and-effect networks are totally redundant at this-here level of operation. Overall cause and effect—the big picture—is imbedded directly in the universe. Not the universe we see and describe, not the universe of physics, but the universe itself.

10

The Primordial Field

. . . .

Metabiology and the Molecular Apparatus

> And Hector was in Ilium, far below,/And fought, and
> saw it not—but there it stood!
> —Matthew Arnold, "Palladium," 1867

*The biosphere furnishes material channels for outing
the universe's hidden plenum.* . . . .

Embryogenesis is the outcome of natural selection, gravity, and
thermodynamics—molecular sorting, wind and water, carbon and
calcium, chemical reactions, matter and energy predation, and procreation. We know that. But it is also, more profoundly, a signature
of how matter goes inside itself to "mean," how meaning is embodied and signified, how the invisible presence in nature federates.

Embryonic folding, convolving, and intorting are how matter
expresses its character, gets its extent and complexity located, makes
a blotch in space-time, and inhabits the lucid dream of having a body.

Life is neither DNA nor cells as such. Life is the cosmic organizing
matron that expresses itself in cell formation and serialization. It represents the fabric of the universe, what it has inside it and *what it is,*

far more profoundly than suns or electrons do, for the embryo is the universe writing itself on its own body. If the universe couldn't make embryos, it would be an entirely different universe and would mean an entirely different thing. It would be a rattly, self-igniting cavern, each separate explosion as massive and meaningless as the last. Stars and atoms would rule an abyss; nothing would happen; nothing would become; nothing would care.

Instead, the universe rushes toward consciousness—or vice versa. Doggedly it fissions and twists into sentience as if expressing its birthright.

Embryos are the only parchment that stars and atoms have that says what they truly are, that has a "within," that finds its way into its own mute depths and turns itself inside-out.

. . . . .

Infinity pokes a hole around the firmament through which our world floats. Yet, as vast as it is, it is a fabulous sideshow, playing as the dense, flickering night followed by matinees.

The essence of the universe is to localize and deepen. Its true nap is not vacuum-thin textile with a few solar sinks; it is whorled every place more intricately than Van Gogh's starry vault, a warp and woof that run through field, stream, weed, and molecule.

It is in volatile stones that oscillate, scissor, twist, fold, divide, layer, curl, thicken, and clone that matter confides its one and only secret. The universe has no canon otherwise.

We hear it, but we don't hear it. We see, but we don't see. A kid squirting wet out of a goat's hind surpasses space travel "because it is coming out of the nowhere into here."[1] Every litter of kittens proves it—coiling and uncoiling, ravelling and unravelling, mewing and suckling—all pretty much from nowhere too. They are the universe's kittens as much as they are the cat's kittens. A hive of spores gestating wasps is the universe's hive.

A tiny wisp of string, a cloudy mottle, that begins at fertilization, ends with habitation by an ego of a mortal shape.

• • • • •

Is it not strange beyond telling that each of us gets fabricated in layers, centrifuged in a uterus, wound into a visceral ball, expelled down a bloody canal? There can be no escaping our origin or its implications. Everyone on the street and running to catch a bus, in military training and courtrooms, adjusting the sails on dhows, herding alpacas, stocking the shelves of markets, bowing to pray, blowing into panpipes, sweeping the streets, operating machinery, began in a germ seed, was swaddled in sheets of self-organizing proteins inside another person, fell into geography as a tiny, innocent clown.

This is what we are, what matter is to have made us. And it is not as though anyone has a better plan or another way. We are, to a one, lambs, calves, pups, caterpillars, puggles, chicks. We are urchins, heirs of the beast.

The universe is comprehensible only as a thing that has been folded many times upon itself. . . . .

An embryo pulls the trillionfold galaxies and their planets, comets, and moons, distended almost but not quite forever along space-time's curvature, down (smack!) into the microcosm. The Big Bang—or, more precisely, its antithesis—reenacts itself in each whelp and whippet from planet to planet, star system after star system, throughout the Zone. Every pilgrim—however simple, however complicated— is an imprint from the zodiac inside the sky. Each body, in correspondent stages, *is* its own birth-chart, the houses, signs, and degrees of the cosmos inscribed and actuated in flesh.

Light-millennia distances in all directions gape and constringe down to one compact, involuting germ; and another single germ, etc., one a bee, one a snail, one a possum, one a Pleiadian "wallaby," each consolidating a chord in the macrocosm. Other "bees" and "crabs" and "croakers" matriculate on their own worlds.

• • • • •

The medieval sciences of sympathetic magic, organized resemblance, emulation *(aemulatio)*, analogy, polyvalency, and polycongruence are ultimately more germane to an understanding of the nature of matter than statistical thermodynamics. Entities are held together at core, in sum and substance, by similitude and correspondence (of which gravity is a subset), but not so tightly that none of them can uniquely exist:

> Through ... antipathy, which disperses them, yet draws them with equal force into mutual combat, makes them into murderers and then exposes them to death in their turn, things and animals and all the forms of the world remain what they are....
>
> The identity of things, the fact that they can resemble others and be drawn to them, though without being swallowed up or losing their singularity—this is what is assured by the constant counterbalancing of sympathy and antipathy. It explains how things grow, develop, intermingle, disappear, die, yet endlessly find themselves again; in short how there can be space (which is nevertheless not without landmarks or repetitions, not without havens of similitude) and time (which nevertheless allows the same forms, the same species, the same elements to reappear indefinitely).[2]

The universe is comprehensible only as a thing that has been folded in many times upon itself, that duplicates itself, reflects, opposes itself, and forms chains with itself everywhere.[3] No wonder its chief expression is embryogenic, its polar inclination inside-out.

• • • • •

A blastula, while creasing and tucking molecular sheets, goes 2, 4, 8, 16 ... 256, cleave, corrugate, extend, inroll, divide, curl back over.

Make its mudra with your hands. Align the fingers and knuckles of two fists in balls; slide and rotate them isometrically and non-isometrically like two socketless balls against each other in rough imitation of fission, introversion, the primitive streak, gastrulation, ectoblast, endoderm, the neural groove, somites, etc.

Invaginating tissue is meaning in a fundamental state. While imposing an algebraic boundary on eternity, it liberates the sum of what is concealed from the bottomless substance in which it is submerged.

This is the universe signing itself. . . . .

Chants of lamas and mantras of shamans, howls of wolves and chittering of moles, all approach the resonance of the cosmic tongue, but the embryo—the infinitesimalized screeching of cells—is creation's bluffest voice, how it excavates and inscribes what it remembers, how it gets inside its own fur.

Bleats, screeches, chirks, mewls, barks, and the like are not rude forerunners of human speech; they are evocations of something prior, something also immediate; something so literal, indifferent, and dead serious we cannot actually hear it. It is the sound "I am."

Listen to unearthly shrieks of gulls over the harbor. They are not unearthly because they do not come from earth, are not made of earth; they are unearthly because they drive the call of something beyond Earth through winged phaetons, something original to human language too but lost amidst its conditions. The same thing that arranged the gull primevally in an egg and ravelled it from seething cell clusters into a small, fuzzy gosling that took to the open sky now sounds destiny itself through the pregnant, altered air.

It wants more than the fish those boats have gathered from the deep. It wants the deep. And it also wants the fish. It soars to the center of everything in the galaxy, in all the galaxies, even as it patrols the open skies, searching.

Where else could such fearful symmetry and lamentation come from? From where else could the dark morn sweep down to the sea?

Meaning is intrinsic to "being." . . . .

Despite all their thermomechanics and deep machinery, embryos are phenomenological, meaning-driven shapes. They generate feelings that are beyond bodies, beyond qualities, beyond time, though we could not feel them "without qualities, bodies, and time."[4]

The embryo uniquely (in the material world anyway) crosses the bridge between "nothingness" and "being." Crossing the bridge *is what embryogenesis is.*

Biology as such is hopelessly underequipped to deal with the difference between "is" and "is not." Life is simultaneously too subtle *and* too gross for its techniques and instruments to detect, its theories to label. Its meta-coding is too subtle, while its embodiment in tissue layers and fractal mass is too swift and too dense.

The passage between mind and matter, nullity and form, occurs in such a basic way that it only appears contrarily to be utterly mechanical and happenstance. Morphogenesis operates in a zone that is latent and fueled to jolt differently from metal balls, atoms, and subatomic charges. Yet *they* are its visible fare.

• • • • •

Why assume that tissue is a *tabula rasa,* the base state from which experience is synthesized? Quite apart from belief (or nonbelief) in disembodied intelligence, experience doesn't pop vacantly out of the void, sprouting wings anemone-like. It is organized embryogenically in blastulas and neurulas; it arranges its meanings the way cells do. It is the explicit, concrete outcome of their orientation to the universe's body.

Experience itself is the ultimate inducer of embryogenesis.

There can be no tissue without experience *first*—an overlooked law of creation. Cells and then tissues rub against one another, find their own existences through the concordances of their bodies, and meld into aggregate meaning.

Interaction kindles morphogenesis. Life is subtle because "being" is subtle.

Biology has made an impossible object. . . . .

An enigma lies at the heart of biology, for biology's "things"—ineffable, ineluctable, improbable—don't add up:

"A wild mallard—just arrived from the home of the north wind."[5]

"A snowy owl [who] makes her nest upon steep rocks, or old pine trees, and lays two eggs, which are of pure white."[6]

"A vast pulpy mass, furlongs in length and breadth, of a glancing cream-color, lay floating on the water, innumerable long arms radiating from its centre, and curling and twisting like a nest of anacondas, as if blindly to clutch any hapless object within reach."[7]

"A bat flits/in moonlight/above the plum blossoms."[8]

• • • • •

The Zen master's staff whacks us. "Don't know!" he declares, meaning that each out-breath expresses the joy of clear mind, each in-breath absolute ignorance, the wonder of being awake in a condition of mystery.

The origin of life didn't happen in a remote epoch. It happens now, everywhere on planet Earth, at incomprehensible speed with a precision that defies thought, let alone explanation.

Look at the embryo and you see substance that, from nanosecond to nanosecond, can be neither purely spirit nor purely matter. It is dancing, agitated, roiling, morphing through multiple interiorizing fields; electromagnetizing, transmuting, organizing *itself.*

A turbine of cosmic bioengineering? No way! It is something . . . something else. It speaks through embodiment to something beyond—or deeper than—embodiment.

• • • • •

There is an alternative explanation. As neither Dr. Blechschmidt nor Dr. Sutherland quite intuited (though many of their shared disciples did), the universe exists simultaneously in a primordial vortex that is oscillating multidimensionally and an atomic, molecular field disposing matter through space-time. Galactic and DNA spirals represent the effects of liquid light and gravitationalized molecules resonating at different metaphorical scales. These combine in the passage of primordial vortices into macromolecular shapes. The ninety-seven percent of DNA that seems to have no purpose may be unzipping helices of information from a cosmic coil like trillions of tiny crop circles, slowly pumping new realities into matter, into cells, into consciousness.

Genes are gears and moorings that maintain congruence between the humongous waves of quantum gravity flowing through space-time and the temporal, material subcurrents gathering within those. They form a microcosmic Tree of Life, which is why their triplets are comparable to the trigrams of the *I Ching.*

Randolph Stone, the founder of Polarity Therapy—unimprisoned by materialistic science—saw its intaglio unfolding:

> Life is built on the pattern of the cells ... which in aggregate make up ... being.... The center of any energy field (the neutron or nucleus) is the life and light of the sphere of action around which the energy particles revolve [to] give it a solidified appearance. In each field are locked up mysteries of beauty as supersonic sound waves and transcendental displays of light waves, as chromosomes and rays of color that mortal eye has not yet seen....[9]

Here the glacial pond sparkles under an alpine shadow. Here the spackled snake winds through underbrush.

> Spirit Energy, in its descent from Spirit to matter, creates a field or neutral pole in the middle of its travel, that partakes of the nature of both [positive and negative] poles. And it is the mind which is this neuter pole. This mind, which is slowed-down Spirit Energy, becomes the functioning factor or positive pole which rules the negative pole of matter....[10]

The embryo is the cosmic mold by which physical and chemical forces array and express themselves, but at the same time other entities may be preassembling hardware to transmit the same messages across the heavenly spheres. The prime mover causing cells to cohere, and tissues to complexify in blossoms, is probably outside Einstein's universe, a tesseract ruled by a geometry we don't see. A realm in some other octave conducts molecules and cells in life patterns much as wind-shifting branches cast depth effects on grass.

• • • • •

If we looked at Sol as it really is, we would not be blinded; we would be transformed:

> The Sun is like a hole that has popped out in the energy field of the Galaxy, allowing energy to flow out in a given area; that's why we see it as bright white light.... [T]he Sun is not a nuclear furnace. [It] basically represents cosmic energy, the true energy of Creation.... [I]n higher levels of vibration, you see a lot more than just the pinhole of the Sun; you see all the vibrations that surround it ... concentric spheres of energy, like ripples on a pond, which are connected by a spiraling coil of energy that expands on out to drive the orbits and rotations of the planets and moons.[11]

Here Hermes Trismegistus meets Mr. Crick and Mr. Watson. Here the cerebrospinal breath meets the visage of Jupiter in rambling belts of smog and membranes. Here the jay lands in a gnarled, lichen-covered apple tree, looks about, and takes off again.

• • • • •

We hardly need yetis and UFOs.

The embryo is the circumstance and moment where spirit and matter fuse or, if you don't accept spirit and matter as the terms of this equation, where the unseen hyperdimensional universe touches a palpable mass-object in three dimensions. "How" we don't rightly know, but we waste far too much time mapping genes and valorizing our capacity for outsider knowledge.

In the cathedral of creation the embryogenic field and the primordial field are synonymous. . . . .

Embryology must finally be a millennial science like astrology, a projective topology dwarfing the physics and chemistry out of which its tissues are manufactured. It precedes all other systems of number and form as—lo, a pulsing orb of (to all appearances) shapeless aspic ... presto chango, a salamander!

When viewed in the contexts of sacred geometry or tarot, embryology is a system of esoteric manifestations. An embryo is a cosmic object as much as a bio-artifact; its masks and gates, vessels and concealments are stations of a feral Zohar. In fact, embryos are the only true meta-objects: they are the sacred algebra under life and the parley of bodily existence.

If we saw the embryo as it truly is, we would see spirit's whirling dervish. The gene is a *ta'wil,* a symbolic exegesis fused irrevocably with a molecular cipher,[12] a totally nonmanipulable object, nondual, endemic to both galactic and extragalactic space. Like alchemical antimony or the spirit of an eagle tethered in a kiva, its double helix is sacred fire, dangerous and unconditional. It cannot be touched. It certainly is not a toy.

Nanomachine activity at the level of chloroplasts, mitochondria, and Golgi bodies transmutes photons, water, oxygen, nitrogen, sugars, carbohydrates, and other sacred emanations, masquerading as molecules, into energy by the alchemy of photosynthesis, oxidation, glycolysis, and the like. But the cosmic machine is weaving life and implanting DNA spools wherever the serial harmonics of planets orbiting stars in three dimensions allow raw creationary material to spill from their suns to be tapped. Molecules seem almost the default medium for transbiological waves that become biological in the context of planetary landscapes.

Perhaps creatures are etched and embodied by an invisible *imago mundi* puppeteering alphabets of pre-cyrillic scribbles. Tiny structures channeling excited electrons through labyrinths of protein metabolize oxygen and generate ATP while they convert cosmic energy into the motion and respiration of native bionts on worlds. Astronomers stare right through this sacred landscape and report only the exoteric explosions *(zahir)* accompanying the hydrogen caul. Candescence without boundary or meaning is only the form of the universe: limitless debris and tinder.

We exist in a Hubbell postcard of exotic galaxies proclaiming a reality that our living and breathing just as adamantly deny. Any scientist or politician has forbidden desires *("the life that is moving and calling us...");* he is fuelled by the ambition to transcend his or her thermodynamically prescribed lot. This fundamental schizophrenia imbues all our laws, customs, and machines.

● ● ● ● ●

Despite the dissonance of their scales, the eternity of their separation—from the background sputter of space to the cries of a nestling—cell and star wring an unaccountable harmony that shatters time, playing forever and only once, recoiling against each other's pulse infinitely and instantaneously. A drake to water, nothing less.

The creation charted by observers aiming instruments into its guts from the inside is a faux megacosm, little more than (as the first astronauts visiting the Moon recognized) "magnificent desolation." The stars and their retinues are uninhabitable at our level, inert even if temporarily infested by squirrels and Vikings. Marginally arable badlands, presided over by gravity and nuclear fission, they are tundra of Thor and Odin, bearing no internal design or subtexture that could germinate life. We cannot find a placenta anywhere in the night. Such a universe is physically incapable of creating either DNA molecules or sentient beings.

The Big Bang represents the current spiritual and moral stage of science, not reality and not us.

Astroembryology . . . .

Because *matter itself* is alive, proto-animate stuff has a propensity to twin, layer, and branch. In intergalactic nurseries, amino-acid milk positively bursts with organized design, though stuff does not become embryogenic until it gets out of both the compression and heat of stars and the absolute-zero vacuum of space. Then it *makes itself alive.*

Smaller sets replicate larger sets, all the way from a massive stellar explosion to a single bud of kelp. An unknown algebra ties their coefficients together. Fractal transitional states—expressed as crystals and cyclones—connect quasars to jellyfish in all galaxies.

The first incarnating aspect of the primordial field is its waters. Water is magical. Wherever it lands, it transforms, turns blue and green, bursts with colored petals and pied tunics. Fluids in resonance existed long before life, before planets. In fact, H_2O and other nectars float through space in opalescent shrouds sparkling with unborn worlds.

It is beyond hyperbole to suggest that all fluid systems in the universe resonate at a scalar level with one another; yet, if Jovian moons Europa and Callisto spawned oceans, these spin relatively close to Earth. Perhaps in the same way that DNA codifies alternate proteins and organs for possible future use, hydraulic systems, vibrating between glacial and meningeal seas, transmit potentialities for diverse morphogenetic profiles across galaxies and light years.

Nonlocal reality may not be particularly useful to creatures prowling or swimming on individual worlds, but it unites them in cosmic molecular dance.

We and other life forms support one another's isolate existences.

• • • • •

Many reels and canticos toil in substance or spirit (or both) to move meaning through domains. The esoteric nature of the universe alone will determine how matter arrives at life by various reproductive molecules, source codes, and biodynamic sequences. These will construct bodies (or their equivalents) on their own terms; they will carve vessels for incarnation that represent some aspect of experience.

Just as terrestrial tidepools and spume metamorphose animate forms over eons, so may Neptunian whirlpools and currents or elemental eddies in the oceans of the belt of Orion or thermoclines on volcanically hot and frozen worlds among the Pleiades, each in their own fashion. Or they may not. Either way, it is a conscious universe.

Capturing local sunlight in autocatalytic networks of molecules, biofilms coalesce and hatch metabolic transmissions, intrinsically and creatively, in native weather in uncountable solar systems. We may someday see this verified, in different kinds of "cells," by "bacterial" snowflakes on a Uranian moon or sea squirts swarming around a Europan geyser—perhaps buffalo look-alikes galloping across Alpha Centaurian tundra or electrolyte sunflies on a Mercurylike planet hugging the inner orbit of Aldebaran.

It doesn't matter how big and massive the world, how turbulent and dense, hot or heatless the medium is. Life deepens from coacervates to sponges to brains, from pulsating bells to tingling filaments, from thylakoids to vegetation, embroidering fountains and fences with its sheer materiality.

• •

Thylakoid: the flattened membrane sack inside a chloroplast that converts photons into chemical energy.

• •

Photographed by NASA in 2000, curling for kilometers along the edge of a Martian crevasse and into its depths is a jointed "glass worm," a harbinger (or artifact) of the embryogenic universe.

• • • • •

Hypothetical eel and dolphinlike creatures evolving limbless beneath miles-thick ice on moons, the oceans of which are warmed by friction of nearby gas giants, are unable to peer through the fifteen miles of snowpack above them and also lack phalanges to engineer a hole in it.

Jupiter's own methane sea, vaster than ten Earths, teems similarly: ropes of vegetation; comb jellies attaining sonic speed from the beating of miles-long braids of flagellae; intelligent "fish" the size of terrestrial cities; and assorted brine stars. Sentient Jovian creatures are too large to escape their planet's gravity well and, in any case, like the inhabitants of their moons they cannot build rockets or saucers far underwater with soft flippers and flagella instead of limbs and digits. They do not even know where their waters are, what clouds lie above them, let alone the unseen planetoids and starry night beyond those. They may not even exist.

But, by meditating upon the waters with their kilometers-long nervous systems and ganglia, they understand the nature of creation as well as we do. They dead-reckon the true galactic center, according to a geography we would not recognize. So did shamans of the Pleistocene, summoning totems from the field of stars and addressing spirit forms of the animals they had slain.

University-educated land creatures on small Sol-3 worlds, vain and anthropocentric though they be, are not the only sages or scientists in the universe. Hampered by layers of "cloud" and "rock" they do not even begin to detect, they preach false absolution.

. . . . .

Consciousness is a one, wherever it occurs. It can always recognize and communicate with other consciousness, even with creatures on planets in remote galaxies if a true-time channel (an ansible) were ever opened. Mind will apprehend itself anywhere, under the most alien conditions, because there is only one mind—man and fish, monkey and octopus, staring at each other, even now on distant Earths.

Everything, from the splanchnic petals and musky hives of aquatic plants, diatomaceous fowl, and gigantic furred ants on worlds in galaxies billions of light years apart, share genesis in the same spat, the atomic germ, tinier and bluer than a robin's egg. Every creature, every papilla and papoose comes from one litter, suckled once, body against body, at the only well. There is no separation or strangeness in the universe—no E.T.s or purple people-eaters to invade from elsewhere, from outside what we are.

Though we probably can never travel to other galaxies, hypothetically we can discourse and break bread with all sorts of exotic creatures on myriad worlds to infinity. In fact, we are already there, as consciousness. That is what astral travel and hyperspace are all about.

Our minds are stars because, undeclared, they condense in the same gravitational fields as stars.

• • • • •

There is too much and too fine dust and debris in an acre of the Moon or Mars ever to catalogue, track, or understand in trillions of lifetimes—to say nothing of all the worlds and systems of worlds in the cosmos. Something else must hold everything together at core, must bring the vagrants home by a force beyond divisibility, an adjacency superseding the annals of space-time. Something else must confer order and meaning.

Knowledge about the universe is not a series of measurements and landmarks drawn from telescopes and electron microscopes. We don't have to travel to Neptune or another galaxy to see what "this" is. Life has a bead on its own nature, on the path of its coming to be, that is neither enhanced nor brought nearer an ultimate unified field theory by objective physical science of the sort valorized these last few centuries on Earth. After all, we are looking through an obstacle so many times thicker than Jovian-system ice that a few more miles of smoky crystal blotting out the starry firmament are a trivial impediment.

We can't abort life. . . . .

The Bible is not the word of God. If you're God, you write a whole lot better than that. The embryo is the word of God. God doesn't talk to Man any other way.

Darwin is only the first stage in the modern qabalization of creation, the transposition of God out of scripture into nature, where by hermetic law he alone abides.

The Bible (for what it's worth) is the word of the ancestors, the deification of a line of paternal elders going back through Winston Churchill and Cotton Mather to Mohammed and Moses to the King of the Apes. Those who call the Bible the word of God wouldn't know the word of God if they heard it, but they want to establish a precedent of punitive authority and get the right to impose it on everyone else. There is no better way to enforce agendas than to proclaim them the oath of God (see pp. 55–60 and 271).

The three great mysteries and touchstones of true Christianity are embryogenic: the Trinity (Father, Son, and Holy Ghost), the Immaculate Conception, and the Resurrection—that is, they all require the involution and transition of spirit through matter; they suggest that ghosts don't just dress up in hauberks and body-suits but incarnate via the same vesicular membranes as everyone else. Even the Son of God, when corporealized and made human on a world, is conceived as a vulnerable, semi-stable life form. He is cell-based; he bleeds. His DNA is also bread and wine. He is *in* creation, not ruling it with a stick or missile shield. He is alive in every cell and trickle of metabolism.

"A plant may not talk," Peruvian shaman Pablo Amaringo tells his questioner, "but there is a spirit in it that is conscious, that sees everything, which is the soul of the plant, its essence, what makes it alive.... [O]ne can sometimes hear how trees cry when they are going to be cut down. They know beforehand, and they cry. And the spirits have to go to other places, because their physical part, their house, is destroyed."[13]

Sad enough! Yet the will of the supernal force incarnating beings overrides any act of vandalism, technological hubris, or pilfered turtle eggs. Pesticided mosquitoes, oil-covered albatrosses, mutilated pelicans, and incinerated ants go swiftly to their destinies, transformed by the absolute nature of life and death. Those who destroy them are left behind to suffer the karma of their "sins."

Do not mindlessly crush innocent birds and spiders, for it is your own mind you are destroying. It took the universe the identical time and care to make them, the same billions of meticulous eons, feather by feather and flap by flap, as it did to make you. They are just as divine. They are just as authentic and real. They have as valid dibs on planetary space and freedom.

When anti-abortionists rage over the murder of unborn souls, they miss the complexity of embodiment. All embodiment is at risk, at stake, all the time. We can't know the problematic circumstances or hidden destiny of any individual sperm, egg, embryo, unused zygote, *in vitro* ovum, human clone, or discarded genetic material.

We can't abort life. We only abort matter.

We are not spectators. . . . .

Look at how the dawn spreads its waves of luminous hue—pink, magenta, and orange—across a cloud-streaked mural. Tree branches rustle. Birds take flight. The machinery of civilization stirs, a car, and then another.

We are in a manifestation. Nothing could be more clear. The manifestation is not our measurement or diagnosis of it. The manifestation is how everything is here, no matter what we call it, no matter what anything in it imagines itself to be.

We can only understand the universe by how we come to be inside phenomena. We suffuse through swiftly-altering morsels of an egg. Our morning is yolk. The belt of Orion is visible above the lights of local aircraft crossing the zenith. Its stars are so much further and so much larger than the shell of our world. Yet both beacons sparkle at the same relative size.

• • • • •

On a planet of grasses teeming with midges, worms, screeching beasts and fowl, human beings (consciousness) awoke from a trance of indeterminate length, as in a fairy tale told by the Brothers Grimm or a creation myth spoken by Dakota bards: a prince morphing from a frog, a medicine chief passing through coyote masks.

The dream of misty waters is the bath from which royal gold saturates in nutlike layers is *la lune* in which lotuses uncurl from other lotuses under petalled suns. Brine of nucleic seeds, we bud at a sub-molecular level from transmutation of metals, at a cellular level from agglutinations of molecules, in the cosmos at large from dissolution of hydrogen gauze through itself. The fetus in womb passes through all these guises en route to body-mind.

In embryonic cloaks we travel deep and dark as celebrants approaching a fair on foot. Getting closer (more membranes), we begin to hear the collective din. Closer yet (neurons synapsing), we discern separate shouts and musicians' songs.

Then we are among them.

• • • • •

Lived experience transcends any conjectured basis of mind or conditional existence. Getting made is a journey of passing into three or more dimensions and four or more quadrants through a series of radical projections that ultimately sew us into bodies. We do not pick ourselves up incrementally like Humpty-Dumpty getting back on the wall. We do not enter from outside; we create context solely by our becoming.

We are not spectators at an airport watching vehicles spit and climb. We are ... all aboard.

The Cosmic Language ....

DNA may have been discovered in cell nuclei by Francis Crick and James Watson, but a doubly twining serpent or paired helix has been an insignia inside us since the dawn of consciousness. When scientists ambush chromosomal artifacts below the threshold of light, they are reclaiming a coin of the oldest gods. Its emblem manifests across our planet in ouroboros snakes, coiling dragons, serpent boats, and elves descending on a twisted rope ladder between heaven and earth. The royal scepter of Mesopotamia (two serpents rotating helically around each other to form a magical staff) is the Egyptian hieroglyph for flax and the caduceus under which the medicine guild was established.

Across the indigenous Amazon, similar motifs abound. Peruvian shaman Pablo Amaringo's paintings of landscapes visited after imbibing hallucinogenic ayahuasca reveal not only illuminated vines, giant snakes wearing crowns, and inhabited planets parading just above the rain forest but credible triple helices of collagen, DNA spools, chromosomes in transition, and the embryonic network of an axon with its neurites.[14] Amaringo adds, "... the hair, the eyes, the ears are full of beings. You see all this when ayahuasca is strong."[15] They are full, that is, of protean life forms.

The rocks, masks, and figurines of old Africa, the markings on the sides of drums, native American glyphs on buffalo hide, aboriginal Australian rock chalkings of turtles, kangaroos, and emus all reveal other "photo-micrographic" fragments. The decorative shield surrounding a Walbiri Rainbow Snake shows something like early prophase and late anaphase of mitosis and a startlingly precise coiled helix.

As communiqués pass precognitively through the universe, at their core is not only DNA but a pagan cipher that gives rise to DNA. Its template is not a nucleic acid, but "an aperiodic crystal that traps and transports electrons with efficiency and that emits photons (in other words, electromagnetic waves) at ultra-weak levels...."[16] Capable of storing and transmitting almost limitless metabiological information in complex uncertainty states, it equally underwrites photophosphorylation, glycolysis, neural synapses, and syntactic strings.

According to anthropologist Jeremy Narby, a master code is generative of both DNA and consciousness. Vested in the submolecules of life forms, a languagelike strand specifies amino acids. These brew hallucinogens inside vines, mushrooms, cacti, some fish and insects, etc. That is, the text became life even as it became consciousness.

The relationship between morphogenic and hallucinogenic molecules mirrors the link between DNA and human symbols. It was by the latter that DNA was discovered microscopically, represented mathematically, and provisionally "decoded." It was by the former that the poisonous snake and vine were snared and cultivated by microscopeless shamans and priests.

Images we view in ordinary circumstances are not "things" either, of course. They are facsimiles, currents and cascades of neurons coursing along membranes. Excited photons bombarding the retina of the eye and tympanic cavity of the ear, amplified along molecular gradients, are then kindled into holograms. Their collective input is sorted and resurrected in ganglia according to an electrochemical convention of "what's out there." Consensus reality has evolved through many phases, water-based and land-based, of long-extinct organisms.

By a masterstroke of bioelectric energy, plant spirits by-pass all such vibrations. They speak directly into molecules of brains, transmitting information that precedes culture, that is both internal (inside cognition itself) and extrinsic (emanating in chlorophyll-based life forms). Two tongues (one biogenic, one totemic) conduct signals congruently into life forms and apparitions of life forms:

> The spirits one sees in hallucinations are three-dimensional, sound-emitting images, and they speak a language made of three-dimensional, sound-emitting images. In other words, they are made of their own language, like DNA.[17]

The facets DNA goes through, coiling and uncoiling, threading and unthreading, tangling and twisting, and the complex manner in which its immense length and existential fineness are stored in minute capsules suggest the supernatural techniques by which wise pixies appear and disappear and arise spontaneously from humble weeds.

How bridges of molecular and geographical reality—widely separated channels of consciousness—are traversed is a mystery, but no more a mystery than how they were navigated on the prior occasion; e.g. the assemblage of creatures from molecules by code. Information at different scales, in different mediums, converges. One pagan cipher (that of heredity) becomes lodged in another (the ascending networks by which the brain translates signals into images and information); both are scored in Kabuki-like renderings inside nucleic structures that simultaneously transmit archetypal messages and biological designs.

Transcultural correspondences like these are either fanciful and coincidental or they are clues to the origin of meaning itself.

· · · · ·

Icaros—the power songs of ayahuasca—arise spontaneously. Students, without being given words, suddenly intone the same magic choruses their predecessors likewise mouthed without a memorized text—many of them without even knowing the language. In fact, it is the unrehearsed singing of an *icaro* that cues the shamanic teacher as to where his student is, in what part of the universe he is travelling, how far he has come, what nonhuman intelligence is guiding him.

When Narby asked Carlos Perez Shuma for an explanation of the recurring giant snakes of his visions, the ayahuasquero suggested that he bring his camera with him the next time so he could take their picture and analyze them at his leisure. Narby objected that such renditions would not show up on film, but Carlos persisted in his playful metaphor: "Yes they would ... because their colors are so bright."[18]

Ayahuasca landscapes, more real-seeming than dreams, shimmer in holographic panoramas in the mind. Pigmented objects saturated almost to phosphorescence are illuminated in the brain as if coming from its own sense organs.

Bird-headed and beaked humanoids, gabby fish, kings borne atop coiled serpents, jewelled spirit palaces in which enchantments are taught, dragon-prowed ships descending to refuel from from ponds tended by mermaids and snakes, seraphim drawn toward the charm of a full moon—like the winged and speckled denizens of astral space, these entities propel themselves through their own indigenous realms; yet they engage with humans on Earth in the telepathic *lingua franca* of creation. They don't just simulate alien panoramas in our nervous system; they *are* citizens of alien worlds, in faraway star systems, on some other frequency or plane, or in some other time frame. Or at least we must act as if they are, for no other premise legitimates their outsider knowledge or would presuade us to take them seriously.

Manifesting during trances and ceremonies, as plainly and explicitly as if they were real people, these beings address shamans and impart knowledge (though probably not without their own agenda). Their holographic excitations through botanicals become revelations about the nature of the universe, human society, plants, animals, heavenly bodies, pharmacy, and other serious matters. Local species are named and rituals are choreographed.

The recipe for the ayahuasca brew was itself deeded to shamans in visions induced by other plants. Even to an astute herbalist this formula is not obvious: it requires leaves of a bush containing hallucinogenic dimethyltryptamine cooked with a separate vine neutralizing a stomach enzyme that would otherwise block dimethyltryptamine from flooding the brain.

• • • • •

The whole UFO question goes nowhere when considered as merely visits by extraterrestrial spacecraft. The phenomena of Greys and cattle mutilators—aliens kidnapping Earthlings and stealing animal organs—is its *reductio ad absurdum,* arising perhaps from disinformation, perhaps from people's projections of nonordinary experiences onto Hollywood-like scripts (I am willing to be convinced otherwise). There is no revelation to be had from almond-eyed visitors, Pleiadians, or other aliens—no interplanetary conspiracy or benign conquest of us by a more advanced technocracy. To that degree we are on our own. The universe is neither linear nor operatic.

We do not understand the nature of time, space, relativity, or dimensionality well enough to see even what is happening on Earth. That is why we imagine uninvited guests coming from impossible places across distances no biological entities could travel without either living millions of years or violating laws of physics as we understand them. Wormholes tunnelling from this solar system to Andromeda and other galaxies or faster-than-light drives might change the equation, but they are, at present, facile plot devices allowing writers to conduct star treks. It is even possible that sentient forms could be beamed as information between solar systems or galaxies, or across dimensions, or employ quantum mechanics writ large to manifest suddenly on the other side of the universe. Such was the implication of Jeff Bridges' "starman" and Kevin Spacey's Prot in recent cinema.

Aliens, DNA spirits, and other occupants of hyperspace may not need ships or logistical operations. They may already be here in epidemic numbers.

Each year in fields of barley, corn, wheat, rapeseed, and the like, layers of circular geometry appear in exponentially complexifying fractal sets. Is this a planet-wide series of hoaxes or an explicit set of communications? Aren't we curious, SETI? A society that pours millions of dollars into radio-telescopes to seek extraterrestrial messages while ignoring crop circles and other possible telekinetic phenomena clearly isn't interested in how nature is structured or where we actually are.

The scientific elders are not invested in truth; while posing as scrupulously empirical, they use metaphors of shallow technoscience to hide from the universe.

We might even rethink the meaning of cells. Though local to the Earth's biosphere, a cell might also be a kind of transistor linking all biological fields across the universe. While a unique style of cellular bud may originate probabilistically in each planet's habitat and climate, the genesis of "cell" itself—a transformational integer of creaturehood—may lie deeper inside matter, inside nature. It may be a basic property of the universe as "being," even as the atom is a basic property of the universe as matter.

• • • • •

Nucleic hexagrams are membranous compact disks bearing (first) the Milky Way's archive of information, (second) pure sentience beyond the mortality of single races, planets, and beings, and (third) us—any "us," any "where," any consciousness. To know the alien riddle, we don't require a cannister left by E.T.s in an Egyptian Sphinx or a digital file cached in a half-feline, half-simian mesa on Mars. The record of the cosmic community is *inside* the biosphere.

In consciousness, there is no difference between another dimension and another planet, between beings manifesting at a different vibration from us and beings across outer space on an oceanic blimp orbiting another star within this vibration. Not only do biology and shamanism ultimately converge but, at the miniscule angle-arc running to the Earth from the center of the Galaxy, they are the same message beamed through illuminated colloidal beings, such as us, such as them.

Life on Earth is working out a primary cosmic code by all the means at its disposal—cellular and symbolic, morphological and semantic. No wonder astrological charts give uncannily lucid disclosures of destiny. At that distance—at that transgalactic declivity—our cells, our languages, our lives, our constellations, and our entire history are overlapping octaves of a single note.

We can't heal anything without petitioning the underlying ciphers that form and sustain us. . . . .

When asked by movement teacher Emilie Conrad to paint her primordial field, a paralyzed polio patient drew a series of homuncular eggs, their components shackled. Then, after she began breathing micro-movements into her limbs, nerve activity gradually returned.

Health (like life) is a pact of cell collaboration based in laws of complexity transcending both cells and organs while encompassing their principles of organization. At foundation, bodies are never just organelles and guts, never isolated, tarnishable tissues; they are metabolic, transpositional fields with visceral and phenomenological aspects. They are minds embodied. There are not even organs finally—in crabs as in ferns—only nodes conducting energies, matterless waves intersecting to form matter.

We are fortunate that our bodies are not as unilinear and mechanical as the medical establishment would have it. If we were ordinary machines, entropy would undo us quickly and summarily and we could not be healed of infections any more effectively than a crumbling sand castle. Viral invasions, tumors, and other malignancies would be everyday events rather than relatively unusual occurrences, and we would not live very long or be too happy while alive. We would probably not even get born.

As it is, we not only expect to be robust but consider disease a mistake or miscue. "Why did this happen to *me*?" the afflicted patient asks without realizing that the real question should be: "Why is this not happening all the time to everyone? Why is everyone's yarn not coming undone? Why do I exist at all? Why have I been so healthy, why have my diseases all been self-limiting till now?"

See how fast a dead thing corrodes, how one by one its cells cease being cells, how its molecules flee their enchantment in droves. Some parts of the body hold their shape longer than others, depending on how deeply they were imbued with the canopy of life rather than its vital principle. More ceramic, they persist as fossils.

Yet that vital principle, so absolute and basic, so uniquely critical to holding it all together, is identifiable nowhere. What ignites and enforces organized activity? What keeps it assembling intricate superstructures, one over another? What is the kernel at the center of the coordination of molecular events leading to cell cohesion and tissue function? What is the partner of hereditary biodynamics? What departs with the last breath? If it is not a spirit choosing an organic-carbon method of incarnation and assembling a molecular vehicle around itself, inhabiting it from within at every vacuole and interstice, then it does one hell of an impersonation of it.

· · · · ·

We do not know how complex metabolic, visceral systems cohere molecularly out of cellular configurations and function so well so long, how highly organized movements of cells and chemical cycles within membranes, billions concurrently, democratically serve one identity and plan.

A possum. A lobster. A maple tree. These are mysteries. They are not just undiagnosed. Their very terms of existence are incomprehensible. On what anvil could such clever nanotechnologies, such "fearful symmetry" have been forged. By "what dread grasp?"

When practitioners of medicine tend toward over-concretization and specification of sites, they miss the sourceless synergy holding life together. So many different stitches, such diverse levels of function and interaction could not be made to resonate in chorus, act in sync, and generate profound, integrated wholes just by solitary outputs. There must be a pulsatory, force too, an invariant jazz that transmits information by fluctuating throughout our bodies and harmonizing the layers of our existence.

We exist because of an imperceptible central principle, not through ligatures of gross fabrics. That is why diseases often cannot be tied to proximal causes. How could they be if existence itself, if life's breath cannot be proximally explained, is lit by an invisible taper?

That is also why meticulously conducted pharmaceuticals and surgeries have repercussions that are sometimes as pathological as— or more pathological than—the malaise they are ordained to treat. How can coarse drugs pluck subtle uncertainty effects and their electrolytes in cells and tissues? How can a knife sunder and splice a ghost or a congery of flowing energy? How can material science engage the nonmaterial core luting an animate gestalt? How can an organism subject only to linear vectors in planes of molecules self-organize in the first place?

The Gaia principle (see p. 65) extends to every organism. A life form, a cell, an organism self-regulates in miniature, as the Earth does in collectivity.

Life was never confined within chemocentric networks; if it were, nothing would have evolved. Foreshadowings of creatures would have emerged and dissipated like clouds in empty firmament, without any bestiary being written, any garden being sown.

• • • • •

Intelligence is not gigabytes of information but coordination and self-organization. It is not magnitude but pattern, denomination. That is why beings come into pattern-designs amid chaos, why templates persist and complexify over epochs—why evolutionarily vertebrates rearranged themselves as rebuses of new orders, passing from fish to mammals. Specificity itself is a form of millennial medicine.

In the emergence of complex systems with the capacity to regulate their fluxes to satisfy their own needs, intrinsic intelligence is essential. Enzymes are critical for multiple biochemical reactions underlying metabolism; genes are critical for organismal stability and perpetuation of species plan; but the things we experience as pulsations and thoughts are just as necessary in converting genes, enzymes, and large proteins into autonomous motivated entities. Much as enzymes and genetic messages inhabit molecular space and pour through precellular and cellular grids, so do raw, unsignified thoughts—the distributed content and modulated flow making up primal intelligence and generating the coordinated connectivity of living reaction networks. This is cell talk (see p. x).

It is a hidden communion where mind enters basins of attraction and systematizes (for all intent and purposes) infinitely dense chemical-molecular switchboards, but thought is there at some level from the beginning and always—before it is thinkable, before it is thought as we know it—nonlinear, impalpable and inextricable, disperse and indivisible, hierarchical enough to run a mammal but quantal enough to be present in a one-celled animalcule without leaving a mark.

How to conduct archetypal intelligence and specify not only its target but its character, energy output, and pathway (so that it gets to cells in a therapeutic fashion) is the premise of all healing systems, in fact the premise of any life. And this is the case whether those messages are directed consciously and medically or cascade under the threshold of waking mind along with the pagan chatter of cells.

A doctor attending to just tissue structures or an unlucky pathology in them misjudges the overall effect at hand. Something is sustaining the elusive reality of an organism, powerfully so, more powerfully perhaps than all the separate mechanical causes and effects of biophysics—and it is that "something" which must be addressed by medicine to get that organism healthy and fully alive again.

When a Taoist or Ayurvedic physician takes a patient's pulse, she is sensing that person's as well as the disease's vibration. She is reading the embryogenic field. She may attempt to introduce herb-based molecules or bioelectric needles at the same vibration. Getting on the disease's wavelength, getting the organism to recognize its own state of being, is much like getting the blastula to recognize the organism into which it is projecting. It is an act of pure healing that transcends the chemico-mechanical paradigm, the ordinary machine of the medical establishment. A vast, wireless noncomputer grid organizes cell activity.

Metapharmacological grains of homeopathic tinctures are presumed to transmit curative power by imprinting information in molecules of water. The design contained in each remedy's nanodose (composed of configured electrons) is hologrammaticized as a "similar"—a nonpathologized replica to a mirror-image disease in the body-mind. Like an embryogenic organizer or splice variant, it activates new patterns at the disease's exact vibration.

The spark that conferred life in the first place must be fanned again and conducted through dormant nodes within the body. Information transcending any syllabary or thermochemical hierarchy must be induced to unleash its meanings and convince tissues made of molecules that they are still alive and still want to live for the good, old reason—the same one as in the first place.

• • • • •

Technological medicine, while also an epiphenomenon of life, is so highly purged, abstracted, and rationalized beyond actual pulsations of protoplasm that it can operate only concretely and mechanically from the standpoint of ritual objectification. Doctors are trained as expensive, hi-tech engineers. Inanimate devices and drugs must do everything on their own. They cannot invoke much cell talk. They must, in fact, replace its primal language, imposing their own literal prostheses in its stead. Of course, their sheer horsepower and material refinement are incredible; they are the acme of intellectual will and acuity applied industrially. At their best, they get done categorically and kinetically what healers must do by sixth sense and magic.

Medical science and education are trapped in verification of not only the results but the means. The trouble is that the "means" are buried in infinitely dense, emergent series of nonlinear phase states that we recognize mainly as differentiation itself.

• • • • •

As long as life is its own labyrinth, intuitive approaches to health, cure, and maximizing function will rival the big guns of science. Each tissue complex reports in symbols, biorhythms, and other mediums not necessarily classifiable, on its relative health or pathology and how free it is to shift in its homeostatic field, how resonant with adjacent tissues. No restriction or disease exists in isolation; everything in body/mind is in depth, igniting from depth, expressing itself in depth, and transposing through infinite probability states.

This is what we feel we are—the one thing we could be and also be us. This explains the infinite depth of the universe in which we find ourselves. What we experience within body-mind is what we see in matter or emblazoned across the window of night.

Doctors, like shamans communicating with cosmic serpents, must intercept and communicate in vital codes; after all, they are enlisting cells to reprise the primal act of embryogenesis, the recruitment of loose molecules into interacting networks of pulsating organs metabolizing one another and communicating globally at scales stretching across cellular space from the nano to the visceral—nothing more, nothing less. Whether they acknowledge it or not—whether they even recognize it—they must invoke, arouse, and transmute the originary germinal field.

A notion as mystical and spurious as "vital force" provides as good an explanation of healing as any extant medical metaphor—for, unlike gravity, heat, and information, it is indebted to no designant, no hard locus, no metaphor or paradigm at less than full system depth. How else to explain the effects of prayers, voodoo, charms, sacred ceremonies, and shamanic sleights of hand? Radionics, Reiki, homeopathy, Body Electronics, and therapeutic color, sound, scent, touch, and even placebos apparently incite and synergize deep, sometimes miraculous cures. Ancient medicine complexes (Ayurvedic, Taoist, Mayan, Tibetan), refined over millennia, have empirically arrived at unknown precepts of biological organization. Call them vital or whatever; they somehow get into the traffic of enzymes, energies, and electrolytes. They go native and speak "cell."

Codes flow back and forth between sentience and matter because the cell is the basis of mind, before and after the fact. It is no wonder that the cell recognizes the mind; it *is* the mind. Every thought that arises from DNA helices to pulsate through H_2O crystals and alchemical salts silently impregnates tissues with unconscious meaning, the only meaning there is. That is how tissue is made.

The arbiters of science may authorize solely molecular effects, but these vital and constitutional systems have (guess what!) *exquisite* molecular effects. They induce unnamed subentities that function much like enzymes and subcodes in regulating homeostatic supersystems. They do not work in chemo brigades but by diacritically placed catalysts—after all, it is easy to overdose at a level of Golgi bodies and microfilaments.

· · · · ·

Each healing modality patches a faster-than-light hypertext between the types of alphabets indexed in symbolic fields and the older runes that lodge in tissue nodes and the nuclei of cells. Healing breaks into the psychosomatic feedback loops by which the energetic basis of each organ maintains afferent communications simultaneously with its own Golgi bodies and lysosomes, the pathways of its axons and dendrons, and system-wide discharges of enzymatic communication. A skillful palpator can actually feel this language change, the cure take effect, as heat, trembling, and texture fluctuations under his fingers.

Of course cells understand human language and intentional touch. Of course language and touch invoke fields of cells. Words and acts are an ongoing, evolving form of proto-cell talk. Mitochondrial alphabets, foreshadowing Phoenician and Iroquois ones, presaged oriole and dolphin too. In that sense, prayers and palpations—like caws and miaows—carry unadulterated vibrations. They come directly from the universe, without semantic interference.

Decisions in us are made by carbon rings and proteins fluctuating in quantum states of phenomenality and incident, transferring their intelligence into shape-shifting assemblages and psychosomatic communiqués semi-omniscient in their own spheres. The keepers of voodoo, medicine bundles, ritual songs, sonic stones, aroma essences, and the like—many of whom belonged to primitive, prehistoric cultures—grasped (at a noncognitive level) the meaning and principle of life. Long ago they stumbled upon the cycles, pendula, and signals that organize animate systems; thus, they were able to shoot their commands, nonliteral petitions, and even their hoaxes and trickeries respectfully into cell talk and get results. They became healing clowns, quacks, Bear doctors, kachinas.

In the end they moved not tissues or organs or cells, or even neuroses, but morphogenetic iconographies—meta-codes fluttering from one tier of manifestation to another. Whatever the system, whatever the method, the healer's vibratory field interacts with the patient's field; the disease is summoned symbolically and biophysically. Its mystery responds to the mystery being proposed to it.

Invocations, spells, mandalas, sand paintings, tissue releases, and other linguistic signals and conjurations are already the visceral discharge of life, the effects of cellular processes. Shamanic replications of the turkey's gobble, the wolf's howl can be cures. That is, they are shortcuts, trails along the nonequilibrium highway into fluctuating chaos fields. They are fodder for twenty-second-century science, the meta-machine, after the Age of Materialism.

• • • • •

Ancient static diseases that permeated the prehuman genome now float in unstable, inertial harmonies beneath a temporary alphabet of genetic surfaces and polyglot patrols of immunity medallions. Everything else, viral and bacterial, thinks it should also be us. And why not? It's all just promiscuous molecules.

Cell nuclei with their relics of endogenous retroviruses and lysogenic phages craft appropriate proteins to solve our emerging crises of bio-identity. Stimulated from esoteric quarters of the morpho-genetic field and perhaps junk DNA, these medallions speak the Old Tongue, alternating so as to change frequencies and sequences of RNA alphabets.

Any organ can hypothetically project itself back into the prelinguistic stems from which its homunculi arose. Stem cells are *in us,* at every level of our potentiality, our hibernating crystals of DNA—no need to assault fetuses and troll blastulas for seeds. We make raw materials as we need them. It may only be a matter of getting our blocks and silences to release, for us to be able to tell our body at the right level what is needed, for cells to become embryogenic again.

Medicine must transcend not only the commoditization of organs but the literalization of enzymes and chromosomes. It must generate stem cells by stimulating ribosomes and spindles to talk to one another, eliciting healthy out of diseased states, regenerating tissue, filling spaces neglected by waves of morphogenesis. This will be the true post-cybernetic shamanic alchemy, conducting palpations and herbal potencies into cardiac tissue, lung, intestinal lining; stuff made of organs themselves.

Outside agents and instruments are usurpations and exaggerations of function. Cells are our once and future makers. Otherwise, asked William Blake, "what [deadly] art/Could twist the sinews of thy heart?/And when thy heart began to beat,/What dread hand? & what dread feet?" Immortal Bird indeed!

Yet it is a kind of New Age provincialism to think that any energy field, any invocation, any conducting of breath, any internal-like olla-podrida will do. We are made from a specific, deeply sheared, topsy-turned hominid design. In getting ourselves back "into shape" we must work intricately and exactly back through the same labyrinth, adhering in our meditation, yoga, prayers, exercises, herbs, or guided touch to the template of our manufacture, the precise (if near invisible) coil of our limbs, spine, viscera, fascia, and gut, the pretzel of interlocking symmetries and asymmetries that underlies the Da Vinci star, the exalted rat that we are.

• • • • •

The science of iridology, presaging a replication of the zones and nodes of the body in the kaleidoscope of the eye, is not so much an ocular palmistry as a clue to the sympathetic transmission of field states between crystals of different sizes and shapes, a link also back to the multipotentiality of the hologram of the blastula. Viscera reflecting in the colors, flecks, and fields of the irides are windows into the luminous heart of the central nervous system—the states and stations its wiring penetrates. Tissue of the whole is projected axiomatically into each zone. As the cell is the organism in miniature, each organ likewise contains a replica of its energy nexus. Liver, kidney, and heart lesions leave specks and other telltale signs in the iris. The body is actually a quantum leap beyond rainbow holograms and pyrochrome or lenticular reflections.

The ear is an atlas too, fashioned of fetal spirals, representing the viscera in a span of concretized ripples near the brain, the fret of each organ tattooing a site in its coils and folds.

The tongue is another diagnostic map; states of relative vitality and obstruction in organs are stained in its pink landscape. Ayurvedic physicians may dispense a metal cleansing rod to scrape off any clouds or cauls before they seep back to cells.

The crisscrossing lines of the palm are a kind of indigenous MRI. The whole face is a medical chart too.

All the organs are a sign of the folding-in of the universe in bodies. Resemblance and signature lie along every seam, opening to the haruspex's gaze.

Even mind is a hyalescent, clairvoyant projection.

• • • • •

Lying atop Mount Desert Island on a granite chunk of South Bubble rooted thousands of feet down, its nub bowing my spine over its back, I feel the quick magnetic pulse of that much rock zinging through me in transcendental x-ray, the dark succussion of stone in my back like submolecular sonar. The limitless laser sun penetrates my chest and belly.

As much as any surgery or drug, stone is a healer, if aligned crystal to crystal to a body and received in full, blank bursts of ancient correction. The sun exploding out of its hole in many dimensions is the most concentrated elixir of all.

Creamy-seeming clouds spiral and break in vanishing galaxies across an azure luminosity; aspen and oak leaves flutter; glitter crosses the disturbed texture of Jordan Pond as if swarms of insects. I feel myself not a concrete thing, not even a glorified jellyfish marooned on a rock, but a field of light and oscillating matter, wrapped in a scarecrow of shirt, jeans, and sneakers, my mind swimming as soft and hard as its capture on stone. My semblance of protoplasm is imbued from above by agitated photons, from below by tiny crystals, echoing through trillionfold matrices, the lost labyrinth of labyrinths that lets any of this be.

11

Meaning and Destiny

• •

The Relation of Consciousness to Matter

> You cannot cut off being from being: it does not scatter
> itself into a universe and then reunify.
>   —Parmenides, 515–460 B.C.

"Being" remains the essential paradox
that science confronts. . . . .

At the start of the third millennium our assumption is that we are
well on the way to knowing everything, that our condition is comprehensible, subject to scientific substantiation.

Not many decades ago Albert Einstein mused that it was miraculous *anything* was comprehensible. Now we think we should be
able to *invent anything*. We surmise it is all machinery or a supercomputer of some sort, down to a nano-level.

Plants, animals, and humans don't come about mysteriously and
existentially any longer, getting born—they are "made," created by
information. We should be able to read their plan, follow it, and
back-engineer them too. We should be able to manufacture and patent
autonomous entities, robots with agendas, humanoids as real as

humans. Reducing creature existence to algorithms and heuristics, we seek to reintegrate them into complete agents capable of making decisions in novel situations.[1]

Mind is presumed to be bioelectrical mucilage, neural salvos of parallel reciprocal fibers connecting independently originating grids and igniting them into a hallucinogenic mirage. The more cells there are, the more complex the organism should be; the more synapses, the closer to consciousness. Existence is quantitative, hierarchical flow.

Trapped in the labyrinth of artificial intelligence, we accept the appearance, the imitation of consciousness, as consciousness.

If mind is mingled solely of molecules, what anneals their separate sputter-sparks and floods their interiors with light? What fans the ember that gives creatures existence? Where is that ember located? What makes consciousness real, makes it one thing in an ant, another in a dolphin? *¿De dónde son las piezas?*

How many membranous filaments do we need to wake up matter—91,750,211? a few more? many more? When there are enough, does a golem suddenly speak for itself as if "real"?

•••••

Despite the fact that most scientists no longer detect any difference between nature and technology, there are the obvious flaws that nature doesn't use machinery, and no machinery (however trillion-fold and subminiaturized its operating system) is discernibly at the basis of life or mind. Plus, machines—even ones that talk back and play chess—are not free agents; they are projections of our existence. They simulate habitual behavioral patterns and data outputs, rearranging generic incidents into familiar syntaxes; they ape the speech algorithms of the masses, thus sound like personalities. Yet they do not replace their own molecules or keep remaking themselves. They are not illuminated from within by awareness. They do not bear inseparable, universal consciousness, whatever that is, wherever it comes from. They do not generate independent proprioceptions or qualia. They are like the celebrated parrot which convinced the court that it understood—because it could speak—French.

A map of the brain or nervous system in no way corresponds to a map of consciousness. Circuitry stores information and makes convincing robots and toys, but it does not make life.

⬤ ⬤ ⬤ ⬤ ⬤

Even a relatively uncreative invertebrate, a cricket or sea urchin, is not just a mechanism—an untenanted configuration of synapses, engendering electric grids and enacting the chemicals of intent. It has an experience of its being; it knows itself. Its autonomy is existential, not vacant. Look at the enthusiasm in its plunge, the desperation of its retreat. These have the fluidity and irrevocability of intelligence, not the sleekly harmonized sputter of programmed circuits. They are initiative and wonder in bodies.

Where do creatures come from? Upon the demise of their bodies, where do they go? If an essence of them precedes life and survives death, how does it shuttle between states? Where is that essence when it is not alive? How do chunnels open between whole dimensions? What gives them *their* landscapes? "In what distant deeps or skies/Burnt the fires of thine eyes." (William Blake)

Who says, "I am!"—not only among men and women but rodents, crustaceans, spiders?

There is a reason "I am" cannot be spoken by machines.

⬤ ⬤ ⬤ ⬤ ⬤

Though we pretend to overwhelm the paradox of consciousness with aggregates of synapses, how can cognition be dependent on a brain that is itself dependent on cognition for its own phenomenal existence? Where do the brain's conscious and unconscious aspects actually reside; how are they distinct from each other; how do they interact? How could they be hardwired if the conscious aspects of the mind are incalculable and unconscious mind is bottomless?

Everything animate, even a lichen or diatom, has mindedness. Nothing inanimate does.

We cannot explain consciousness—how it got here, how it made a home in matter, what its meanings mean. Clearly, neurons accrete into ganglia and ganglia coalesce in cerebral fields—these systems evolve autocatalytically, chemicomechanically, nonteleologically. Yet, pure mind does not arise from or accrue through infrastructure; it does not invent itself or stoke its own thought effects, no matter how thick and consolidated the wiring and how much time is allowed for synapses to flirt with their own algorithms. Nor do the ingredients for mind reside componentially or collectively in cells.

It is unclear why any number—even a near infinite set—of empty shells, spools of wire, or vacant chambers melded and strung together in gradients should develop consciousness. A television, yes; a voice synthesizer, yes; a computer, yes; logic strings, yes; robotized phalanges, yes—but mind itself, no. Circuits are, for all intent and purposes, no better suited than ping-pong balls for fashioning thoughts. Nothing times nothing is still nothing; nothing to the power of nothing does not increase. Silicon circuits are of the same order as plastic around air. If consciousness and "being" are not *in* objects to begin with, those objects cannot invent or breed states of mind by rearrangement. They cannot manufacture mind out of stone. Only the fertilized egg can do that, can put billions of virgin filaments, strands, synpases in order within a densely spiralling, self-generating mass-construct.

Nature makes raw meta-machines that are fundamentally and syllogistically divergent from artificially assembled, mechanically performing devices—though, in principle, both operate in transparently identical nexuses of cause and effect. They should each have quantifiable energy input and output, closed circuitry, and an ascertainable inventory of functional and moving parts transparent to microscopy. No phantom switches, intangible generators and currents, hidden source codes, or invisible threads and networks—let alone minute goblins and jinnis—should be needed to run either type.

That a biont is alive and an engine inanimate does not remit the former's thermodynamic loops or biochemistry. Yet a living machine means an entirely different thing, has properties of a different entelechy. A mere germ cell, i.e., any generic cell at its root, emits infrastructures—morsel by morsel, layer by layer, system by system—into fleshy, epidermally enwrapped torsos with ganglia and limbs. It convenes and charts sets of entangled viscera, dense neuromusculature, enzyme-signalling grids, neural pathways, outlet sensory nodes, transmutative canals, sequence-specific maps, and synchronies generating independent organisms and their modes of sentience, while directing, maturing, and maintaining their gargantuan enterprises as well.

All of this technology is imbedded in a deceptively spare microstack of tubules, filaments, fibers, spindles, and axonemes that, to all appearances, is as primeval and rudimentary as a volvox or amoeba. The intelligence driving mammalian composition, breeding mammalian mind, governing resilient biological factories and their products, and impelling creature talents and acts is inscribed in troglodyte protozoa that function exactly like computer microchips and yet so unlike them that scientists will never arrive at artificial "being" by their present electrical and cybernetic model. There is no such thing, even theoretically, as adding the right genes to the right chemicals, because the system is dynamic and holographic from the onset—a now-inaccessible state.

Where exactly does an embryo tap into a labyrinth conferring life? How does it draw cohesive animism and intelligence? How does it start a new game each time, using the same heathen atoms, the same impartial landscape, the same physical rules?

What does the egg know that science doesn't? What moves does it have beneath its surface of ordinary molecular signals and networks that causes matter to change its very nature? What does it feint, abracadabra, that biotechnologists do not catch while they track its every step with devices far more discerning and particular than the audience at a magic show, fooled by mere sleights of hand? How did it get so clever and efficient by its own algorithm? Where do emergent properties come from if not *intrinsic* complexity?

If a vital factor enters nature by way of the particles of the fertilized egg, do they organize themselves from a blueprint inherited incrementally, or does life use molecules only secondarily as a way of manifesting? If the latter (or even the former), could scientists find what it is about cells and embryogenic designs that establishes mind-edness and then build or attract that thing to their own hierarchies of chemicals and circuits? Could they start with raw materials and make a worm?

If the vital factor originates *sui generis* outside matter, then all attempts to inculcate it artificially will fail. If distant tomtoms send ripples of sound through air, no manipulations of molecules can deploy anything but a static imitation of their timbre and beat. There will always be a difference between the Moon and mere wave machines. It is a matter not only of depth and scope but angle of attraction. Life has a self-correcting accuracy, a stability that tran-scends its own itinerary and plan.

If consciousness is not a collective, hierarchicalized by-product of neuronal, ganglionic networking and zillionfold cumulative synaps-ing, then it can *only* be an indivisible principle, materially affixing to cells but not reducible to them.

And its actual source must dwarf technology in the same way that all the stars and galaxies dwarf human history. We cannot count our way to complexity and meaning any more than we can count our way to prosperity, happiness, justice, or enlightenment. Some things just *are;* their existence is qualitative. They have to do with what creation is, where it is sourced and where it is headed, not just the slab of time-space-awareness on which we happen to dwell.

There is nothing that is not dependent on us for its representation. . . . .

Our most dominant illusion is that we live in a place that is visible, illumined. That is a joke.

Most of this world is latent and incorporeal. Virtually nothing that is here is actually here. It is lodged, all of it, in utter, bottomless tenebrosity. The real fire burns inside *la máscara.*

The stuff that is lit up, drenched by sun-star, neon, and ultra-sound, is nothing much, a glittering colossal surface, a vast effulgent façade. Beneath the clatter of civilization and razzle-dazzle of cultural melodrama stretches an exponentially more immense, obscure underbelly: the ungaugable tolerance levels of behemoth machineries; the silent, stark architectures of cities; the grief of children; the fantasies, internal chatter, and precognitive intentions of urban masses; the pain of animals butchered alive, hanging from hooks in factories; the secret flow between currencies and accounts; the atavistic desires and withheld malignancies of cells; the unfinished business of the dead; the deep gaps inside everything, between everything and everything else—the presentient planet.

Though the carapace sparkles luminously, it is big light roaming loose. The Hubbell discloses hydrogen like so much luminous wool. But the real cordage is beyond magnitude, prior to meaning; definitionless, contextless, unbeheld.

⁕ ⁕ ⁕ ⁕ ⁕

Scientists use optic mirrors floating above the atmosphere plus a radio chamber at the South Pole to tune into the faint flicker and hiss of the cosmos before the Big Bang. Peering fourteen billion years through a medley of stellar, quasi-stellar, and background static at a realm emptied of galaxies, stars, or planets, they approach the curtain of mind, of atomic reality itself. Matter seems to dissolve into something else entirely. It looks like acute, consuming heat followed or preceded by a titanic explosion and dispersal but, given the site and perspective from which we view it now, that is a mirage, a hoax played on matter by itself.

Only five percent of what they find can be deemed matter at all, made up of protons, electrons, and neutrons, the stuff of our bodies, rocks, fire, birds on branches. All the rest—ninety-five percent— is not that. It is not close to anything we recognize. It may not be anything we can *ever* recognize.

Thirty percent of today's universe is now deemed "dark matter," while a whopping sixty-five percent comprises an accelerative, repulsive force, a scion of anti-gravity operating under the trope of "dark energy." The language should not mislead you. These are utterly unconscious, unmanifested things. Dark matter is invisible and can't be detected, disclosed, or measured. Dark energy isn't matter; it isn't energy in the ordinary sense either.

And behind them is the true "dark," which writes all mythology and mortality itself.

• • • • •

Closer to home, the proud gleaming of street lamps, videos, and windows is ridiculously flimsy. A strong wind could spit them out in a moment. Even daylight is a mere vibration; even history....

And you think this place is lit?

• • • • •

What is crucial and determinate cascades below the thought-streams of philosophers, politicians, generals, executives, and (of course) the rest of us.

Every creature participates in constructing existence by subliminal choices and ancient predispositions. Worlds are collaborations of consciousnesses—not just human, but plant and animal, even Golgi body and thylakoid. Daily reality is comprised mostly of the breathing of mitochondria and Krebs cycles, the photosynthesis of chloroplasts, the collective subconscious thoughts and unacted desires of sentient beings, the unminded thoughts in octillions of worms and insects, the incipient intelligences of viruses and bacteria, the dormancy of atoms and molecules yet to become cells, cells perhaps once.

What else could reality be?

Money and molecules are identical emblems. . . . .

Nature is not what appears to spill through a peephole into our jurisdiction. A mere crack in a brick wall allows us to glimpse but a facet of the fathomless unknown. We see the shroud of one band of creation. On the traffic that passes, astronomy and biology impose the statistical mechanics of idealized particles in kinetic states, the flow of time, energy, matter, information—ripples in the ontological veil.

Astrophysics is an advanced form of a very primitive system of measurement and knowledge. It measures only the manifestation we are inside of, the shedded skin of reality. It sees what it allows itself to see and, while defending its candidacy, sees very little even of that—very little of nature, very little of itself.

From the planet on which we now exist, the ancients awoke long before science and perceived something totally different.

"If all existing things were smoke," wrote Heraclitus at the dawn of the Western world, "it is by smell that we would distinguish them."[2]

* * * * *

Appeals to materiality are culture-bound metaphors in a society addicted to static capitalization, in which value is printed on paper and distributed as truth. We have patented molecules, dollars, organs, and other tender in order to liege people to predetermined valuations and feign an exit to the cosmological dilemma we are in. We think that somebody, maybe not us, is in charge, just as we think that something gives us rules. So we speak for that somebody (or something) and evade a responsibility that is ours alone.

Atoms and bank accounts serve the same capital agenda. They control the masses by the ceaseless transformation of matter into commodity-totems manipulated in creature minds for the indoctrination and disenfranchisement of citizens.

* * * * *

Our laws are the laws of a dream: circumstantial, adventitious. They mistake our present incarnation in a physical context as evidence that existence could only be carnal, that reality stops where matter stops.

We do not pause to consider that we are squeezing multidimensional data streams into experiments we certify with no regard for how and where those streams actually originate, how they take on sole proprioception and conceptual form in beings who apprehend them. Confusing meaning after the fact with *a priori* meaning, we are akin to Helen Keller, both deaf and blind, describing the color, speed, and hum of boats moving from the Statue of Liberty toward the horizon. Only by appropriating common language patterns can she hide the fact that she experiences something entirely different, something that can only be hot or cold, leaden or feathery: hieroglyphs along her carapace.

Beware problems that are created in language and then solved through maneuvers of phonemes and sememes within that same language. Most true objects are more remote and inexplicable than the debris pouring from the most distant galaxies, masquerading as the universe at the birth of time. Other, equally strange photons arrive from more local fields and flowers, but they too are ghosts. That is, they exist only insofar as they are configured into shape and meaning inside us.

Physical representations are superstitions. . . . .

The development of linear perspective by the fifteenth-century Florentine artists Filippo Bruelleschi and Leone Alberti has subtly deluded even sophisticated people into believing that a mere artisan's device, a convention of representation, *is* the world. We now depict linear topology and its perspective not only as reality but a facsimile to which all activities in the universe must conform. Scientists valorize models of atoms, molecules, and other artifacts as if at some level they were real things—real matter, real atoms, real gravity, real mass. Not that there are any *other* objects to get at in their stead.

We invent electrons, quarks, tissues, organisms, and their domains by relentless photorealism. A cell, which has dynamic rather than naked existence, is stylized into a picture or diagram of a "cell." We transpose bodies into radiation images of bodies. Even the wet, electrified brain, consensus-designated locus of all representations, has become a Gyro Gooseloose cartoon.

Given the depth of the problem, we have settled on an absurdly cheap resolution.

• • • • •

Conversely, when we admit "stuff" is mostly nonexistent, more energy than matter, we assume we have honored both the ineffable if transient life/mind event *and* the universe's essential materialism. Not so. Our real problem is physicalism, the belief that everything has a locus and trajectory somewhere in our map called "universe."

By deeming phenomena to be "only" representations of electrical flow along neurons, scientists then transpose their materialism into a new phantom domain—the physics of thought. "Although it is indeed a piece of self deception to think of the rainbow as existing apart from the human mind," writes Frank Broucek in *Regaining Consciousness: Awakening from the Nightmare of Scientific Materialism,* "it is also a piece of self deception to think that it exists only in the mind."[3]

While science and philosophy boast that we have long ago transcended the seventeenth-century Cartesian mind-body dialectic with our Wittgensteinian and Heisenbergian propositions, in truth our entire epistemology is a bogus resolution of Descartes. We are entangled everywhere in mind-body fallacies.

We have no explanation for mind that does not either reduce it, in whole or in parts, to body or idealize it as epiphenomenal to body. Mind remains the piece outside even when the puzzle is one hundred percent filled in.

We are now trying to find our way out of the devious proposition that mind is "a combination of conscious and non-conscious elements which are connected by a central control system. We are ignorant of the existence of such a control system by which we trigger the release of our deep wisdom and mental resources. Given our dazzling success in exploring the physical universe and our abysmal failure in understanding the human mind, [it is as though] we must instinctively avoid such knowledge.... This characteristic ... prevents us from grasping what controls, as opposed to what just describes, our mind."[4]

But this diagnosis itself is the cue that we barking up and down the wrong trail. The unsearched forest emits howls and other cries far more eloquent. My god, they are even calling us by name.

*The universe is utterly dense and utterly transparent at
every point.* . . . .

We worship the efficacy of numbers, and we employ their rubrics
to map every layer and morsel of reality. But numbers are series of
conventions, "indeterminate at a fundamental level,"[5] their ontological status in limbo. Perhaps the mathematical properties of gravity, black holes, and waterfalls have existences independent of us,
but clearly such existences are not what we encounter, and they certainly cannot be rendered by quadratic or other equations. Abstract
numerical objects—isosceles triangles and moebius strips, abscissas
and sigmoid curves—are proxies too, albeit elegant, spare ones, like
music, close to the source of thought. Charles Stein points out:

> Particles and forces are not paydirt if the laws that specify
> them are special cases of something else. But how we should
> think about that "something else" depends on what mathe-
> matics actually is. There may be a vast array of mathemati-
> cally describable universes, say, among which our universe
> is a product of evolution; each universe might present itself
> equipped with variant, mathematically specifiable laws. But
> it makes a difference whether or not mathematics is a descrip-
> tive language constructed by us for modeling empirical data;
> for how can what is beyond this universe be nothing but an
> artifact of something constructed from within it?[6]

We keep making the same mistake of misplaced concreteness in
different ways. We dismiss cognition as a basic fact, while creating
a purely nonconscious, algebraic universe out of the products of
consciousness, out of an illuminated condition.

The rational numbers are infinitely compressed everywhere
because between any integer and its neighbor—any two natural
numbers—lie limitless ratios and repeating decimals: one infinity
inside another inside another. The natural numbers and the negative

ones they cast in their mirror are a translucid continuity. The irrational, imaginary, and complex numbers are jagged and diaphanous, strewn through a universe without actual points in space. Random numbers, i.e., most of everything that happens, are mere passing tallies distinguished by the fact that they cannot be generated by formulas less unwieldy than themselves.

These squiggles reflect not the universe but our relationship to it by neurons drawing their count from the digitalization of gravity and heat. Einstein's space-time grid, Stephen Hawking's star spill, and the multiverses of *Star Trek*, *Star Wars*, and other fantastic lands gyre and gimble in the wabe.

* * * * *

Looking inside stuff at a microscopic or digital level for what is most basic or real tricks us every time. We will always detect something because there is something everywhere. It is no wonder that, once we contrived powerful enough equipment to delve inside, to enlarge the invisible a billionfold or more, we found atoms inside of molecules, electrons inside of atoms, quarks inside of electrons, mitochondria and genes inside of cells, etc.

The universe is utterly dense and utterly transparent at every point. Open a portal and there is another portal inside, open that portal ... and so on, forever.

Embryogenesis is superconscious radiant energy flowing into molecularized world-stuff. . . . .

Why have stuff at all? Why bother? Why not just void—no time, no space; no matter, no thought; no here, no there?

A moment's consideration answers this question: the existence of the universe is the aftermath, not the ground of thought—and not thought as we imperfectly measure it. From mind springs matter to incarnate its properties. We find something everywhere because mind is everywhere.

The universe is not solely matter or vacuum. The universe is knowledge, utter.

Before the Big Bang, before matter, there was consciousness. Molecules congealed from the same original flow as we do, as our thoughts do. Time and space are dimensions ultimately of mind. We dwell in a temporal, spatial projection.

The type of universe we have is determined at its root by "being," not thermodynamics. Thermodynamics arises from being, from consciousness flowing across a landscape in search of its own true nature.

Nature is not just random movements of inert matter. Matter is an explicit effect of mind, the medium in which mind is enacted.

The rustle and rattle of event and stuff—including hurricanes, sunspots, waves against coastlines, and mighty rivers—happen because thought is omnipresent and restless.

A footstep on the branch of a fallen tree makes a snap no one heard that we call the Big Bang.

• • • • •

Experiments show that water molecules form either harmonious or discordant crystals depending on the types of music played to them—symmetrical ones when the concert is Bach, disorganized ones when it is heavy metal or rap. Even the emotions of people handling the vials affect the gels inside. Love and anger transmit molecularly quite different landscapes.[7]

These events provide a hint of the missing science of mind over matter. Despite no physical evidence for telekinesis, our reality becomes whatever we think it is, not all at once (in some spontaneous transmogrification), but over epochs. We impose subplots unconsciously on a mysterious, frangible event.

Technology is not just the result of incremental applications of thermodynamics and properties of materials to the landscape in which creatures find themselves. It is the precise meeting ground of mind and matter evolving through Stone, Bronze, Machine, and Atomic Ages.

It is not just convenient and fortuitous that we were born in a universe than could be chipped, sharpened, glued, melted, alloyed, wired, combusted, transitorized, and networked; it is the only sort of universe in which we can exist.

The cauliflower cities of civilization and their industrial artifacts and traffic are not a prodigy bursting rough and presto from industrialized rudiments. They are the inevitable descent of primal energy vibrations into density and turbidity. They invent the world of which they are the seeming product—minute after minute, hour after hour, day after day, year after year, lifetime after lifetime.

. . . without ever turning on the light. . . . .

The problem with physics in its infancy is that experimenters do not realize that forces of nature yield conveniently to man-made laws and mathematicizations only because they are also forces of mind. Atoms, sub-atomic particles, and DNA helices are shadows of consciousness, which is why we find and manipulate them so readily, using a technology we invent out of the stir and patter of our thoughts. Cars, planes, radios, and computers materialize almost magically out of consciousness even as they are simultaneously constructed of minerals and petrochemicals. They were on this planet somewhere at its genesis—dormant during the Ice Ages, latent in the pumps and war wagons of ancient Assyria and Persia. As we rushed (and rush) pellmell toward mechanism, the shadow of the machine precedes the machine.

How else were the stones of the Egyptian Pyramids and Celtic star monuments transported and hoisted into place, did Peruvian shamans trance-glimpse the serpent mounds of North America, did the Crucifixion imbue Western civilization with a triune God, did this planet become linked up by cell phones and fibers? How else was a Pliocene landscape transformed into farms, fishing fleets, factories, and cities?

· · · · ·

Researchers keep finding new stuff, mechanically, molecularly, experimentally. The potential inventory from cyclotrons, silicon circuits, electron microscopes, and the like has a long way to go before it exhausts itself, but it will run into mind everywhere it seeks, even in unconceived dimensions, even as it tries to flee mind's innate contamination. It is possible that there is in fact no actual universe—just a shifting universelike collaboration in which thought forms occupy imaginal space.

In order for science to work, for technology to take us where we are truly going, able to go, it must find where consciousnessness meets matter, where thoughts themselves converge with atoms, where cells convene on reality. Otherwise, it is not simply that we won't solve the riddle of nature or make our grand unified field theory; it is that our machines won't work, won't be good enough or run long and powerfully enough. We will be helpless against fundamental crises of resource depletion, industrial pollution, energy itself, and human violence. We will fail. We will fail without ever meeting ourselves. We will fail without ever turning on the light. We will become extinct while still strangers to one another. And we will never see all the others who are standing right beside us.

Thought creates biospheres. . . . .

Embryos are the sheerest, most acrobatic molecular processes that allow nature to get at itself, at its primal atoms, to "think" its own existence, to seep into space-time, to undergo metamorphoses. These will never be unravelled by science because they are both inside and outside physical effects. They are at the same level as the nature of being itself—pure process, pure reality, and (apologies to Whitehead) pure novelty.

Modes of morphogenesis—big mind assembling vessels for itself out of its own matter—matriculate everywhere that thoughts meld climes of atoms suitable for bodies to inhabit. All life processes are thought working on and partially solving its own conundrums, making designs that allow it to explore the realms of its own existentiality. The universe needs bionts on planets to become whole.

Consciousness is still, ever and once, finding its way through matter, finding its way through sentient beings everywhere to itself.

In fact, the missing ingredient that congeals bodies, that holds them together, that keeps their beads and morsels from unravelling, their molecules from running amok is the living current of thought binding every piece of matter inside a biological field to the shape of that field. Consciousness is a glue holding forms under entropy together (because consciousness is also entropy, is unconscious too).

We are alive because matter is alive. We talk because cells talk so much faster and deeper, without the burden of words. We are conscious because the universe is conscious. It should be no surprise when pigs and donkeys come prancing out of the barn.

. . . . .

Creatures on all worlds derive from one omniscient source. The varieties of embryogenic events known to us here on Earth or proceeding elsewhere among the stars are instigations provided by matter for its own evolution. They represent thought using available mass-forms and tissue-metaphors to export itself, to continue to spread across zones of already-spreading intrinsicality, tilling blastulas, gastrulas, and neurulas while generating each unique biosphere and noosphere out of its own bodylike, bodiless nature.

Thought needs embodiment to become Spinoza or Buddha or Elvis or even algae. Without mindedness at the heart of the universe none of these life forms would exist, here or anywhere. The ratios of a snail's shell and the sori and bifurcations of a fern are as much "mind" as Ten Commandments or a Bhagavad Gita. A spider egg, ravelling and unravelling, slowly becoming a creature, is consciousness, the same consciousness as we are, crawling out of a hive of its own strings, uncoiling into space and time.

Put it this way: If thought wants to swim in Jovian oceans, life will be there, breathing ammonia, not only because Darwin insists upon it, but because mind incarnates it. Likewise, if thought wants to infest the hot sulphurous clouds of Venus or boiling muds beneath its surface, airborne microbes and underground tendrils will spring forth, molecule by molecule, chewing metal.

Where the river is taking us is totally unknown. . . . . . .

The myth of our creation leads ever to a familiar grail: Robert Silverberg conducting Lord Valentine to the castle of Majipoor, Philip José Farmer seeking the lord of universes, Roger Zelazny in Amber, C.S. Lewis in Narnia, Olaf Stapledon and his Starmaker. . . . "from creation to creation, until he would come face to face with the being whose brain-child it was.

"But what if that maker of universes was a madman? Or an imposter? Or a super-criminal hiding from the wrath of his own superiors?"[8]

The surprise is: the creator of universes is us—us in exile, us with amnesia, us summoned by a resonant horn, us passing through the back of a closet into a sunny day in another dimension . . . us constructing our own origin tale.

. . . . .

Thought wants to have at itself because it knows even pain is better than nothing. Thought wants to enact something, to feel something. Thought wants to find itself in whatever there is, find in itself whatever there is, too. That is why this is so damn deep and lacerated.

Emotions in their elemental form make tissue out of atoms in order to have structures through which to sense and extend themselves—blood into arteries, sugars in the veins of leaves. Life forms arise from the physics of desire. Creatures are timeless essence quoiled into narratives. If an animal is a machine moving through the world, it is a machine for experience, an extremely subtle one.

Mindedness is not only requisite but absolute—the fount of unpersonified intelligence from which the syrup of matter flows. No wonder matter returns to it, always. No wonder matter finds it everywhere and gives it back its things, genes among them.

Despite the nightingale's haunting song and the sunset's radiance, "nature [Alfred North Whitehead] is a dull affair, soundless, scentless, colorless; merely the hurrying of material, endlessly, meaninglessly."[9]

A jukebox spooling ballads in a local bar, lilting melodies of *Danny Boy* and *Old Man River* across the glen, it is in us that the vast illuminated creation takes on meaning and trembles with hope "... among the fallen apples on the ground ... wasps rolling drunk with the juice, or creeping about the little caves in each fruit...."[10]

> Mind Energy is the finest form of matter.... In the human body, Mind Energy flows over the brain and the nervous systems and becomes animated Intelligence.... MIND ENERGY IS A NEUTER meeting ground between Soul and Matter.... He who can control his own mind in its entire fields of function can control the universe....[11]

Without this personified connection to its own creationary source the universe is an out-of-control mess, and we are headed for either catastrophe or oblivion or both. Already our reality is deteriorating at exactly the speed and resonance of the relentless negative information, con games, and commercialized debris our supercivilization is churning out—sometimes even in the guise of spiritual imprimatur.

Unless we acknowledge our role in creation, we are likely to ravage nature as mere products rather than a fragile shimmering in our hearts.

Matter reflects the conflict of mind's untamed grasp at its own bottomless depths. . . . .

That, for all its divine labor, the universe is not a delightful ingot of paradise, or anything close, and is filled with crooks, sadists, terrorists, pimps, and murderers, and bedevils itself every which way, is neither a defect nor a harbinger of future burnishment. It is a fact of the rugged nature of unquieted mind, of thought entering its own gnarled terrains and irregular pockets as it unfolds and displays its startling innards.

Look at it exploding crudely, exuberantly, in galaxies and stars, playing demolition derby with rocks and fireballs.

Look at its molten landscapes on Mercury and Io, the icy desolation of Pluto, Charon, and innumerable worlds like them, the glaciers of Callisto and Antarctica, the *mares* of the Moon and canyons of Mars. What about the deserts of Africa, the wind-swept snowy ranges of Afghanistan and Nepal, the density and shear plane of granite? These are magnificent untamed zones of original molecularity, original mind.

• • • • •

There is a tremendous amount of murk in matter to go through. Murk arises from the extension of a force brighter than light trying to open molecular space to itself. Murk is deeper than memory, which is why we forget.

We don't have any choice but to celebrate the majestic displays and horrific assaults on our integrity—the devastating, ecstatic, excruciating transformations of life. The many layers of shifting luminosity and shadow we pass through as we plunge and navigate our way down the boundaryless sea of "being"—these are inevitable, inescapable. Tragedy and comedy follow, as fish and vassals of all shapes and textures probe amorphous, swift-moving elements.

For those who want to move immediately to biblical or other prophecy, to channel Ra, the Pleiadians, or the Council of Elohim, I would point out that the world is in the way. The world got here for a reason, even if it is "no reason," and it is not going to budge and allow our projections access to their own pure source. Rampant debasement, contrariness, and obdurateness are not illusions. There is magic everywhere, but wishful thinking as such won't get us there. We have to abide by the rules, our rules, because if mind and matter didn't need such harsh, impregnable terrain, such treacherous quicksand everywhere, such long, notched chromosomal wires, such abyssal waves of cause and effect, they wouldn't be here.

We have to play the hand we were dealt. We can't rush things.

• • • • •

By dead reckoning, in the rain, my umbrella blown and then popped out, amid lunch crowds, I know that this is a ragged realm, a stormy, wild place, always on the brink of tearing itself apart. Though at times lucent and calm, its weather on good behavior, it is not mellow. Its itinerary is not a safe one. Its premise is neither bliss nor prosperity. It has no vacation or lunch break.

It is an opening out—in every sunset, thunder and lightning belch, tornado, battlefield, prison, and mangled bus—of the illimitable, unfathomed depth inside it: the cracked egg that will continue to crack and spill until everything is here. That is why the fox is tearing apart the quail, right through its guts to its bones, why the boat is dragging metal plows through habitats of countless creatures who are devouring one another, why fire is igniting acres of forest populated with deer, fowl, and nesting bugs, flames shooting hundreds of feet up into the planet's sky while turning bodies to ash.

These are the thoughts of a cosmic child, the child we collectively are.

· · · · ·

Creatures are modes of knowledge only, and knowledge continues to vomit out a universe unknown mostly to itself, that it itself is creating out of its own massless and eternal omneity—the true expanding equation issuing from the Big Bang.

Rough circumstances from the get-go . . . so a rough ride.

"Only fools and rascals could bear to live. . . ." . . . .

When Darwin set forth his proposal for the origin of species and (by proxy) the inception of life, the implications were poorly understood. It was presumed that man and woman could still maintain their divine sphere on Earth, quartered outside the monotonous, malevolent sway of matter. Nature was vulgar and detestable, while humanity remained inviolate and sublime. The predatory combat of beasts leading to the survival of only the least merciful (or lustiest) of them was—at best—remotely then a human problem. Various conscienceless armies had disturbed mankind's sanctuary, but they were uncivilized ruffians and pagans—huns, visigoths, comanches; they did not obliterate Zion. That was left to the high-brow constabularies of the twentieth century; they alone created a "cosmos devoid of directionality and overall significance ... indifferent to our suffering."[12] Their notions brought to our gates far more terrifying vandals than any jihad or Khan.

Novelist Peter Matthiessen sets the futile prophecy of a "heavenly city on Earth" in a late nineteenth-century context. At issue, in the mind of outlaw E. J. Watson, a decade or so after *Origin of Species,* is the status of a giant alligator: "[a] huge motionless thing across the river. It was simply *there,* oblivious, indifferent, and that indifference was more terrible than [any] literary notion of malevolence could ever be."[13] Darwin's antipathetic alligator machine had replaced not only God but the Devil.

Thomas Hardy, writing at approximately the moment the fictional Watson is pondering reptile nature, reenacts the dizzying brink upon which Western society teetered, having just begun to turn away from its idealization of life toward an apprehension that everyone, not just factory workers and peasants, but church-goers and nobles too, is alone in the universe. Nature cares nothing for them; their situation is utterly hopeless: "... happiness was but the occasional episode in a general drama of pain."[14]

Darwinian selection suffuses the crypto-landscape of Egdon Heath with inscrutable vegetation and wildlife, dwarfing humanity:

> [M]ummied heathbells of the past summer, originally tender and purple, now washed colourless by Michaelmas rains, and dried to dead skins by October suns.... Huge flies, ignorant of larders and wire-netting, and quite in a savage state, buzzed about him without knowing that he was a man.... Snakes glided in their most brilliant blue and yellow guise, it being the season immediately following the shedding of their old skins.... Litters of young rabbits came out from their forms to sun themselves upon hillocks, the hot beams blazing through the delicate tissue of each thin-fleshed ear,

and firing it to a blood-red transparency in which the veins could be seen.[15]

All random, all exquisite; all relentless, all inevitable! The furze-cutter, once a Paris businessman, now "seemed to be of no more account in life than an insect ... a mere parasite of the heath, fretting its surface in his daily labour as a moth frets a garment...."[16]

Nature is a pervasive, implacable clock ("the united products of infinitesimal vegetable causes"[17]), capable of beauty only in the chance appearances of its common operations. Snakeskins, rabbit flesh, "the intermittent husky notes of male grasshoppers,"[18] the indigo and gold marsh flowers, however gossamer, however poignant, are inert and impassive, to what *they* are and what *we* are.

• • • • •

In Hardy's account of *Tess of the d'Urbervilles*, the heroine's young brother Abraham inquires of her, as they guide their wagon down a country road just before the crack of dawn, if the stars overhead are really worlds. Displaying knowledge that had permeated uneducated gossip by then (1891), she answers in the affirmative.

"All like ours?" the child asks.

"I don't know; but I think so. They sometimes seem to be like the apples on our stubbard tree. Most of them are splendid and sound—a few blighted."

"Which do we live on—a splendid one or a blighted one?"

Tess's quick, unwavering response mirrors a fashionable prejudice: "A blighted one."

"'Tis very unlucky that we didn't pitch on a sound one, when there are so many more of them...! How would it have been if we had pitched on a sound one?"

"Well, father wouldn't have coughed and creeped about as he does, and wouldn't have got too tipsy to go on this journey; and mother wouldn't have been always washing, and never getting finished."[19]

The crisis of our location in the universe comes down to the basics (as Karl Marx demonstrated) of who does whose wash on any given planet. If this is a blighted world, everything will be blighted here; everyone will be in servitude. "... [W]hy so often the coarse appropriates the finer...," Hardy sighs, "many thousand years of analytical philosophy have failed to explain...."[20]

"The truth," he concludes in another novel, "seems to be that a long line of disillusive centuries has permanently displaced the Hellenic idea of life, or whatever it may be called. What the Greeks only suspected we know well; what their Aeschylus imagined our nursery children feel. That old-fashioned revelling in the general situation grows less and less possible as we uncover the defects of natural laws, and see the quandary that man is in by their operation."[21]

A quandary indeed! Some thirty years later, George Bernard Shaw gets to the heart of the matter. Humanity has been duped by nihilism masquerading as physical law:

Natural selection, he declares, is "a blasphemy, possible to many for whom Nature is nothing but a casual aggregation of inert and dead matter, but eternally impossible to the spirits and souls of the righteous.... [It] has no moral significance: it deals with that part of evolution which has no purpose, no intelligence, and might more appropriately be called accidental selection, or better still, Unnatural Selection, since nothing is more unnatural than an accident. If it could be proved that the whole universe had been produced by such Selection, only fools and rascals could bear to live."[22]

Now, eighty years after the publication of those words, we have exactly that, a civilization of fools and rascals, or in which fools and rascals hold all positions of power, of sagacity and science, and most citizens accept them as leaders and aspire to their status and goals.

Natural selection has been ingested, internalized, assimilated, and honed into a global ethos. In the process we have lost the roots of our own consciousness, the meaning of our sojourn in this amazing place. The so-called physical universe has been scoured of any pretense to morality or sacredness and "is simply too terrifying and depressing," laments one molecular biologist, "to be borne...."[23]

Each object has responsibility only to the physics of its zombi manifestation.

> [Tess's] narrative ended; even its re-assertions and second-
> ary explanations were done....
> The fire in the grate looked impish—demoniacally funny,
> as if it did not care in the least about her strait. The fender
> grinned idly, as if it too did not care. The light from the water-
> bottle was merely engaged in a chromatic problem. All mate-
> rial objects around announced their irresponsibility with
> terrible iteration. And yet nothing had changed since the
> moments when he had been kissing her; or rather, nothing
> in the substance of things. But the essence of things had
> changed.[24]

Her wagon is smashed. She is kidnapped and raped. Her one
erotic interlude terminates lovelessly. The whole tragic tale will conclude in desolation among the menhirs at Stonehenge where she is
captured by the police and executed, a Wahhabi court likewise ston-
ing a girl for the crimes of another, not even crimes, pummelling her
until she is dead.

All natural objects now mock mankind. Did we really think we
were special? Did we think we were capable of love? Did we think
we were alive?

> When she ceased, the auricular impressions from their pre-
> vious endearments seemed to hustle away into the corners
> of their brains, repeating themselves as echoes from a time
> of supremely purblind foolishness.[25]

The heavenly city—Zion, Camelot, Shangri-la—is now merely a
relic of a naive phase of human history. The proposition that life is an
expression of the laws of thermodynamics running circuits has led
unfailingly to societies of robots mining minerals, cyborgs turning
on and off switches, customers hoarding goods. Our species once

confabulated bare, innocent technologies, whittling pebble tools and shillelaghs, forging bronze, sculpting kayaks, catapults, and carts—useful, aesthetic devices, independent of human destiny. Now, having declared *ourselves* machines, we have become machines. Finding no difference between our acts and those of predators, we have achieved a world of total predation.

Shooting political prisoners, dissecting them on the spot, and selling organs hot out of their bodies for currency can pass as acceptable behavior only when one regards life as commodity, commodity as suitable capital, and matter as lacking any real consciousness or moral implication.

> "Tis because we be on a blighted star, and not a sound one, isn't it, Tess?"[26]

"Welcome to the desert of the real!" . . . .

Because we don't include consciousness in our ethics and technology, because we don't recognize ourselves from within as "being," but instead degrade existence from without as matter (or, more tellingly, degrade matter and ourselves as hollow and circumstantial), our whole approach to reality, all our world-views and belief systems, are skewed. Now we can't even find thought in nature. We can't find us. We can't make a sustainable planet. And we lose our dead when they pass.

We don't have to be overly moralistic about it, yet an incalculable negativity must fuel all the concepts behind materialism. It is a negativity that rids nature of spirit, purpose, or meaning—that condemns metals to be forged into planes and cars. The rabbit and fox are cruelly, even brutally skinned, the sea squirt centrifuged into gunk, so that we can create motherboards, engines, and cyclotrons. That is, we must tear down everything to its molecular components and then reorder them according to the imagined needs of our technocracy.

• • • • •

Our tools and toys zip by faster than our nervous systems and tissue layers comprehend or can assimilate. Nanotechnology is on the putative event horizon—commerce and war conducted in virtual fog by atoms.

"...are there beings (& are there beings)," warns the poet Robert Kelly, "who step on us as lightly as we tread ants?/[T]hat is the hideous question someone is always asking/Egypt after Egypt."[27]

No doubt there are, and they smell us and catch us in their webs from countless dimensions away. And yet we are heedless, even arrogant; we take no precautions and do not cover our scent. In fact, we entrap ourselves in the very tedium of doom. Our thought forms lead us to carnage, as if we had bought a ticket beforehand.

. . . . .

We have fallen into worshipping appearances. To a degree that makes mere veneration of a golden calf look like child's play, we have replaced God with physical and algebraic effigies far more powerful than any Baal.

In atoms and cells we are committing blasphemy and idolatry at an unprecedented level.

Witness the mood disorders, petty neuroses, and madnesses among scientists and doctors as well as the populace at large. This is the price of sacrilege. We are not yet trained to live in an algorithm; we do not know how to blend molecular into sacred space.

What scientist has the courage of a twelfth-century samurai, the simple faith of a Dalai Lama (or even an average Zen student supported in empty body-mind by the algorithmic blue of the sky)?

. . . . .

The movement out of sacred space into matter's bottomless catacombs has presented a chillingly clear glimpse into things—clearer than humankind can grasp. It is not that the message of science is itself so dire. We are who we were before its disclosure. ("Nothing has changed ... but the essence of things has changed."[28]) Nor can we (this far along) close Pandora's box or reverse our destiny. It's just that we weren't ready for the "truth," so we remain stubbornly stuck, unresigned to our new station, unwilling to embrace or redeem it.

We see a heartless, predatory leviathan embodied in nuclear, nucleic fire. Its presiding officials are a sniper, a slumlord, an interrogator with electrodes, a prison guard, a hanging judge, a mafia boss, a sonofabitch, a pit bull, a great white shark....

And there is no one else—no one out there and no one down here.

• • • • •

The gluttony and ruthlessness in nature are terrifying to us because they are incidental and circumstantial. They cannot be tamed by law and custom, for they *are* law. Generic and deep, it begins with single cells or less in planetary oceans. Raw swimming nodes, dissociated from any shared consciousness, ravenous beyond the possibility of compromise or reconciliation, gnaw into each other, cut open and chew apart their fellow creatures. They don't understand collective consciousness, mercy, or empathy, and anyway, they have no choice. From the creation of life it is as though nothing existed but hunger, fury, and fear. Blind beasts, though possessed of eyes, snap at blind beasts.

It came ashore with a vengeance, reenacting itself in tidepools, hives and anthills, along littoral beds, at watering holes, across tundras, and throughout the open sky (snake and frog, owl and wren, hawk and mallard, tarantula and hare, feline and mole, bubba holding a knife to the poor goat's neck).

Consciousness would not have gotten beyond pure cell cannibalism and tissue carnality without penetrating deeper into matter, hence its spiralling augmentations of cerebral ganglia. From fish to frogs and geese, a faint recognition of our circumstance and mission is accruing. The capacity for love is present even in dumb beasts. But gray matter alone won't do it from here. As consciousness, we must continue to work its transformation from within, using literally nothing, or nothing more (because more cells and capacity are precisely the wrong direction now). We must walk on air. We must do it for the famous reason: because it's there. Because we're here.

Zen mind travels through fathomless space, recognizes its own reflection, and acts. There is no other reason. There is no other motive. There is no other desire, but there is at least *that* desire.

• • • • •

The conduct of lamas and other honest priests is crucial. What other hope have we? What else dares to oppose tooth and claw as a statement of the innate "being" of the universe? What causes charity, empathy, morality, generosity? The mother giraffe defending her baby ... the whale using his own body to shield his mate from the harpoon.... If we are really different, if we do more than snarl, hoard, kill, and commit mayhem, then we have to plant those other roots as deep in mind and matter.

Unless we stand in consciousness, *for* consciousness, *as* consciousness, unless we experience the unhaltable spread of compassion *inside* our stuff, we are going to be thrown to the jackals and insects or even worse. We are going to be tossed naked back into the star machine.

• • • • •

We miss the double meaning. We don't recognize that the universe is transparent in order to allow us to find ourselves. We think it is transparent because it is made of particles and bits and plunder and there is nothing real or meaningful in it. We don't understand that meaning must, in fact, be this spare if it is going to handle our transit from state to state—or, as the Tibetan philosophers have understood for centuries, from one bardo, one bridge between, to another.

This world of life, like the worlds after lives, is a bardo too. Essence is in moments of transit, not places as such.

We are visitors in the universe, getting to share with each other the incredible radiance inside matter. As painful and irreconcilable as the conditions (mostly) are—from the disembowelment of all creatures to our absolute and complete vulnerability and the devastating theft of everything ever gained, even desire, even knowledge, even hope—

they are apparently the only conditions that will support meaning and destiny, that will take us to the heart of things and also bring us home.

I think that life is brave. It takes an unbelievable amount of courage to be an insect and test the voracious fields of summer, to carry strange armor as part of oneself and do battle with tireless invaders. It takes courage to be a rabbit, stalked from birth to death; to be a hyena or coyote or hawk, to have to maintain the high standards of ferocity and the purity of the kill, also from birth to death. Fish are brave just to be fish, to dwell in soft packages in watery currents. Every domain of nature reveals the dignity, unquenchable will, and valor of life forms, whether as humble as a worm or rapacious as a piranha. They are all doing as well as they can against the inexorable force of gravity and the chemical bonds that make them up, the queer bodies, instincts, and lifestyles they are meted by a deity or despot who will not take no for an answer. It is all the beetle can do to get its cab over the next bump.

The lion *would* lie down with the lamb if it were allowed. But nothing is allowed except this, so the lion and lamb make peace at a far more profound level by completing their assigned acts.

• • • • •

There is not a creature it isn't hard to be. A moth, a guinea pig, a snake, a koala or thrush, a penguin on the southernmost ice, a seal robbed of its mate and children by fur traders, a lone bear in the wilderness with its imperative, the oaks and spruces that make up much of that wilderness. These are all valiant shapes to struggle with, heroic bodies to steer through the cold hells of creation, intrepid actions to attempt on stormy worlds, always ignorant, always uninitiated, always itinerant, innocent. Even the cruelest of predators boastfully howling over its quarry is blameless to the end.

And humans are the most audacious and foolhardy of all, for tangling with the savage mind itself, with the totems and paradoxes of a universe that forever evades knowing.

• • • • •

Consciousness had to find itself. It began in an original miasma, more deeply subliminal than unconsciousness, on the nether side of entropy, concealed beneath dark energy as well. It was not in the universe at all—because it had not *made* a universe.

Matter in its most primeval form trickled out of even more primal stuff. It blossomed—electron-like, cell-like, lotus-like, embryo-like, tree-like—petal by petal, sheet by sheet, coil by coil, syllable by syllable into the thing we call "nature": inanimate nature, alive nature.

It needed hunger. It needed rapacity, theft. It needed paramecia and bacteria. It needed sharks and spiders, a reptile brain and teeth.

Knowledge began as desire, a distinction between self and other. Wisdom and compassion incubated in acts of stalking, in cannibalism. Language had to initiate from strife, pillage, and rape. The Word was a wound. Mind itself was aggression, assault. What meeker mouse dares interrupt the cosmic Om?

Only when self begins to witness, through its own consciousness, consciousness outside itself, only then does the real universe begin. The weary warrior gets to enter the temple and wash his/her hands in holy water. Matter is liberated; the map becomes the territory; the molecule becomes the Word. Only then is the path behind/before us clear—the way the universe has to travel in order to disinter and redeem itself.

A noncarnivorous universe probably could not advance beyond a vegetable, a giant galactic strand of kelp. Sad but true. When we get to the next stage (in a few billion years, hopefully somewhat sooner), we will understand why we rode there on the beast, wearing the beast's own body.

• • • • •

By the standard of angels our civilization is an unsightly disaster, a bare step beyond pure predation, far more Satan than God.

By the standard of the sea, by the law of the jungle, it is a gallant and promising start at reclaiming matter from mindless carnivorousness and abstract violence, at preserving consciousness for some future leap or transformation. It is the faint rustle of a remote yet coming jubilee.

Right now we feel both: what we arise from, what we are turning into. The mournful requiems of this state permeate every tune and human sound—if not their actual words and melodies, then the passage of air through their horns, the moans of molecules colliding here, the relative mixing of substance and vacancy, churning light and darkness, absence and presence, between epiphany and despair.

. . . . .

Here in the West we keep practicing losing our faith to see if we can find it again—bigtime on a global scale. This is a deadly game, a merging of Hide-and-Go-Seek with Russian Roulette and Chicken. For we spend so much time trying to demonstrate we aren't peeking and have really kept our eyes shut that we may be losing our ability even to know what faith is and what it is like to find it. We have been keeping ourselves blindfolded with such compulsion for so long we may forget that taking the blindfold off is also part of the game. And not just at the last possible moment.

Faith is not a luxury, and it is not something we can afford to let slip away while trying to meet the universe eye to eye without blinking. It is the only thing that keeps lives, that keeps whole civilizations afloat in the great dark.

When God speaks in Job, it may not be God, Creator of Time and the Universe, but our own unconscious voice—unacknowledged, unheeded—certainly back then, how much more so now! Our voice calls to us in the guise of the most omniscient demiurge we can imagine, to tell us that the majority of our nature lies lost within, beyond worship, beyond hubris, beyond knowledge.

It is God because it is not us. It comes from a being so far above us as to make us cower and hide from ourselves. He tells us, as we tell ourselves in His voice, that so long as we pretend to honor Him, to obey His law, to mete His justice, while making ourselves His crusaders of a defective and fatally wounded universe in alias of His better one, that long will we be homeless mendicants, barred from the temple.

As long as we kill in God's name, as long as we take it in vain, as long as we run from our own shadow and deny to Him who we are, as long as we blame random tyrants and comets and infidels for our downfall, our isolation, for taking away our birthright and happiness, that long will God rage inside us; that long will we punish ourselves and then condemn our acts in His name.

Science doesn't begin to address the problem of nature, but the anonymous author of Job, tells it like it is. Does this rapper ever lay it down!

> Who is this who darkens counsel/By words without knowledge...?/Where were you when I laid the foundations of the earth?/Tell me, if you have understanding who determined its measurements...?/Or who stretched the line upon it?/To what were its foundations fastened?/Or who laid its cornerstone,/When the morning stars sang together..../Or who shut in the sea with doors,/When it burst forth/and issued from the womb;/when I made the clouds its garment,/And thick darkness its swaddling band;/When I fixed My limit for it,/And set bars and doors...?/Have you entered the springs of the sea?/Or have you walked in search of the depths?/Have the gates of death/been revealed to you...?/Can you bind the cluster of Pleiades,/Or loose the belt of Orion...?/Or can you guide the Great Bear with its cubs?/Do you know the ordinances of heaven?[29]

We must now face the real limits, the actual bars and doors, the lines and ordinances of the universe, its foundations and cornerstones, its measures and springs.

Our soul yearns for release from a self-imposed prison, this life sentence, centuries before the atom, gene, or galaxy were disturbed in their briar. It is man calling out to man, woman to woman, not God but our own damaged, still intact psyche trying to bring us home.

Whoever doesn't know how to look unflinchingly into the eye

of the tiger and at the same time feel the tiger's bottomless heart is doomed to anxiety, phobia, depression, insecurity, and insomnia. Hearing only its empty, nihilistic roar, we flock like gazelles into passing ideologies.

• • • • •

Yes, this universe will eventually collapse and disintegrate. Yes, the Sun will erupt in a nova and then fag out. No way anything here (or anywhere else) survives. No way anything has ultimate value or moral significance. No way out.

Yet, given how mysterious this whole situation is and how little we really know about it, how out of context the present cosmos or our existence in it is (i.e., lacking any objective compass), we can hardly claim to know anything about the next moment, let alone billions of years in the future or past.

If we all woke up tomorrow, twenty feet tall, made of light, and sailing through a landscape of shimmering psychic hues, we would accept it just as we accept morning here or any other trance or bardo. Certainly *this* Sun and *this* universe and *this* world are entropies running down, arid husks in the unmaking. Our molecules and our children's children's molecules are going to be immolated along with them. Where do we make our stand?

The answer is that meaning is bottomless and can take on almost any identity, any political or theophanic form, including ones that will eventually incinerate, redeem, or even rescue us.

Maybe the universe *is* a bier of death, an engine of obliteration. But we are here by a hair's breadth anyway, and where we come from and where the river is taking us are totally unknown.

No use in basing a society or its morality upon events so far in the future that virtually *anything* can happen between now and then.

• • • • •

We don't get it. Karma is not something to study and obey; karma is something to feel, in the depth of our very existence. One must behave well not because the universe is keeping score and there are rewards and punishments. We must behave well because it is the right thing, because how we behave is who we are. How we behave is what the universe is, what the universe of ourselves and our children will become.

If we trick ourselves into reducing our existence to molecules and cells, then the meaning of our life will evaporate too. We may be winning the secular game but we are blowing the only one of consequence because, having made ourselves petty tyrants over matter and over our acts, we have become mere hirelings and henchmen, totally myopic in relation to the epic event of which we are the source, the guardian, and also—when we let ourselves sing—the very song.

The death of God in the West, while doing no injury to God himself/herself, has devastated humankind.

Form does not differ from emptiness. . . . .

Why is there something rather than nothing? "That is the hideous question. . . ."

But if there were nothing, this would all still be the same, us here as "nothing" rather than as "something." In fact we are *nothing*.

Neo-Darwinians argue that life is random and mind illusory, while physicists have proven that matter is *nada* beyond energy and mass, that things are mostly empty space.

Emptiness is what matter is and not some metaphor for it. Cells arise from nowhere and disappear. All that exists at core is entropy: matter and energy. And matter, come to think of it, is little more than energy (particle waves) gathering in dense enough bevies that other particle waves cannot pass through them. There is nothing that is not a transient wave-form generating other transient wave-forms.

We are little more than phantoms—that is, brief laser-like shells for molecules that hold fields together for a time. Then our conscripted monads snap the illusion and eke back into the cosmos.

It *is* nothing, all nothing—but not in that way.

Something or nothing is not the universe's question, nor is it any question at all.

In Dzogchen exegesis, a thought arises while the mind is observing itself. The mind seeks the source of that thought, where it originates, how it lasts and extends itself, and where it goes. Yet the recognition of a thought causes it to dissipate and vanish; then there is nothing at all—no source, no origination, no abiding, no place it goes. "We find there is nothing, which is why the essence is said to be emptiness."[30]

Far beneath the instincts and unconscious thoughts of sentient beings, underpinning the layered waves of embryos, the universe billows in total latency. What we experience as life is but the conditional manifestation of certain of its aspects. Yet such sentience holds together even the simplest creatures.

Rising from the tall grass, a newborn foal extends its head to its mother to be licked, its eyes searching hers for an explanation, for solace, for reassurance.

Buddha nature sinks through matter, creating the illusion of a world. All reality is a defilement of our pure essence which confronts infinity by a realm of no time or space, no duality. Down here, "I" emerges as a form of dampness, hardness, fragmenting, splintering into many "I's," each manufacturing worlds.

Embryogenesis is what happens as Buddha nature swirls through materializing thought forms. Latency becomes shape and design.

We are alive now, but that itself is forever. How else can we comprehend the eternities before our birth and after our death? We may not have always been us, but we were always consciousness, and "us" was always proximate to our passage in consciousness. Now we are doing us here; then we will be doing us before and after here. But it will be changed forever by the intimate, radiant manifestation we know as the big L, life.

"... *till we die, scooby-doo....*"

None of the big human events break into the true reality—not birth, death, love, or murder. As major as these are, they evanesce and pass. The world moves on, absorbing horrendous slaughters and terrorist attacks, conceding brutality and devastation, rationalizing or denying the torture and killing of children, surviving unendurable extinction and loss. The planet reconstructs itself daily, almost as if nothing happened, because a more powerful wind drives it. It ignores our petty, passing conflicts and concerns, our temporary attachments, as it conducts the unfolding of consciousness and matter.

We cannot even keep our attention on the horror of the world's debacles. Prayers are intoned; absolutions are made. A by-stander's coat is laid over the victim's head. The corpses are removed; the show goes on. (By a year later, the events of 9-11 were referred to on the Beltway as "the flying Arabian circus" and our leaders had moved on to their "shock and awe" campaign and tax-incentive mirages).

Life prolongation makes no sense, for the body is a temporary palace of disintegrating crystals, its molecules coming and going, replacing one another. Nothing is ultimately ours except maybe the grid itself, which exists nowhere.

Pure existence can't alight in a place where things (including us) are mostly space and orbiting charges.

Billionaires of the future may try to stay embodied longer than the span that mortality grants them (by cryonics, stem-cell injections, harvesting of new body parts, even stealing bodies), but it will always be orbiting crystals.

Everyone has to die sometime. Immortality is a sterile, futile ambition.

In fact, no one should have to experience so much history in a body anyway. To have to be here forever would be hell.

· · · · ·

But that doesn't mean nothing here is real. Everything is real, even if its ultimate reality doesn't reside in its lapsing configurations.

This world is no mere figment or passing fantasy. Lives are in fact incredible things. We wake and go to sleep, wake and go to sleep some 250,000 to 350,000 times, all in a single day. We see clouds, meet people, play, attend parties, make love, laugh, splash, hide, tumble, go to school, change shape, dine, sob, give birth, travel, read books, have fevers, remember childhood, mourn, pray, embarrass ourselves, sit under the stars drinking mead or beer, forget what was once so important, and die, and they are all the same thing.

We spend our duration with this one job, managing the life of this other person, deciding from moment to moment what he or she should do, when to eat, what to eat, when to rest, where to go, whether to do this or that. Even a mouse must manage a mouse, a spider a spider. It is chemicodynamic, but its acts are real. They are the only real acts, the only indelible events in the anatomy of the universe.

Ego and identity—personal memory—are ephemeral and expendable, while consciousness itself is beyond history or self. It is what makes dew on the grass, bright dandelion heads and chalices of morning clover. It lies outside the galactic clusters we view through corridors of eternity. It is first love, lake water against flesh, wind sorting cloud cream through prism blue of worlds uncounted, untold. It is able to kick a ball, acquire knowledge, lapse into total despair, regenerate; it has a sense of competing to win, of ultimate meaning.

"Billy, I'll tell you, buddy. You're a long time dead. You'd better enjoy every day until then."[31]

It is tough to get a body. It is also tough to get out of here alive— that is, on our terms, unindicted, unsullied by implacable surgeons operating the many jaws and bureaucracies of death—unscathed by sorcerers commandeering spirit portals as dragons and eagles.

• • • • •

Despite the impermanence of bodies, we don't just get tossed around universes willy-nilly, and we also don't get asked our preferences. Lifetimes can't be skipped or refused; every life is an aspect of the cosmos. No lives are superfluous. All must be stumbled through.

That is what the universe is: "... *one last time from Freddie's joint...."*

Being alive is the original visionary experience to which other sprees and nightmares defer.

Our existence is shaped like a rose, not out on a stem but internally, a subtlety of petal and curve—a sheathing as thick as a galaxy and as gracefully etched as a cell, a roselike layering of shapes and aromas, celebrations and lamentations. Once we embrace that rose, we can go anywhere, can dispatch energy and compassion to any site.

Nirvana is finally samsara, samsara nirvana. The material realm is plumb spirit, the spiritual realm a literally en*light*ened state of material.

· · · · ·

Minnows wriggle at lake's edge, sun casting perfect black, undulating runes of each one, as they propel themselves with itchy sinews, darting among rocks. Not many rotations ago of a planet charmed to the migration of a sun-star, they were mere atoms collecting in the ancient body of a mature fish—stones that were not stones, stones that germinated. They arose Aladdin-like from transubstantiations of jellied balls. Now they are swimmers. A child discovers them, tiny scooting things shimmying back and forth in the wonder of their medium. A woman sees them through her child's memory, different minnows in a different pattern in a different body of water, covered with lily pads, the same dance, the same sun, the same helices, the same golden lotuses; just as solemn, equally in the moment, as exquisite and irrevocable, and as transitory.

This is truly strange. This is unbelievably wonderful. This is so fragile and dangerous that we should do anything we can to get to the bottom of it, the bottom of ourselves, before it is too late.

· · · · ·

Do you see that the universe has made living houses out of its own molecules and flakes, woven them around and around inside themselves, inside *itself*, in slumbering layers, as owls and antelopes, as hungry ravens, fuzzy larvae crawling on weeds, as cats slinking through underbrush, whales plumeting in the watery abyss, as impetuous skunks and vibrating blue-winged needles, as sunflowers and periwinkles, all out of the mutating opacity of eggs and seeds? The universe has given room for states of consciousness and identity, action-forms and meditation shapes—all (in fact) that nature will allow.

Sentient figurements come into existence through their cell and karma essence—inexplicably, spontaneously—recognize themselves for an instant, and vanish. There are no eyes, no ears, no mouth, no tongue—no consciousness either. As the Heart Sutra puts it: no death, but also (since nothing lives or dies anyway) no extinction of it.

* * * * *

Those privileged to be here do not realize what the universe goes through to set them floating in their trances, in SUVs, talking on cell phones. Looking down from skyscrapers at taxis shooting along streets, teeming in crowds to bazaars, they have all but forgotten what they knew in being born—that it is a prayer, a view. It was not deeded. It was not obvious. It was not perpetrated by God the Creator, and it did not just molecularly accrete from umbra of inanimate mass.

It is pure experience, and it can go anywhere it wants, anywhere it doesn't get in its own way. It can ride explosive charges in tanks, blasting at the enemy, or it can journey, a pilgrim in grace, and bring medicine to the poor. It can sit under a tree and witness the gold ingot at the horizon spilling its wave-lengths through violet interference. It can dance the hop; it can cast spells.

We are here, at last. Consciousness is not an illusion. We are rooted in it grave and hard. And it is our job, in fact our only job, to sense it, to reach toward it, to find and touch aspects of it and bring them into ourselves into the world.

* * * * *

The creation of this plane of reality is a gift. To travel the road of life and death in a body, in the animated and electrified suit of the gods, is an honor.

Thanks!

. . . . .

The science of the future must not only address the presence of mind in matter but translate the extraordinarily subtle effects of what has been lost in unconsciousness or death into explicit effects and consequences.

There is an unexamined law whereby the more deeply something real is buried, forgotten, or destroyed, the more profoundly and fatefully does it arise elsewhere and impose its unresolved karma and meaning on its world. It can return as an event, a force, a species, or even a machine. This is represented in astrology by the birth chart and in Tibetan Buddhism as the passage between bardos.

Likewise, the further we go from somewhere important and the less we are actually present there, the more globally and unconsciously leveraged our absence/presence becomes, not only in the place but in ourselves.

This is how the laws of embryogenesis work. It is how, if someone leaves the Earth and all its crises and attachments and goes to another world, even another solar system, he brings those crises and meanings with him—essentialized, synopsized, and potentiated—even as he leaves himself back home to the same degree and in the same way. This is what holds the universe together and why, until we evolve beyond this stage, nothing different will ever happen. Antipathy will rule empathy and sympathy, from holy lands to ocean depths.

But this is also what gives consciousness a future, the universe itself a future, into eternity, beyond any contraction or singularity. Antipathy must be resolved. Antipathy must be resolved by our making real emotional connections with sentient beings and the stuff of our own lives, by feeling unerringly through the dense and tangled fabric of embodied matter, not by machinery, not by physical force or skills, not by intellectual or psychic power, but by ourselves alone—by deepening recognition of our flawed perfection, by unabashed candor, by unconditional love. Everything must be transformed, without losing *anything*, which is the only way it will finally recognize itself, the only way it will stick.

Consciousness will win, not because of any particular thought or insight, but because acts of incomprehensible cruelty and violence (e.g., Iraqui soldiers tying dynamite around detainees and blowing them apart; Nazi soldiers herding families into gas chambers—or worse; Contras pushing prisoners out of planes; African militias hatcheting off children's hands; calves and chickens butchered alive into cuts of meat; suicide attacks yet to come) are reaching blindly toward acts of selfless compassion and service. Probably the single-most reason for this whole messy, bloody creation is to disturb the ghost's dreamless sleep and get it to be real. To escape the trap of its own facile omniscience, to sparkle in its own view like cobble after rain, to assemble itself from its own sheets of stuff and ceaselessly emergent layers, to thrash with actual fins in its own icy depths, consciousness must cut its teeth on everything that could still be unconscious—everything. By being capable of both bottomless terror and bottomless love, by knowing both so well that they are equally irrefutable—in fact *only* by being capable of both, ruthlessly and inexorably—does consciousness transcend both, transcend also its own seeming insubstantiality, and become a thing.

This provides the barest tidbit of safety in our troubled state, but it is our sanctuary in the universe at large.

· · · · ·

The wonder is not the megawattage of galactic light in the universe, the incomprehensible vastness of suns, sun after sun, each conflagration exploding quadrillions of times the size of the poor Earth … but that that same universe is epitomized inside us, that we occur inside matter, that we are made of light.

Consciousness is not threatened by the death knells of planets, stars, or even universes. Consciousness is not adumbrated by the vastness of space-time or the complexity of matter, not then, not now, not ever.

The universe in all its dire mystery and bewildering plethora, in its terrifying, antipathetic grandeur, is a precise invariant replica, a mask of the collective unconscious, a projection of all that is thought and unthought, made and unmade, created and destroyed. It is us. It is our single hope of becoming, our only home.

The licentious anarchy of nature, its ferocity and privation are what make existence possible, real, even joyful, perhaps immortal—they alone provide enough depth for it to happen at all. Anything less would bottom out long before we awoke from enchanted dreams of nothing.

In San Cristobal villagers gather beneath a glittering heaven of kachinas, to acknowledge that creation is not only above but within them. Slowly at first, then a gathering wave, they send their chorus spiralling outward: *"Praise God for the light within me./Praise God; let love abide."*

* * * * *

It hardly matters if we are atoms and molecules and our lives are not really real. What is real will survive somehow, even through obliteration of whole universes. What is illusion will be discarded. We don't have to worry about this one.

Notes

· ·

General

Embryos, Galaxies, and Sentient Beings was begun after the publication of my book *Embryogenesis: Species, Gender, and Identity* in April 2000, and completed in April 2003. It is a sequel, at once the précis and the extension of *Embryogenesis.* It is also a totally different book written in open space virtually without reference to its predecessor.

Embryogenesis (Berkeley: North Atlantic Books, 2000) is nearly a thousand pages long, with a glossary, hundreds of black-and-white line illustrations, and color plates. My goal was to explore biological and developmental mysteries in depth. Because that book already existed, I had no need to repeat its level of detail here.

The ratio of expository information between the two works differs from topic to topic. Various embryogenic stages (like meiosis, formation of egg, sperm, blastula, gastrula, and neurula), the development of particular organs (like the ear, genitals, brain, blood, etc.), plus a range of other topics (including gender and sexuality, primate and hominid evolution, ontogeny and phylogeny, the biological basis of death, creation cosmology, theories of reincarnation, animal consciousness, etc.) are covered in a paragraph up to a few pages in this book, whereas whole sections or chapters are devoted to them in *Embryogenesis.* On the other hand, a number of themes that were only introduced in *Embryogenesis* are discussed here in greater depth: the embryo as the universe writing itself on its own body, the effects of runification and subminiaturization over epochal time, the relation between thermodynamics and information, nanomachines and molecular technology, the discontinuity between natural selection and morphogenesis, the relationship between molecular and cellular activity, the limits of genetic determinism, the gap between empirical and phenomenological definitions of being, the impact of commoditization on ethical and epistemological systems, and the application of embryogenic principles to extraterrestrial worlds (exo-embryology)—to name a few. Additionally, I have addressed a few new topics: the aftermath of the deciphering of the genome, the impact of the 9-11 terrorist attacks and the militant Islamic (and fundamentalist) critique of Western science, Thomas Hardy's nihilistic Darwinism, and the relationship between DNA and ayahuasca.

Embryos, Galaxies, and Sentient Beings explores two works with which I was not familiar at the time of writing *Embryogenesis:* Stuart Kauffman's meta-thermodynamic principles underlying biological systems and Erich Blechschmidt's model of morphogenesis by topokinesis (see the references below).

This book also revisits the underlying themes of the first two volumes of my trilogy that concluded with *Embryogenesis*—the shamanic and paraphysical basis of healing in *Planet Medicine* and the deconstruction of the Big Bang in *The Night Sky.*

In the notes below, I have listed the chapters in *Embryogenesis* that roughly correspond to material in this book. The only chapter in *Embryogenesis* with no material here is "Transsexuality, Intersexuality, and the Cultural Basis of Gender."

Introduction/Acknowledgments/Dedication
See also Preface and "Spiritual Embryogenesis" in *Embryogenesis.*

1. Robert Sheckley, *Untouched by Human Hands* (New York: Ballantine Books, Inc., 1954), p. 54.

2. H. R. Voth (translator), "Hopi Origin Myth," in *Io* No. 4, *Alchemy Issue,* Ann Arbor, Michigan, 1967, p. 145.

3. Robert Duncan, *The First Decade: Selected Poems 1940–1950* (London: Fulcrum Press, 1968), author's preface on front flap.

4. Charles Olson, "Enyalion," in Richard Grossinger (editor) *Ecology and Consciousness: Traditional Wisdom on the Environment* (Berkeley: North Atlantic Books, 1992), p. 214.

5. Charles Olson, *The Maximus Poems,* edited by George F. Butterick (Berkeley: University of California Press, 1983), p. 130.

6. ibid., p. 184

7. Duncan, p. 9.

8. ibid, p. 62.

9. Robert Kelly, *The Alchemist to Mercury,* collected and edited by Jed Rasula (Richmond, California: North Atlantic Books, 1981), p. 3.

10. Edward Dorn, *Some Business Recently Transacted in the White World* (West Newbury, Massachusetts: Frontier Press, 1971), p. 82.

11. Duncan, author's preface on front flap.

12. Charles Stein, "Assisi" in *Io,* No. 2, Amherst, Massachusetts, February 1966, pp. 64–65.

The words to the "Maine Stein Song" were composed by Lincoln Colcord.

The Big Kahuna (2000) was written by Roger Rueff and directed by John Swanbeck.

Chapter One
What is Life?
Evolution, Thermodynamics, and Complexity
See also the Preface, Chapter One entitled "Embryogenesis," and "Chaos, Fractals, and Deep Structure" in *Embryogenesis.*

General references for this chapter: Chris Lucas, "Self-Organizing Systems (SOS) Frequently Asked Questions" (USENET Newsgroup comp.theory.self-org-sys, version 2.8, May 2002) and James Gleick, *Chaos: Making a New Science* (New York: Penguin Books, 1987).

1. François Jacob, *La logique du vivant: Une histoire de l'hérédité* (Paris: Gallimard, 1974), p. 320; translated and quoted by Jeremy Narby in *The Cosmic Serpent* (New York: Jeremy P. Tarcher/Putnam, 1998), p. 134.

2. E. O. Wilson, *Sociobiology* (Cambridge, Massachusetts: Bellknap Press, 1975); quoted in Mae-Wan Ho, *Genetic Engineering—Dream or Nightmare: The Brave New World of Bad Science and Big Business* (Bath, England: Gateway Books, 1998), p. 173.

3. Jonathan Franzen, "My Father's Brain," *New Yorker,* September 10, 2001, p. 81.

4. Loren Eiseley, *The Immense Journey* (New York: Vintage Books, 1946), p. 210.

5. Edward Dorn, "This is the way I hear the Momentum," in *Io, Ethnoastronomy Issue,* No. 6, Summer 1969, Ann Arbor, p. 110.

6. Robert Pollack, *The Faith of Biology and the Biology of Faith:*

Order, Meaning, and Free Will in Modern Medical Science (New York: Columbia University Press, 2000); quoted in Frederick Crews, "Saving Us From Darwin, Part II," *The New York Review of Books,* October 18, 2001, p. 52.

7. Kevin Padian and Alan D. Gishlick, "The Talented Mr. Wells: A review of *Icons of Evolution: Science or Myth? Why Much of What We Teach About Evolution is Wrong* by Jonathan Wells (Washington D.C.: Regnery Publishing, 2000), *The Quarterly Review of Biology,* Vol. 77, No. 1, March 2002, p. 36.

8. Charles Darwin; quoted in Frederick Crews, "Saving Us From Darwin," *The New York Review of Books,* October 4, 2001, p. 26.

9. Barry Commoner, "Unravelling the DNA myth: the spurious foundation of genetic engineering," *Harper's Magazine,* February 2002, pp. 40–41.

10. I am grateful to Harold Dowse for helping guide me through entropy and Maxwell's gedanexperiment.

11. ibid. Thanks again.

12. Charlotte P. Mangum and Peter W. Hochaka, "New Directions in Comparative Physiology and Biochemistry: Mechanisms, Adaptation, and Evolution," *Physiological Zoology,* Vol. 71, No. 5, 1998, p. 476.

13. ibid., p. 480.

14. ibid., p. 476.

15. Niall Shanks and Karl H. Joplin, "Redundant Complexity: A Critical Analysis of Intelligent Design in Biology," *Science,* Vol. 66, June 1969, p. 272.

16. Gleick, p. 133.

17. ibid., p. 134.

18. Lucas, section 2.8.

19. Shanks and Joplin, p. 273.

20. ibid., p. 274.

21. ibid.

22. Stuart Kauffman, *Investigations* (Oxford, England: Oxford University Press, 2000), p. 144.

23. ibid., p. 125.

24. ibid., p. 162.

25. ibid., p. 125.

26. ibid., p. 2.

27. ibid., p. 3.

28. ibid. pp. 151–152 (phrases in brackets mine).

29. ibid., p. 109.

30. ibid. (phrases in brackets mine).

The epigraph is from Max Planck, *Treatise on Thermodynamics,* translated from the German by Alexander Ogg (New York: Dover Publications, 1926). p. 1.

The illustration of Maxwell's demon on p. 21 is by Oliver Chin.

Thanks to Jess O'Brien for "Pretty trippy...," etc.

Chapter Two
Is There a Plan?
Creationism, Cultural Relativism, and Paraphysics
See also the Preface and Chapter One, "Embryogenesis," in *Embryogenesis.*

1. Emile Durkheim, *The Elementary Forms of the Religious Life,* translated from the French by Joseph Ward Swain (London: George Allen & Unwin Ltd., 1915), pp. 358, 375.

2. Michael Behe, quoted in Nina Shapiro, "The New Creationists: Seattle's Discovery Institute leads a national movement challenging Darwinism," www.seattleweekly.com, 2001.

3. Michael Behe, *Darwin's Black Box* (New York, Touchstone/Simon and Schuster, 1998), pp. 39–40.

4. Charles Darwin, *Origin of Species,* Sixth Edition from 1872 (New York: New York University Press, 1988), p. 154.

5. Richard Dawkins, *River out of Eden* (New York: Basic Books, 1995), p. 83.

6. I am grateful to Harold Dowse for this distinction and insight.

7. Niall Shanks and Karl H. Joplin, "Redundant Complexity: A Critical Analysis of Intelligent Design in Biology," *Science,* Vol. 66, June 1969, pp. 276–277.

8. Michael Behe, quoted in Shanks and Joplin, p. 271.

9. Charlotte P. Mangum and Peter W. Hochaka, "New Directions in Comparative Physiology and Biochemistry: Mechanisms, Adapta-

tion, and Evolution," *Physiological Zoology,* Vol. 71, No. 5, 1998, p. 476.

10. Frederick Crews, "Saving Us From Darwin," *The New York Review of Books,* October 4, 2001, p. 27.

11. ibid, p. 26.

12. Alvin Plantinga, "Saving Us From Darwin: An Exchange," *The New York Review of Books,* November 29, 2001, p. 63.

13. Shanks and Joplin, p. 281.

14. Steven D. Schafersman, "Michael Behe and Intelligent Design on National Public Radio 'Talk of the Nation,'" March 2002, www.freeinquiry.com/behe-npr.html.

15. ibid.

16. ibid.

17. Kevin Padian and Alan D. Gishlick, "The Talented Mr. Wells: A review of *Icons of Evolution: Science or Myth? Why Much of What We Teach About Evolution is Wrong* by Jonathan Wells (Washington D.C.: Regnery Publishing, 2000), *The Quarterly Review of Biology,* Vol. 77, No. 1, March 2002, p. 36.

18. ibid.

19. ibid.

20. ibid.

21. Boris F. Rice, Sr., "Letters," *Scientific American,* Vol. 287, No. 5, November 2002, p. 14.

22. "The Wedge Project," quoted by James Still, www.infidels.org.secular_web/feature/1999/wedge.html.

23. Charles Darwin; quoted in Roger Shattuck, "Saving Us From Darwin: An Exchange," *The New York Review of Books,* November 29, 2001, p. 63.

24. Charles Darwin; quoted in Alvin Plantinga, "Saving Us From Darwin: An Exchange," *The New York Review of Books,* November 29, 2001, p. 63.

25. Stephen Jay Gould; quoted in Alvin Plantinga, "Saving Us From Darwin: An Exchange," *The New York Review of Books,* November 29, 2001, p. 63.

26. Noel Charlton, "Gaia Theory," www.lancs.ac.uk/users/philosophy/mave/guide/gaiath~1.htm, no date given.

27. Durkheim, p. 375.

28. ibid.

29. Donna Haraway, "The Promises of Monsters: A Regenerative Politics for Inappropriate/d Others"; in Lawrence Grossberg, Cary Nelson, and Paula Treichler, *Cultural Studies* (New York: Routledge, 1992), p. 304.

30. ibid, p. 312.

31. Jule Eisenbud, *Parapsychology and the Unconscious* (Berkeley: North Atlantic Books, 1983), pp. 111–129.

32. Jean Millay, *Multidimensional Mind: Remote Viewing in Hyperspace* (Berkeley: North Atlantic Books, 1999).

33. Gary Snyder, "On Geography: An Interview Conducted by Richard Grossinger and David Wilk, 9 November 1971," in Richard Grossinger (editor), *Ecology and Consciousness: Traditional Wisdom on the Environment* (Berkeley: North Atlantic Books, 1992), p. 145.

34. Mao Tse-tung; quoted in Ian Buruma and Avishai Margalit, "Occidentalism," *The New York Review of Books,* January 17, 2002, p. 5.

35. Omar Khayyam Moore, "Divination—A New Perspective, *American Anthropologist* LIX (1957), pp. 69–74.

36. I am grateful to Harold Dowse for this explanation.

37. "Do Hells Exist? A Discourse by His Holiness Living Buddha Lian-sheng at Ling Shen Ching Tze Temple in Redmond on October 16th, 1999," translated by Janny Chow, in *The Purple Lotus Journal,* March 2000, Issue No. 12, San Bruno, California, p. 8.

Heraclitus' epigraph comes from Philip Wheelwright (editor), *The Presocratics* (Indianapolis: The Odyssey Press, 1966), p. 73. Dates of birth and death are approximate.

The stormy petrel and caribou come from Knud Rasmussen, *Intellectual Culture of the Iglulik Indians, Report of the Fifth Thule Expedition,* Copenhagen, 1929.

Anthroposophist Ellias Lonsdale laid the groundwork for my discussion of negative identity. I am also grateful to him for "Christ died so that people would be as he was" and the preceding aphorisms.

Chapter Three
Biogenesis and Cosmogenesis
General reference for this chapter: Jeremy Narby, *The Cosmic Serpent* (New York: Jeremy P. Tarcher/Putnam, 1998). See also the chapters "The Original Earth," "The Materials of Life," "The First Beings," "The Cell," "The Genetic Code," and "Sperm and Egg" in *Embryogenesis.*

1. Stuart Kauffman, *Investigations* (Oxford, England: Oxford University Press, 2000), p. 143.

2. ibid., p. x.

3. Rudolf Virchow, *Disease, Life, and Man: Selected Essays,* translated by Lelland J. Rather (Stanford, California: Stanford University Press, 1958), pp. 84 and 106; *Die Cellularpathologie in ihrer Begruendung auf physiologische under pathologische Gewebelehre/Zwanzig Vorlesungen gehalten waehrend der Monate Feburar, Maerze, und April, 1958, im pathologischen Institute zu Berlin* (Berlin: August Hirschwald, 1858), p. 3; translated (where necessary) and quoted together in Harris Livermore Coulter, *Divided Legacy, A History of the Schism in Medical Thought, Volume II—Progress and Regress: J. B. Van Helmont to Claude Bernard* (Washington, D.C.: Wehawken Book Company, 1977), p. 621.

4. Narby, p. 110.

5. ibid., p. 86.

6. ibid., p. 88.

7. Erwin Schrödinger, *What is Life?* (Cambridge, England: Cambridge University Press, 1944).

8. Kauffman, p. 5.

9. ibid., p. 91.

10. The results of this contest, held by the *Washington Post,* were circulated on the Internet on 8 February 2001.

11. Kauffman, p. 129.

12. ibid.

13. ibid., pp. 142–143.

14. Charles Stein, "I Dwell in Possibility," a review of *Investigations* by Stuart Kauffman, *Nature Biotechnology,* June 2001, Volume 19, No. 6, p. 506.

The epigraph comes from Samuel L. Clemens (Mark Twain), *Adven-*

tures of Huckleberry Finn, originally published in 1884 (Cambridge, Massachusetts: The Riverside Press, 1958), p. 101.

My Jovian landscape was inspired by Ben Bova's science-fiction novel *Jupiter* (New York: TOR, 2001). This book combines an embarrassingly adolescent cast of stick figures in a disingenuous anti-spiritual—disguised as anti-fundamentalist—plot, set nonetheless in an ingenious landscape of an inhabited Jupiter on which there occurs a convincing episode of interplanetary interspecies communication.

Chapter Four
The Principles of Biological Design
General references for this chapter: D'Arcy Thompson, *On Growth and Form* [1917] (Cambridge, England: Cambridge University Press, 1966); Erich Blechschmidt, *The Beginnings of Human Life,* translated by Transemantics, Inc. (New York, Heidelberg, and Berlin: Springer Verlag, 1977); Erich Blechschmidt and R. F. Gasser, *Biokinetics and Biodynamics of Human Differentiation* (Springfield, Illinois: Charles C. Thomas, 1978); Philip C. Hanawalt and Robert H. Haynes (editors), *The Chemical Basis of Life: An Introduction to Molecular and Cell Biology* (San Francisco: W. H. Freeman and Company, 1973). See also the chapters "The Original Earth," "The Materials of Life," "The First Beings," "The Cell," "The Genetic Code," "Chaos, Fractals, and Deep Structure," and "Ontogeny and Phylogeny" in *Embryogenesis.*

1. James Gleick, *Chaos: Making a New Science* (New York: Penguin Books, 1987). p. 221.

2. ibid., pp. 221–222 (quote within quote from French mathematician Adrien Douady).

3. Charles Olson, *Human Universe and Other Essays,* edited by Donald Allen (New York: Grove Press, 1967), p. 122. This is from the essay "Equal, That is, to the Real Itself."

4. Herman Melville, *Moby-Dick or, The Whale,* originally published in 1851 (Berkeley: Arion Press/University of California Press, 1979), p. 320.

5. Olson, *op. cit.*

6. Herman Melville, quoted in Olson, p. 118.

7. Blechschmidt and Gasser, p. 6.

8. Blechschmidt, p. 19.

9. ibid., p. 11.

10. ibid., pp. 36–37.

11. Richard Strohman, "Manuevering in the Complex Path from Genotype to Phenotype," *Science,* Vol. 296, 26 April 2002, p. 701–702.

12. Kauffman, p. 5.

13. Blechschmidt, p. 2.

14. Blechschmidt and Gasser, p. 114.

15. ibid., p. 75.

16. Blechschmidt, p. 97.

17. Blechschmidt and Gasser, p. 122.

18. Blechschmidt, p. 97.

19. ibid., p. 23.

Paracelsus' epigraph is from Arthur Edward Waite (editor), *The Hermetic and Alchemical Writings of Paracelsus: Volume II, Hermetic Medicine and Hermetic Philosophy* (Berkeley, California: Shambhala Publications, 1976), p. 279.

Chapter Five
The Dynamics of the Biosphere

General references for this chapter: "Nanotech: The Science of the Small Gets Down to Business," *Scientific American,* Special Issue, September 2001; Michael A. Nielsen, "Rules for a Complex Quantum World," *Scientific American,* Vol. 287, No. 5, November 2002, pp. 68–75; Bruce Alberts, Dennis Bray, Julian Lewis, Martin Raff, Keith Roberts, and James D. Watson, *Molecular Biology of the Cell* (New York: Garland Publishing, Inc., 1989); Jeremy Narby, *The Cosmic Serpent* (New York: Jeremy P. Tarcher/Putnam, 1998); and Stuart A. Newman, "Generic physical mechanisms of tissue morphogenesis: A common basis for development and evolution," *Journal of Evolutionary Biology,* No. 7, 1994, pp. 467–488. See also the Preface and the chapters "Morphogenesis," "Biological Fields," "Sperm and Egg," "Ontogeny and Phylogeny," and "The Origin of Sexuality and Gender" in *Embryogenesis.*

1. Nielsen, p. 71.

2. ibid., p. 73 [quote rearranged by me].

3. ibid., p. 72 [quote rearranged by me].

4. ibid., p. 71.

5. Charles Olson, *The Maximus Poems,* edited by George F. Butterick (Berkeley: University of California Press, 1983), p. 184.

6. Charles Olson, "Enyalion," in Richard Grossinger (editor) *Ecology and Consciousness: Traditional Wisdom on the Environment* (Berkeley: North Atlantic Books, 1992), p. 214.

7. Nielsen, p. 71.

8. Herman Melville, *Moby-Dick or, The Whale,* originally published in 1851 (Berkeley: Arion Press/University of California Press, 1979), p. 348.

9. J. Madeleine Nash, "Inside the Womb," *Time,* 11 November 2002, p. 74.

10. Barry Commoner, "Unravelling the DNA myth: the spurious foundation of genetic engineering," *Harper's Magazine,* February 2002, p. 47.

11. ibid.

12. Charlotte P. Mangum and Peter W. Hochaka, "New Directions in Comparative Physiology and Biochemistry: Mechanisms, Adaptation, and Evolution," *Physiological Zoology,* Vol. 71, No. 5, 1998), p. 473.

13. ibid. [I have replaced some commas with semi-colons in this and the previous entry to make them easier to read within the punctuation conformities of this book.]

14. *Holy Bible: The New King James Version* (Nashville: Thomas Nelson Publishers, 1979), p. 525.

15. Paul Valéry; quoted in Adam Phillips, *Terrors and Experts* (Cambridge, Massachusetts: Harvard University Press, 1995), p. 13.

16. Tom Abate, "Genome discovery shocks scientists: genetic blueprint contains far fewer genes than thought—DNA's importance downplayed" (*San Francisco Chronicle,* February 11, 2001), pp. A16–A17.

17. J. Madeleine Nash, "When life exploded," *Time,* 5 December 1995, p. 68.

18. Kauffman, p. 209.

19. Kevin Padian and Alan D. Gishlick, "The Talented Mr. Wells: A review of *Icons of Evolution: Science or Myth? Why Much of What We Teach About Evolution is Wrong* by Jonathan Wells (Washington D.C.:

Regnery Publishing, 2000), *The Quarterly Review of Biology,* Vol. 77, No. 1, March 2002, p. 35.

20. Stuart Kauffman, *Investigations* (Oxford, England: Oxford University Press, 2000), p. 209.

21. Nielsen, p. 71.

22. See note 8 in this chapter.

23. Harold Dowse provided me with the description of Slowpoke.

24. Maurice Merleau-Ponty, quoted by Anita Shreve in the epigraph of *Where or When* (New York: Harcourt, Inc., 1993).

The epigraph comes from Herman Melville, *Moby-Dick or, The Whale,* originally published in 1851 (Berkeley: Arion Press/University of California Press, 1979), p. 11.

Chapter Six
The Limits of Genetic Determinism

General references for this chapter: Mae-Wan Ho, *Genetic Engineering—Dream or Nightmare: The Brave New World of Bad Science and Big Business* (Bath, England: Gateway Books, 1998) and Barry Commoner, "Unravelling the DNA myth: the spurious foundation of genetic engineering," *Harper's Magazine,* February 2002. See also the chapters "The Genetic Code," "Chaos, Fractals, and Deep Structure," and "Biotechnology" in *Embryogenesis.*

1. *Holy Bible: The New King James Version* (Nashville: Thomas Nelson Publishers, 1979), p. 525.

2. Ho, p. 23.

3. ibid.

4. Barry Commoner, "Unravelling the DNA myth," p. 40.

5. ibid., p. 41.

6. ibid.

7. Tom Abate, "Genome discovery shocks scientists: genetic blueprint contains far fewer genes than thought—DNA's importance downplayed" (*San Francisco Chronicle,* February 11, 2001), pp. A16–A17.

8. ibid.

9. Richard Lewontin, "After the Genome, What Then?", *The New York Review of Books,* July 19, 2001, pp. 36–37.

10. Richard Strohman, "Maneuvering in the Complex Path from Genotype to Phenotype," *Science,* Vol. 296, 26 April 2002, p. 703.

11. Ronald Bailey, "Is Biologist Barry Commoner a Mutant?" reasononline, http://reason.com/rb/rbo12002.shtml, January 30, 2002.

12. Barry Commoner, "Failure of the Watson-Crick Theory as a Chemical Explanation of Inheritance," *Nature,* Vol. 220, October 26, 1968, p. 339.

13. Bailey, *op. cit.*

14. Thomas S. Kuhn, *The Structure of Scientific Revolutions* (Chicago: The University of Chicago Press, 1962), p. 158.

15. Harold Dowse arranged this chronology.

16. Luis W. Alvarez, Walter Alvarez, Frank Asaro, Helen V. Michel, "Extraterrestrial Cause for the Cretaceous-Tertiary Extinction," *Science,* Vol. 208, No. 4448, June 6, 1980.

17. Stuart Kauffman, *Investigations* (Oxford, England: Oxford University Press, 2000), p. 185.

18. Niall Shanks and Karl H. Joplin, "Redundant Complexity: A Critical Analysis of Intelligent Design in Biology," *Science,* Vol. 66, June 1969, p. 280.

19. Barry Commoner, "Unravelling the DNA myth," p 44.

20. Commoner, "Failure of the Watson-Crick Theory as a Chemical Explanation of Inheritance," *Nature,* Vol. 220, October 26, 1968, p. 337.

21. Commoner, "Unravelling the DNA myth," pp. 43-44.

22. Ludwig Wittgenstein, quoted in Richard Strohman, "Manuevering in the Complex Path from Genotype to Phenotype," *Science,* Vol. 296, 26 April 2002, p. 703.

23. Commoner, "Unravelling the DNA myth," p. 43.

24. Richard Strohman, "Maneuvering in the Complex Path from Genotype to Phenotype," *Science,* Vol. 296, 26 April 2002, p. 701.

25. ibid, p. 702 (slightly rearranged to fit context).

26. Roberto P. Stock and Harvey Bialy, "The sigmoidal curve of cancer," in press, *Nature Biotechnology,* 2003 [quote slightly altered by me for grammar].

27. David Rasnick and Peter H. Duesberg, "How aneuploidy affects metabolic control and causes cancer," *Journal of Biochemistry,* Vol. 340, 1999, p. 621.

28. Stock and Bialy, 2003.

29. David Rasnick, "Aneuploid theory explains tumor formation, the absence of immune surveillance, and the failure of chemotherapy," *Cancer Genetics and Cytogenetics,* Vol. 135, 2002, p. 66.

30. ibid, p. 69.

31. Stock and Bialy, 2003.

32. "Food for Our Future, Food and Biotechnology" (London: Food and Drink Federation, 1995), p. 5; quoted in Ho, p. 130.

33. Richard Strohman, "Maneuvering in the Complex Path from Genotype to Phenotype," *Science,* Vol. 296, 26 April 2002, p. 703.

34. ibid., p. 701.

35. ibid., p. 702.

36. ibid.

37. ibid., p. 701.

38. Commoner, "Failure of the Watson-Crick Theory as a Chemical Explanation of Inheritance," *Nature,* Vol. 220, October 26, 1968, p. 340.

39. Strohman, p. 702.

40. Richard Strohman, unpublished response to Frederick Crews, "Saving Us From Darwin," *The New York Review of Books,* October 4 and 18, 2001.

41. Richard Strohman, "Maneuvering in the Complex Path from Genotype to Phenotype," *Science,* Vol. 296, 26 April 2002, p. 701.

42. Harvey Bialy, personal communciation, 2002.

43. Stock and Bialy 2003.

44. Jacques Derrida, *Of Grammatology,* translated by Gayatri Chakravorty Spivak (Baltimore: Johns Hopkins University Press, 1974), pp. 64–72.

45. C. G. Jung, *Psychology and Alchemy,* translated by R. F. C. Hull (London: Routledge and Kegan Paul, 1953), p. 37.

46. Eugene P. Odom, *Fundamentals of Ecology* (Philadelphia: W. B. Saunders Company, 1959), p. 30.

47. Dougal Dixon, *After Man: A Zoology of the Future* (New York: St. Martins Press, 1981).

48. Commoner, "Unravelling the DNA myth," pp. 42–43.

49. Ho, pp. 15-16.

50. Scott F. Gilbert, "Ecological developmental biology: Develop-

mental biology meets the real world," *Developmental Biology,* Vol. 233, 2001, p. 8

51. ibid., p. 4.
52. ibid., p. 5.
53. ibid., p. 4.
54. ibid., p. 5
55. ibid.
56. ibid.
57. Pierre Teilhard de Chardin, *The Phenomenon of Man,* translated from the French by Bernard Wall (New York: Harper & Row, 1959).

Heraclitus' epigraph comes from Philip Wheelwright (editor), *The Presocratics* (Indianapolis: The Odyssey Press, 1966), p. 73. Dates of birth and death are approximate.

Chapter Seven
The Wahhabi Critique of Darwinian Materialism
General references for this chapter (excluding the 9-11 gap): Mae-Wan Ho, *Genetic Engineering—Dream or Nightmare: The Brave New World of Bad Science and Big Business* (Bath, England: Gateway Books, 1998); Bruce Alberts, Dennis Bray, Julian Lewis, Martin Raff, Keith Roberts, and James D. Watson, *Molecular Biology of the Cell* (New York: Garland Publishing, Inc., 1989); and Barry Commoner, "Unravelling the DNA myth: the spurious foundation of genetic engineering," *Harper's Magazine,* February 2002. See also the chapters "Biotechnology" and "Self and Desire" in *Embryogenesis.*

1. Commoner, p. 45.
2. ibid.
3. ibid., p. 46.
4. ibid., p. 47.
5. ibid., p. 46.
6. ibid.
7. Ronald Bailey, "Is Biologist Barry Commoner a Mutant?" reasononline, http://reason.com/rb/rb012002.shtml), January 30, 2002.
8. Commoner, p. 46.
9. ibid.

10. Paul Foster Case, *The Tarot: A Key to the Wisdom of the Ages* (New York: Macoy Publishing Company, 1947), p. 162.

11. ibid.

12. Osama bin Laden; quoted in Ian Buruma and Avishai Margalit, "Occidentalism," *The New York Review of Books*, January 17, 2002, p. 6.

13. Generally reported from websites associated with al Qaeda.

14. Osama bin Laden; quoted in Buruma and Margalit, p. 5.

15. Buruma and Margalit, p. 6.

16. This statement is informally reported on the Internet as being from Metin Bobaroglu, "a pharmacist from Istanbul, Turkey, with a deep interest in Sufism and philosophy." It is unknown whether al Qaeda hijacked these exact words, but the purpose for which they were intended is diametrically opposite political martyrdom.

17. Mao Tse-tung; quoted in *Hungry Ghosts: Mao's Secret Famine* (New York: Free Press, 1996), p. 62.

18. Ho, p. 211.

19. ibid., pp. 223–224.

20. Andrei Dmitriyevich Sakharov; quoted in *The New Yorker*, May 21, 2001, p. 38.

21. ibid.

The epigraph is from M. L. Rosenthal (editor), *Selected Poems and Two Plays of William Butler Yeats* (New York: The Macmillan Company, 1962), p. 91.

The image of the Taliban in the Afghan landscape is substantially informed by "Temporal Vertigo: A last road trip through premodern, postmodern Afghanistan" by John Sifton, which appeared in *The New York Times Magazine*, September 30, 2001, pp. 48–51.

I thank Philip Wohlstetter for sharing his insight into the Departments of War and Defense.

I took the notion of opposing global world-views from a PBS radio interview with Benjamin R. Barber, author of *Jihad vs. McWorld: How Globalism and Tribalism Are Reshaping the World* (New York: Ballantine, 1996).

"The dead are weird while we're here having fun" comes from Ellias

Lonsdale, co-author (with his channelled wife) of *The Book of Theanna: In the Lands that Follow Life* (Berkeley, California: Frog, Ltd., 1995).

The battle of McWorld versus Jihad has proceeded through the course of this book into Operation Iraqi Freedom of 2003. McWorld (in the current guise of George W. Bush and the Military-Industrial Judeochristian Complex) said to Saddam, "Disarm or be disarmed." But their own weapons of mass destruction and the nuclear arsenal of the Israelis, in the service of the international flow and elitist concentration of capital, were not judged by the same standards. That is because commoditization is blind to its own ambitions. Saddam and his sons and Chemical Ali and the Baath Party appeared to the West as immeasurably more brutal because their exercise of power and callousness were primitive and direct rather than institutional. They did not use consumption and "the good Life" to distract the masses from their application of power and cultural appropriations. They went with excision of tongues, torture of children in front of parents, warheads packed with Sarin, and the like. That kind of stuff is far too medieval for our lobbyists and spin doctors. In the sort of media democracy we have turned into, appliers of raw power prefer distractions of pleasures and goods, the pretence of popularity polls.

The irony is that the invasion of Iraq is so much about oil that President Bush actually believes it is *not* about oil. McWorld does not recognize that it too destroys villages and families and incurably pollutes natural and cultural landscapes. It hardly ever uses weapons of mass destruction directly, but they are a deterrent behind which it carries out its global hegemonization.

The "coalition" bulldozed the living trying to surrender with the dead while airing Centcom sound bites and feel-good reality TV. Give plastic arms to one poster child, but incinerate evidence of thousands of others. The torture of the West is in the lie, in class-based judicial systems and death rows, in constant, bottomless manufacture of desire, road rage, alienation, abstract terror, mock guilt, jaded innocence, and dispensable grief. This is the globalized evil that gruff mullahs and ayatollahs shield parishioners against, yet replacing the West's tyrannies, prisons, and embargos with their own bogus theocracy and its menial laws, phobic constabularies, and baroque executions.

Jihad is fighting a long guerrilla war against McWorld. It has inherited the role of the Communist resistance and redefined it in Islamic terms. Yet it is seeking to eradicate not only the feudalism and greed of McWorld but the pluralism and democracy of its institutions. It exterminates those in its path while shamelessly calling on Allah, a god of peace, to bless its mission and agenda—to impose a ruthless pseudo-Islamic Republic tougher and more tyrannical than the old Soviet gulag.

False prophecy is a dangerous business—stealing in the name of the lord. As Normal Mailer writes in *The New York Review of Books* (March 27, 2003): "[W]e can never know for certain where our prayers are likely to go, nor from whom the answers will come. Just when we think we are at our nearest to God, we could be assisting the Devil....

"If I were George W. Bush's karmic defense attorney, I would argue that his best chance to avoid conviction as a purveyor of false morality would be to pray for a hung jury in the afterworld" (pp. 52–53).

Saddam Hussein, Osama bin Laden, and their colleagues stand no chance at all of a hung jury. They have clearly sided with a dualistic universe that pits Islam against Satan right down to the atomic core.

Meanwhile Dante has his own circle of hell for those in "mega-corporations [who] do their best to appropriate our thwarted dreams with their elephantiastical conceits" (p. 53).

The ultimate blowback!

Chapter Eight
Topokinesis

General references for this chapter: D'Arcy Thompson, *On Growth and Form* [1917] (Cambridge, England: Cambridge University Press, 1966); Erich Blechschmidt, *The Beginnings of Human Life*, translated by Transemantics, Inc. (New York, Heidelberg, and Berlin: Springer Verlag, 1977); Erich Blechschmidt and R. F. Gasser, *Biokinetics and Biodynamics of Human Differentiation* (Springfield, Illinois: Charles C. Thomas, 1978); (illustrations) Erich Blechschmidt, *Vom Ei zum Embryo* (Stuttgart, Germany: Deutsche Verlags-Anstalt, 1968). See also the chapters "Fertilization," "The Blastula," "Gastrulation," "The Origin of the Nervous System," and "Neurulation and the Human Brain" in *Embryogenesis*.

1. Blechschmidt, pp. 24–25.
2. ibid., p. 16.
3. Blechschmidt and Gasser, p. xiii.
4. Blechschmidt, p. 16.
5. ibid., p. 25.
6. ibid., pp. 18–19.
7. ibid., p. 19.
8. ibid., pp. 20–21.
9. ibid., p. 49.
10. ibid., p. 43.
11. Blechschmidt and Gasser, p. 6.
12. Blechschmidt, p. 100.
13. ibid., p. 103.
14. ibid., p. 105.
15. Blechschmidt and Glasser, p. 96.
16. ibid.
17. Blechschmidt, p. 107.
18. ibid., p. 103.
19. Stuart Kauffman, *Investigations* (Oxford, England: Oxford University Press, 2000), p. 212.

The illustration on page 307 comes from Erich Blechschmidt and R. F. Gasser, *Biokinetics and Biodynamics of Human Differentiation* (Springfield, Illinois: Charles C. Thomas, 1978). The illustrations on pages 297 and 299 come from Erich Blechschmidt, *Von Ei zum Embryo* (Stuttgart: Deutsche Verlags-Anstalt GmbH, 1968.) Courtesy of Dr. Traute Blechschmidt.

Empedocles' epigraph comes from Philip Wheelwright (editor), *The Presocratics* (Indianapolis: The Odyssey Press, 1966), p. 132. Dates of birth and death are approximate.

My source for the discussion of William Sutherland and the Breath of Life is Franklyn Sills, *Craniosacral Biodynamics (Volume I): The Breath of Life, Biodynamics, and Fundamental Skills* (Berkeley: North Atlantic Books, 2001).

I thank Renate Stendahl for her translations of Erich Blechschmidt's captions from *Von Ei zum Embryo*.

Chapter Nine
Tissue Motifs and Body Plans
General references for this chapter: Erich Blechschmidt, *The Begin-nings of Human Life,* translated by Transemantics, Inc. (New York, Heidelberg, and Berlin: Springer Verlag, 1977); Erich Blechschmidt and R. F. Gasser, *Biokinetics and Biodynamics of Human Differentiation* (Springfield, Illinois: Charles C. Thomas, 1978); and Stuart A. Newman, "Generic physical mechanisms of tissue morphogenesis: A common basis for development and evolution," *Journal of Evolutionary Biology,* #7, 1994, pp. 467–488; (illustrations) Erich Blechschmidt,*Vom Ei zum Embryo* (Stuttgart, Germany: Deutsche Verlags-Anstalt, 1968). See also the chapters "Ontogeny and Phylogeny," "Organogenesis," "The Mus-culoskeletal and Hematopoietic Systems," "Birth Trauma," and "Heal-ing" in *Embryogenesis.*

Parts of the section on osteopathy were co-written into the seventh edition of *Planet Medicine: Modalities* (Berkeley, California: North Atlantic Books, 2002). I took early drafts of several pages and adapted them differently to each book.

1. Blechschmidt and Gasser, p. 82.

2. Thomas Hardy, *The Return of the Native* [1878] (New York: New American Library, 1959), p. 72.

3. Blechschmidt, p. 102.

4. ibid.

5. Blechschmidt and Gasser, pp. 142–143.

6. ibid., p. 142.

7. ibid., pp. 102–103.

8. Blechschmidt, p. 88.

9. ibid., p. 102.

10. J. Madeleine Nash, "Inside the Womb," *Time,* 11 November, 2002, p. 71.

11. Blechschmidt, pp. 24–25.

12. Erich Blechschmidt, *The Ontogenetic Basis of Human Anatomy: The Biomechanical Approach to Human Development from Concep-tion to Adulthood,* translated by Brian Freeman (Berkeley: North Atlantic Books, tenative 2004 publication); from *Anatomie und Onto-genese des Menschen* (Heidelberg: Quelle & Meyer, 1978).

13. ibid.

14. ibid.

15. Blechschmidt and Gasser, p. 35.

16. Blechschmidt, *The Ontogenetic Basis of Human Anatomy.*

17. Blechschmidt and Gasser, pp. 33–34.

18. ibid., p. 18.

The illustrations on pages 325, 343, and 349 come from Erich Blech-schmidt and R. F. Gasser, *Biokinetics and Biodynamics of Human Dif-ferentiation* (Springfield, Illinois: Charles C. Thomas, 1978). The illustrations on pages 328, 331, 337, 341, 346, and 357 come from Erich Blechschmidt, *Von Ei zum Embryo* (Stuttgart: Deutsche Verlags-Anstalt GmbH, 1968.) Courtesy of Dr. Traute Blechschmidt.

Heraclitus' epigraph comes from Philip Wheelwright (editor), *The Presocratics* (Indianapolis: The Odyssey Press, 1966), p. 78. Dates of birth and death are approximate.

My understanding of the lymph system comes from a lecture by Bruno Chikly at an Upledger Institute seminar in San Francisco in 2001.

I thank Renate Stendahl for her translations of Erich Blechschmidt's captions from *Von Ei zum Embryo.*

Chapter Ten
The Primordial Field

General reference for this chapter: Jeremy Narby, *The Cosmic Serpent* (New York: Jeremy P. Tarcher/Putnam, 1998). See also the Preface and the chapters "Spiritual Embryogenesis," "Cosmogenesis and Mortal-ity," and "Healing" in *Embryogenesis.*

Parts of the section on healing were co-written into the seventh edi-tion of *Planet Medicine: Modalities* (Berkeley, California: North Atlantic Books, 2002). Thus, different versions of the same text appear in each book. See also the Appendix of *Modalities* entitled "Seven Laws of Cure."

1. William Empson; quoted in Adam Phillips, *Terrors and Experts* (Cambridge, Massachusetts: Harvard University Press, 1995), p. vii.

2. Michel Foucault, *The Order of Things: An Archaeology of the Human Sciences,* translated from the French (New York: Pantheon

Books, 1970), pp. 24–25.

3. ibid., pp. 25–26.

4. Maurice Merleau-Ponty, *The Primacy of Perception*, translated from French and edited by James M. Edie (Evanston, Illinois: Northwestern University Press, 1964), p. 27. Edie did not translate the entire book but did translate the phrase quoted.

5. Thomas Hardy, *The Return of the Native* [1878] (New York: New American Library, 1959), p. 93.

6. Zadock Thompson, *Natural History of Vermont* [1853] (Rutland, Vermont: Charles E. Tuttle Company, 1971), p. 65.

7. Herman Melville, *Moby-Dick; or, The Whale* [1851] (Berkeley, California: University of California Press, 1979), pp. 282–283.

8. "Selected Haiku" of Buson (1715–1783), translated by Robert Hass, in Stephen Mitchell (editor), *Bestiary: An Anthology of Poems about Animals* (Berkeley, California: Frog, Ltd., 1996), p. 22.

9. Dr. Randolph Stone, D.C., D.O., *Polarity Therapy, Volume One* (Sebastopol, California: CRCS Publications, 1986), p. 18.

10. ibid., p. 20.

11. David Wilcock, quoted in Wynn Free, *The Reincarnation of Edgar Cayce* (Berkeley, California: Frog, Ltd., tentatively 2003).

12. Henry Corbin, *Creative Imagination in the Sufism of Ibn Arabi*, translated from the French by Ralph Manheim (Princeton: Princeton University Press, 1969), p. 50.

13. Luis Eduardo Luna and Pablo Amaringo, *Ayahuasca Visions: The Religious Iconography of a Peruvian Shaman* (Berkeley, California: North Atlantic Books, 1991), pp. 33–34.

14. Narby, p. 70.

15. Luna and Amaringo, p. 34f.

16. Narby, p. 110.

17. ibid., p. 71.

18. ibid., p. 19.

Emilie Conrad, founder of Continuum, proposed the relationship between the embryogenic field and the primordial field at a workshop in Berkeley in 2001.

The glass worm on Mars is discussed in Richard C. Hoagland, *The*

Monuments of Mars: A City on the Edge of Forever, 2001 edition (Berkeley, California: Frog, Ltd., 2001). A NASA frame of it appears on the back cover.

In Ursula K. Le Guin's science-fiction universe of Hain, the ansible is "a device which produces a message at two points simultaneously. Since this transmission is truly instantaneous, distance is immaterial to the ansible.... The ansible operates according to the Constant of Simultaneity, which is in some ways analogous to gravity." http://husted.com/ hgsf/Ansible.htm.

Pablo Amaringo's paintings appear in *Ayahuasca Visions: The Religious Iconography of a Peruvian Shaman* (see above).

I thank Dr. John E. Upledger for the concept of "cell talk."

The quotes on pp. 394 and 400 are from William Blake (1757–1827), "The Tyger."

Chapter Eleven
Meaning and Destiny

See also the chapters "The Evolution of Intelligence," "Mind," and "Death and Reincarnation" in *Embryogenesis.*

1. Phoebe Sengers, "Fabricated Subjects: Reification, Schizophrenia and Artificial Intelligence," www.desk.nl/~nettime/zkp2/fabricat.html), 2001.

2. Heraclitus, quoted in Philip Wheelwright (editor), *The Presocratics* (Indianapolis: The Odyssey Press, 1966), p. 73.

3. Frank Broucek, *Regaining Consciousness: Awakening from the Nightmare of Scientific Materialism,* unpublished manuscript.

4. Kerwin Lebeis, lazyape@home.com, *Brain & Mind Brainstorming,* www.epub.org.br/cm/n04/comment1_i.htm, December 30, 2000.

5. Charles Stein, "I Dwell in Possibility," a review of *Investigations* by Stuart Kauffman, *Nature Biotechnology,* June 2001, Volume 19, No. 6, p. 506.

6. ibid.

7. John Douillard, "Ayurveda and Reaching Mainstream America," talk at Ayurveda International Symposium, California Association of Ayurvedic Medicine, Berkeley, California, May 4, 2002.

8. Philip José Farmer, *The Maker of Universes: The Enigma of the*

Many-Leveled Cosmos (New York: Ace Books, 1965), back cover.

9. Alfred North Whitehead, *Science and the Modern World* (Toronto: Collier/Macmillan, 1925), p. 54.

10. Thomas Hardy, *The Return of the Native* [1878] (New York: New American Library, 1959), p. 280.

11. Dr. Randolph Stone, D.C., D.O., *Polarity Therapy, Volume One* (Sebastopol, California: CRCS Publications, 1986), p. 54.

12. John Haught, *God After Darwin: A Theology of Evolution* (Boulder, Colorado: Westview, 2001); quoted in Frederick Crews, "Saving Us From Darwin, Part II" *The New York Review of Books*, October 18, 2001, p. 55.

13. Peter Matthiessen, *Bone by Bone* (New York: Random House, 1999), p. 168.

14. Thomas Hardy, *The Mayor of Casterbridge* [1886] (Hertfordshire, England: Wordsworth Classics, 1994), p. 260.

15. Thomas Hardy, *The Return of the Native*, pp. 60, 254.

16. ibid., p. 278.

17. ibid., p. 59.

18. ibid., p. 289.

19. Thomas Hardy, *Tess of the d'Urbervilles* [1891] (New York: Penguin Books, 1978), p. 30.

20. ibid., p. 119.

21. Thomas Hardy, *The Return of the Native*, pp. 171–172.

22. George Bernard Shaw, *Back to Methuselah* (New York: Brentano's, 1921), pp. lxi–lxii.

23. Robert Pollack, *The Faith of Biology and the Biology of Faith: Order, Meaning, and Free Will in Modern Medical Science* (New York: Columbia University Press, 2000); quoted in Frederick Crews, "Saving Us From Darwin, Part II," *The New York Review of Books*, October 18, 2001, p. 52.

24. Thomas Hardy, *Tess of the d'Urbervilles*, p. 297.

25. ibid.

26. ibid., p. 72.

27. Robert Kelly, "First in an Alphabet of Sacred Animals," in Richard Grossinger (editor), *Ecology and Consciousness: Traditional Wisdom on the Environment* (Berkeley: North Atlantic Books, 1992), p. 60.

28. See footnote 24.

29. *Holy Bible: The New King James Version* (Nashville: Thomas Nelson Publishers, 1979), p. 533.

30. Chögyal Nankhai Norbu and Adriano Clemente, *The Supreme Source* (The Kunjed Gyalpo): *The Fundamental Tantra of Dzogchen Semde,* translated from Italian by Andrew Lukianowicz (Ithaca, New York: Snow Lion Publications, 1999), p. 85.

31. Johnny Unitas, the quarterback, to Bill Curry, his center, on a day at practice when the temperature was "about a thousand degrees." This was quoted in *NFL Insider,* Vol. 4, No. 4, February-March, 2003, p. 96.

Parmenides' epigraph is from Philip Wheelwright (editor), *The Presocratics* (Indianapolis: The Odyssey Press, 1966), p. 96. Dates of birth and death are approximate.

The quote on p. 405 is from William Blake (1757–1827), "The Tyger."

My Cytherean landscape was derived from the science-fiction novel *Venus* by Ben Bova (New York: TOR, 2001).

"... Freddie's joint ..." comes from an uncredited sound track at the end of an episode of *The Sopranos* in 2001, the lyrics apparently from a song called "Black Book" by Nils Lofgren.

"... *till we die, scooby-doo....*" comes from "So Much in Love," written by George Williams and Roy Straigis and sung by the Tymes (1963).

Some of the insights in the closing pages are derived from a talk given by Lama Tharchin Rinpoche in San Rafael, California, spring 2001.

The Closing Hymn

I asked jazz musician Joshi Marshall and an old friend Alan Powers to help me with a presentation of the "Light Within Me" mantra. What I first got was:

Praise God for the Light Within Me

But once the chords of people singing together merged, they created (according to Alan) a deeper music implied by the melody line:

Something else entirely came pouring through. The melody rose in harmony to meet the word "… *light*…," then expanded and spread. The tune sank through *"within"* so that the mantra of *"n"* blended with the *"m"* of *"me"* in the near hint of a minor chord. At *"me"* the light suffusing each of them became a tender epiphany of everything we are and could somehow be—the inexpressible outcome of this amazing journey. It carried the brief immanence of lost worlds, the flicker of something here and gone in an instant, something too important to lose but impossible to hold.

Index

as purest form of consciousness,
xxxii
replication as inherent property of,
121
segmentation as inherent property
of, 121
two states of, xxix–xxx, 83
See also matter, embryogenic;
matter, thermodynamic
matter, embryogenic
exo-embryology and, xxx–xxxi, 380
gravity and, 121–122
metamorphosis to, xxix–xxx
morphogenesis as addressing
domain of, 83
potential of, as universal, xxix–xxx
matter, thermodynamic
astrophysics as addressing domain
of, 83
metamorphosis of, to embryonic,
xxix–xxx
Matthiessen, Peter, 424
Maxwell, James Clerk, 20–22
Maxwell's demon, 20–23, 33, 109
MCA (metabolic control analysis),
213, 223n
McVeigh, Timothy, 273
McWorld, 272–276, 463–464
meaning
consciousness and, 430–432
courage and, 432
embryogenesis as generating,
83–84, 356, 369–370
and faith, 434–437
identity of, 436
as intrinsic, nonlinear, extrinsic,
229–232
as intrinsic to being, 373–374, 408
invaginating tissue and, 373
of life, 280

materialism and loss of, 260–261
redemption of, 430–445
mechanics
commodification of life and,
427–428
hybrid default mechanics, 112–114
intelligence and, 403–408
See also materialism
media, 66, 445
medicine
body as source of, 400–401
cell talk and, ix–x, 397–398, 399
codes hidden within other codes
and, 398–401
homeopathy, 165, 396
intelligence and, 394–398
iridology, 400–401
minerals as source of, 401–402
osteopathy, 318–321
over-concretization and, 394, 396,
400
Taoist and Ayurvedic, 395–396
touch, therapeutic, 319, 320, 398
See also disease; health; medicine
and biotechnology
medicine and biotechnology
conditions addressed through,
259–260
patents of indigenous knowledge,
189–190
police acts of, 260
vaccines, 246–247
See also antibiotics; biotechnology;
medicine
medieval sciences, 371–372
meiosis, 283
Melville, Herman, 123–124, 174
membranes, biokinetics and, 281–283,
285, 358, 365–367
memory, materialist view of, 3

Index

ontogeny recapitulating philogeny
and, 134–135
science as reducing all to, xii
statement of, 10–12
vibrating grid of life and, 321
"zeroth" law of, 11
See also entropy; materialism;
science
thermodynamics, second law of,
10–11, 110
thermodynamics, third law of, 11–12
thermogenic functions, 27–28
thought. *See* mind
thylakoids, 381
Tibetan Buddhism, 431, 443
time
being and, 231
DNA complexity and lack of, 173
sympathy and antipathy and, 372
See also deep time and space;
space-time continuum
tissue
alliances among cells weaving, 88–91
as composite of convergent
creatures, 179–180, 181
culture and. *See* culture as shaped
by tissue
as emergent property, xxix
exaptations, 169–170
experience as preceding, 374,
380–381
health outcomes expressed in
terms of, 90
metabolism and structure of. *See*
topokinesis
vascularization of, 286, 292, 322, 325,
335, 336, 340–341
Titan, xxxi
toads, 251
Tolkien, J. R. R., xxiii

tomatoes, transgenic, 258
tongue, 301–302, 401
tool making, 351
topoisomerase, 146–147, 163
topokinesis (metabolism and
structure of tissue)
culture and, 301–302, 308–313, 330,
351–352
defined, 129
embryogenesis and
brain and nervous system,
294–296, 298, 299–300,
302–307, 322, 325, 327–328, 333,
338–341
buoyant and suspensory factors
within body, 315–316, 321–322,
323–324
face and head, 294–302, 333
gastrointestinal system, 334–338
gastrulation, 288–290
genes and, 356–367
germ becoming blastula, 281–285
glands, generally, 321–328
hair, 324
heart/cardiovascular system,
321–322, 325, 327–328, 333–334
interior, formation of, 286–287,
288–292, 315–316, 321–322, 337
limbs, 346–353
limiting tissues and, 316, 322–323,
332–333, 336, 340, 342–343, 358,
365–366
liver, 325, 327–328, 333, 338
lymph system, 328–331
membranes as primary agent,
281–282, 285, 358
musculoskeletal system, 338–346
renal system, 331–333
respiratory organs, 315–316,
327–328, 333–335

Afterthought

● ●

Question: Did consciousness already exist when the first primitive
cells took shape?

Answer: It wouldn't have looked like consciousness.

In a dimensionless web (i.e., evolving life), every so-called regulator, every switch, every partial connection mutably nondeter-
mines/determines every other in series that radiate and ricochet in
every indirection forever—until you get Golgis, mitochondria, cells,
snakes, thoughts, cities, etc. This is not because (by differential selection) trillions upon trillions of molecular acts wire *ex post facto* queues
or because one-way plexuses of near-infinite entanglement snare
themselves in synapses of equally unfathomable extenuation.

No deterministically obedient network or historically (ontogenetically) requisite hierarchy of codes and decisions operates quasi-
cybernetically at *any* level of complexity. In fact, there is no
complexity, no molecular algebra or cytoplasmic circuitry, no nucleic
software or self-organizing protozoan alias, no binary/spliced
labyrinth or machinery that a technoscientist can get into to regu-
late/reconstruct with meticulous sequencing—*nada*. The entire thermochemical, information-retrieval, ergodic enterprise is a projection
simultaneously of Western civilization onto nature and the mystery
text of nature onto us. There are countless other equally convincing, antithetical grammars hidden under matter's sphinxlike smile.
What appears to be deep algorithmic design is latency itself imparting tangibly.

Multidirectional waves spread in trellis-like motifs—e.g. bodies.
Chromosomes are contingent, not sovereign. Plants, animals, and

their kind swap more pagan emulations. That fern could be that crow, that tumid ant. Genes are *in* the trellis; strings of them cascading across time as space (open any hatch and out they pour like the thousand dalmations), but they do not initiate or provide forms. Nature has cobbled a map, but not a trail (nothing like a trail).

What corporate cyborg, while auditing present assets and future profits, tolerates ghosts and chimeras in the rigging? Certainly not self-consciously militarized technoscience. The umbra of false basis, of illusory precept, of nonlinear determinism that the experiment itself metes through natureculture is a perfidious mirage. It signifies not a solemn truth gambit but a neurosis underwritten by error and chicanery. Science is tainted by dissociation, melancholia, self-doubt, while limning a regime in which absolute, godlike confidence is required. The Aboriginal philosophers didn't know either, but they were innocent of hubris, naked under the storm.

Most yogis and lamas, popes and rabbis likewise bethrall themselves and others with karmic myths. On the tundra where universal Marx meets universal Christ, blind armies collide with armies of the blind. Da Free Johns, integral-consciousness visioneers, four- and ten-step self-anointed enders of suffering, jemadars of liberation armies, and other politicians, ayatollahs, and vainglorious preachers probably mean well and mean deep, but whenever the ego and pride of the bossman (bosswoman) bleed through the deeds, the system serves the guru, is not for the disciples' benefit or edification. Listen to his/her actual dictates: Are they compassionate or self-exalting? There is no middle ground. Yes, some sutras are near egoless: *Job*, Dzogchen, the Upanishads, the words of the Fourteenth Dalai Lama, the sonnets of Shakespeare. It is psychospiritual authority as such that is suspect.

Science got this much right: men and women incarnate alone and equal in the hundred eyes of God. There be no castes, no ordinations, no priests.

Western Mountain, July 2003, added in blueline

A graduate of Amherst College, Richard Grossinger received a Ph.D. in anthropology from the University of Michigan, writing an ethnography of fishing in Maine. He is the author of many books, including *Planet Medicine; The Night Sky; Embryogenesis: Species, Gender, and Identity; Homeopathy: The Great Riddle; New Moon;* and *Out of Babylon: Ghosts of Grossinger's*. He and his wife Lindy Hough are the founding publishers of North Atlantic Books in Berkeley, California.